There Are No Such Things as Theories

There Are No Such Things as Theories

STEVEN FRENCH

OXFORD
UNIVERSITY PRESS

OXFORD
UNIVERSITY PRESS

Great Clarendon Street, Oxford, OX2 6DP,
United Kingdom

Oxford University Press is a department of the University of Oxford.
It furthers the University's objective of excellence in research, scholarship,
and education by publishing worldwide. Oxford is a registered trade mark of
Oxford University Press in the UK and in certain other countries

First Edition published in 2020

Impression: 1

Published in the United States of America by Oxford University Press
198 Madison Avenue, New York, NY 10016, United States of America

British Library Cataloguing in Publication Data
Data available

Library of Congress Control Number: 2019950429

ISBN 978–0–19–884815–8

DOI: 10.1093/oso/9780198848158.001.0001

Printed and bound in Great Britain by
Clays Ltd, Elcograf S.p.A.

For my dad, who didn't know much about art (but knew what he liked)

Contents

Preface

What *are* scientific theories and models? Back in the days of my youth, when I was negotiating the shift from physics to the philosophy of science, the answer would have been clear and straightforward: theories are sets of propositions, closed under logical deduction. This answer is explored in Chapter 1, but of course, it then invites the next question: what is a proposition? I touch on possible answers to that, in preparation for later discussions, focussing in particular on fictionalist and eliminativist answers.

By the time I'd taken up my first job, in Brazil, an alternative answer to my central question was gaining prominence: theories are families of models, construed in terms of the set-theoretic formalism. For many people this 'Semantic' or model-theoretic Approach is the new 'hegemony' and in Chapter 2 I outline the bones of it, together with recent criticisms. These can be understood as part of the inevitable push-back, with the pendulum swinging back towards proposition-based accounts. Recent work by Lutz and others has shown how these have the resources to accommodate many of the features of scientific practice that were previously invoked to support the model-based view.

Nevertheless, I continue to maintain that, suitably tweaked, the latter offers the better framework for philosophers of science to represent, *at the meta-level* as it were, those features of practice. Note the shift here, from answering the question above to considering what might be the best perspective from which to philosophically describe and analyse scientific practice. This shift was already laid out in my early work on the partial structures variant of the Semantic Approach with Brazilian logician and philosopher, Newton da Costa (da Costa and French 2003) but let me be clear: although I think that the best way of characterizing theories and models is in terms of partial structures, this is not to say that theories or models *are* such structures.

In particular, as I suggest in Chapter 3, the really significant advantages of this approach emerge when we consider the *representational* role of theories and models in science. Even more importantly, perhaps, at least as far as this book is concerned, this is where we find the first set of comparisons with artworks. Bluntly, and on the first iteration, theories and models are taken to represent in broadly the same manner as certain depictive artworks such as paintings are taken to represent. And indeed, we find various examples of such artworks deployed to both support and undermine certain accounts of representation in science. An obvious question, then, is whether such examples are *appropriate* in this context and I shall suggest that a certain amount of care is required here, whilst also noting that we can find parallels to the Semantic Approach in the philosophy of art.

Thinking of theories and models as like (or perhaps not) paintings when it comes to *representation* then opens up the broader issue whether they may be compared to artworks when it comes to their *ontological status*. Are they, for example, like musical works, not least in not being identifiable with a particular score or inscription or 'performance'? Popper, famously, placed theories along-side pieces of music and other artworks in his 'World 3' of entities that are neither physical nor mental and in the next couple of chapters I consider this view, as extended via Thomasson's account of artworks as 'abstract artefacts'. This allows for further close comparison of theories and models with various kinds of works of art, from comics and literature to photographs and musical compositions, and along the way I discuss the multiple instantiability, or rather, 'discovery', of theories and models as well as their modal flexibility. Ultimately, however, I conclude that not only is the view of such theories and models as abstract artefacts ontologically profligate but major concerns arise when it comes to accommodating the associated heuristic processes.

In Chapter 6, then, I discuss an alternative account that holds that theories and models are *fictions*, and which, again, draws on certain moves and devices in the philosophy of art. This discussion also touches on a number of important issues, such as the role of imagination in conceiving of such fictions, about which more needs to be said than I can manage here. Following Weisberg, however, I worry that this account may not be able to accommodate the kinds of complex models that are a feature of current practice and conclude that it should also be rejected.

That takes us to the view I favour, set out in Chapter 7: eliminativism, in the sense that, as it says on the tin, I believe *there are no such things as theories*. Such a claim generates a number of obvious concerns. First, what are we to make of assertions *about* theories and models? Here I too borrow a device from the philosophy of art by way of metaphysics, namely Cameron's version of truth-maker theory, which he applies to musical works. This allows us to acknowledge that, for example, the claim that 'quantum mechanics is empirically adequate' is *true* but what makes that claim true is not some feature of quantum mechanics regarded as some thing or entity, whether abstract artefact or fiction, but rather some feature of *scientific practice*. This is not to identify theories with such features—the latter simply act as the truth-makers for our claims 'about' the former.

Focussing attention on scientific practice then allows us to accommodate scientists' own views about what they think they are doing. And, in particular, as I maintain in Chapter 8, historical research reveals just how complex and diverse that practice is, and how contestable are the claims about the identity of particular theories and models, whether classical or quantum or whatever.

Finally, in Chapter 9 I return to the issue of representation and the question of how we should understand that, as a relationship, if there is no theory or model that represents. Likewise, how should we take realist claims that theories are true, or approximately so, given this eliminativist stance? The answers of course are revealed by reflecting a little carefully on what we are doing as philosophers of science—that is, we construct various frameworks, such as the Syntactic and Semantic Approaches, in terms of which certain features of scientific practice are characterized and in terms of which we can talk, in strict terms, of truth and representation. Just to be clear: this is not to adopt an anti-realist stance about what science tells us about the world but it is to be a kind of anti-realist about the devices by means of which it does the telling!

In the end what does that matter? Perhaps little in the Grand Scheme of Things but I think it is helpful to show that we can keep our ontological commitments minimal and that there are interesting comparisons to be drawn between not only art and science but the philosophies of art and of science, even if, granted, some care must be taken. And finally, I would hope that we might be encouraged to think a little further about what it is, exactly, that we do as philosophers when we think about science.

Acknowledgements

I am grateful for conversations with a number of people about aspects of this work, most notably Otávio Bueno, George Darby, Victor Durà-Vila, Milena Ivanova, Matthew Kieran, Aaron Meskin, Alice Murphy, Dean Rickles, Juha Saatsi, and, perhaps most significantly of all, Pete Vickers. The core idea of the paper with Pete, which lies at the heart of Chapter 7, was presented to audiences at the annual conference of the *British Society for the Philosophy of Science*, at a work-in-progress seminar here at Leeds, as well as at workshops on aesthetics and science organized by Dean, George, Otávio, and myself and in the philosophy departments of the Universities of Hertfordshire, Miami, and Notre Dame. Chapter 9 is based on a paper that I presented to the *New Thinking about Scientific Realism* conference in Cape Town, South Africa, in 2014, organized by Emma Ruttkamp-Bloem. I can't say in all honesty that I won anyone over in the respective audiences but I'm grateful for their comments all the same! I'd also like to thank the three readers who were asked to give their views of an earlier draft by Oxford University Press—all of you made useful comments but one went 'above and beyond' (and not in the trance music sense). Such a careful and supportive reading was truly exemplary and the book is vastly better for the helpful suggestions that were made.

I'm also grateful to the publishers to reproduce material from:

And finally, a massive shout-out, as always and forever, to Dena and Morgan (and The Ruffian).

1

Theories as Sets of Propositions

Introduction

What is a scientific theory? We can certainly point (figuratively and literally) to any number: the Boveri-Sutton chromosome theory, Ragnar Nurske's balanced growth theory of economics, the molecular orbital theory of molecular structure and of course those old favourites, Newtonian mechanics, Maxwell's theory of electromagnetism, the special theory of relativity, quantum theory... the list goes on. But can we characterize or otherwise pin down what a theory *is* in terms that go beyond simply listing examples? Even if we stay at the level of the list, obvious questions come up: what makes one theory different from another? When it comes to the above examples, the answer to that seems obvious: in these cases the theories are *about* very different things or phenomena and there is simply no issue about confusing the Boveri-Sutton chromosome theory, say, with the special theory of relativity.

But now consider the latter theory: the introduction to Einstein's classic 'annus mirabilis' paper of 1905 (Einstein 1905a) ends with the following passage:

> Like every other electrodynamics, the theory to be developed is based on the kinematics of the rigid body, since assertions of each and any theory concern the relations between rigid bodies (coordinate systems), clocks, and electromagnetic processes. Insufficient regard for this circumstance is at the root of the difficulties with which the electrodynamics of moving bodies must presently grapple.
>
> (ibid., p. 141)

The focus throughout is on the behaviour of (rigid) *rods and clocks*, whereas the theory is now, of course, typically taught using the framework of Minkowski *space-time*. Minkowski's paper, published after Einstein's, makes no mention of rods and clocks but instead talks of 'space-time vectors' and, crucially, of rendering certain symmetries evident (Minkowski 1908)[1] And famously Minkowski went on to insist,

[1] Einstein gets only two mentions in this work, for taking Lorentz's 'hypothesis' that bodies suffer length contraction due to their motion, in accordance with the famous Lorentz transformation, and presenting it as 'a new way of comprehending the time-concept which is forced upon us by observation of natural phenomena' (Minkowski 1908, p. 53). Einstein himself famously dismissed Minkowski's

There Are No Such Things as Theories. Steven French, Oxford University Press (2020). © Steven French.
DOI: 10.1093/oso/9780198848158.001.0001

> The views of space and time which I wish to lay before you have sprung from the soil of experimental physics, and therein lies their strength. They are radical. Henceforth space by itself, and time by itself, are doomed to fade away into mere shadows, and only a kind of union of the two will preserve an independent reality.
>
> (Minkowski 1909, p. 1)

Now, is Minkowski's work a 're-presentation' of Einstein's theory? Ostensibly, they are about two different kinds of thing—clocks and rods on the one hand, and Minkowski space-time on the other. So, should they be considered different theories? If so, how do we accommodate the apparently well-founded belief that in passing from Einstein's 1905 paper to Minkowski's 1908 one, we have a progression within, at least, the same research programme; or indeed, more strongly, that the Minkowskian presentation *is* the special theory of relativity? If they are not considered to be different, however, what are the identity conditions that they satisfy? Indeed, what are the identity conditions for theories, in general?

As another example, consider quantum mechanics. As is well known and much commented upon, the theory as it is used and loved by physicists, chemists, and others consists of a heavy duty mathematical framework, with associated algorithms, and a light touch physical interpretation sufficient to make minimal sense of the predictions and explanations that can then be obtained with regard to various phenomena. What it doesn't have is an agreed, 'full-blooded' interpretation that tells us 'what the world is like' according to quantum theory.[2] There are various options 'on the table'—such as the Ghirardi-Rimini-Weber interpretation, the Everett or 'many worlds' interpretation, the Bohm interpretation (and we'll come back to these in Chapter 8)—but no consensus among either physicists or philosophers of physics as to which one should be adopted. So what is the difference between 'the' quantum theory and its interpretation, taken from any of the above, or others? Is it only a theory when it has an agreed interpretation? Or if we adopt one or another interpretation do we then get a different theory, as some advocates of certain of these interpretations seem to think?

Thinking about quantum theory and its history then raises a further set of questions. Take Bohr: he introduced his 'old' quantum theory across three now-classic papers in 1913 and alternately referred to what he was talking about as a 'theory' and a 'model'. So, what's the difference between a theory and a model? Some scientists in some disciplines, like Bohr, use the terms more or less interchangeably, whereas others think of theories as more general, models as more

work as 'superfluous learnedness' (Pais 1982, p. 152) but then came to recognize its significance, not least in helping to lay the basis for the development of general relativity.

[2] And which we can then accept as true if we are realists or take as telling us what the world could be like, if we are not.

specific. In many if not all of these cases, models are presented in the same terms as theories—via sets of equations, with idealized assumptions, perhaps, or via qualitative descriptions. Thus, the autoregressive conditional heteroskedasticity (ARCH) models that, according to Wikipedia, are used to characterize and model observed time series in econometrics are written down using mathematical notation, appropriately interpreted, in just the way we would expect a highly mathematized theory to be (see http://en.wikipedia.org/wiki/Autoregressive_conditional_heteroskedasticity). But in certain cases, models are clearly different from what we usually consider theories to be: think of the famous Crick and Watson model of DNA, built out of tinplate and wire, which would have your eye out if it fell on you!

Can we answer the question, 'what is a model?', then, with an account that covers all the different types? Can we likewise say what theories are in a way that embraces all kinds, as indicated in the examples given above (or, at least, all such kinds across the relevant and particular science, as suitably demarcated)? And can we manage the implausible and construct a unitary account of both theories and models?

In this chapter and the next, I shall sketch two of the more well-known attempts to give just such an account, before eventually suggesting that we—philosophers of science, scientists, or simply those who are interested in science—need to adopt an entirely different approach to these questions.

The 'Syntactic' Approach to Theories

A well-known view of scientific theories holds that they are simply sets of logico-linguistic propositions, closed under classical logic. Within such a set we may distinguish certain propositions that are granted priority and regarded as axioms of the theory, from which the others may be derived (typically via classical logic) as theorems. A semantics is then supplied via some interpretation which, of course, relates the terms in these propositions to features of the world, whether observable or unobservable. Let me just pause at this point to emphasize: a theory *is*, on this view, just such a set of propositions. Of course, that then raises the further question: what is a proposition? I'll come back to that shortly but let me flesh out some of the details.

On this approach, then, a theory is taken to be a partially interpreted abstract formalism, incorporating a language in terms of which the theory is formulated. Thus, a theory is taken to consist of:

(i) an abstract formalism F;
(ii) a set of theoretical postulates (taken to be axioms) T;
(iii) a set of 'correspondence rules' C.

F incorporates a language *L* in terms of which the theory is formulated and a deductive calculus defined. *L* contains logical and non-logical terms, where the latter are typically taken to be divided into observation terms and theoretical terms; the 'correspondence rules' then function as a kind of dictionary or bridge by relating the former to the latter.

A 'partial interpretation' of the theoretical terms and the sentences of *L* containing them is then provided by the theoretical postulates—which contain only theoretical terms—and the correspondence rules, which correlate the non-logical, theoretical terms with observable phenomena by allowing for the derivation of certain sentences containing observation terms from certain sentences containing theoretical ones. The partiality of the interpretation arises because the theoretical terms are not explicitly defined. This gives a certain degree of structural leeway that allows for the addition of further correspondence rules as science advances, thus extending the interpretation of these terms:

> All the interpretation (in the strict sense of this term, i.e. observational interpretation) that can be given for L_T [the theoretical language] is given in the *C*-rules, and their function is essentially the interpretation of certain sentences containing descriptive terms, and thereby the descriptive terms of V_T [the theoretical vocabulary]....
>
> For L_T we do not claim to have a complete interpretation, but only the indirect and partial interpretation given by the correspondence rules....
>
> ...Before the *C*-rules are given, L_T, with the postulates *T* and the rules of deduction, is an uninterpreted calculus....Then the *C*-rules are added. All they do is, in effect, to permit the derivation of certain sentences of L_O [the observational language] from certain sentences of L_T and vice-versa. They serve indirectly for derivations of conclusions in L_O, e.g. predictions of observable events, from given premises in L_O, e.g. reports of results found by observation, or the determination of the probability of a conclusion in L_O on the basis of given premises in L_O. (Carnap 1956, pp. 46–7)

If *T* is the conjunction of theoretical postulates and *C* the conjunction of the correspondence rules, then a scientific theory is typically taken to be the conjunction '*TC*'. Thus, as already noted, according to the Syntactic Approach, the answer to our core question is: a theory *is* the logico-linguistic entity *TC*.

However, it is obviously not the case that theories, as presented, discussed, critiqued, etc. by scientists are presented in such a form. The above framework can best be thought of as a 'rational reconstruction' (Reichenbach 1938), which yields the form in which thinking processes are communicated to other persons, rather than the form in which they are actually performed. An example here might be the contrast between the actual thought processes of scientists in, for example, arriving at a given hypothesis or theory and what is actually presented in papers, conference talks, etc. (something we'll touch on again in later chapters). Laudan (2007) takes this as indicating that this process should be understood as essentially an *epistemic reconstruction* in which the features of scientific practice to be highlighted and incorporated into it should be those relating to the truth or falsity of the relevant hypotheses.[3] Within that context, *TC* can be seen as part of the reconstruction (Demopoulos 2007), as that which can then be identified as truth-apt. The *C*-rules are then to be understood as principles of epistemic interpretation.

Now, the language in terms of which *TC* is presented may appear artificial but what matters is whether it can illuminate the epistemology of science. Of course, this understanding then enters into tension with the above claim that the theory *is TC*; after all, how can this be the case if *TC* is not what scientists actually hold in their thoughts, in some sense? It is precisely this tension that I shall be exploring in this book and as we'll see, there's an obvious way of dissipating it.

Models in the Syntactic Approach

As Frigg and Hartmann note in their survey essay, 'Within this [the Syntactic] approach, the term model is used in a wider and in a narrower sense. In the wider sense, a model is just a system of semantic rules that interpret the abstract calculus and the study of a model amounts to scrutinizing the semantics of a scientific language. In the narrower sense, a model is an alternative interpretation of a certain calculus' (Frigg and Hartmann 2012). The former sense corresponds to that due to Tarski, whereby a model is just any model in *the 'model theoretic' sense* of the formal calculus and as Lutz remarks, Carnap and other proponents of what came to be known as the received view of theories (see Suppe 1977) certainly made extensive use of *this* kind of model (Lutz 2012, p. 92; see also Mormann 2007).

What about the second, narrower, sense? Well, if we take the central distinction between a theory and a model to be that the latter incorporates some crucial idealizing feature, on the basis of which we know it is not or cannot be true (a distinction that is also disputable), then we can run the same analysis as above: a model can be obtained from a given theory by providing an alternative interpretation of the underlying formal calculus, where the terms of this new interpretation refer to such idealizing features, via which further understanding could

[3] Lutz 2012, p. 105 gives a detailed account of this process in terms of providing a framework for explicating specific scientific theories. From such a perspective, criticisms that it cannot explicate certain features of practice or can explicate modes of reasoning that are not part of such practice simply fall wide of the mark (ibid.). In particular, he claims, it was meant to explicate only those theories that are sufficiently well formalized as to be axiomatizable. This might be thought of as imposing a bit of a restriction on this approach.

be provided by such models. A well-known example would be that of the 'billiard ball' model of a gas, where the formal calculus embodied in Newton's laws of mechanics is interpreted not in terms of gas atoms, but rather the more familiar (to scientists of a certain class anyway!) set of objects knocked about on the billiard table.

On this analysis, then, models are models *for* theories and, basically, are just theories themselves (see da Costa and French 2003, ch. 3):

> a model for a theory *T* is another theory *M* which corresponds to the theory *T* in respect of deductive structure. By correspondence in deductive structure between *M* and *T* is meant that there is a one-one correlation between the concepts of *T* and those of *M* which gives rise to a one-one correlation between the propositions of *T* and those of *M* which is such that if a proposition in *T* logically follows from a set of propositions in *T*, the correlate in *M* of the first proposition in *T* logically follows from the set of correlates in *M* of the propositions of the set in *T*.
>
> Since the deductive structure of *T* is reflected in *M*, a calculus which expresses *T* can also be interpreted as expressing *M*: a theory and a model for it can both be expressed by the same calculus. Thus, an alternative and equivalent explication of model for a theory can be given by saying that a model is another interpretation of the theory's calculus. (Braithwaite 1962, p. 225)[4]

Two questions then arise: does this structural equivalence between theories and models yield an accurate description of the relevant elements of scientific practice? And in the light of our answer to that, what can we say about the ontological status of these elements?

Answering that first question hinges on what one takes those elements of practice to be. In the early years of the Syntactic Approach, the role of models in the cutting-edge physics of the time—namely quantum mechanics and relativity theory—was, at best, downplayed or dismissed as merely heuristic or pedagogical. In the case of quantum mechanics, what subsequently came to be called the Copenhagen interpretation had established its hegemony by the early 1930s (see Cushing 1994) and even if we grant that this was not a monolithic account, it left little room for alternatives at the time. Of course, this is not to deny that scientists themselves would have constructed models within the framework of the theory in order to obtain predictions concerning a specific type of system or just to explore certain features of such systems. Quantum mechanics, as it came to be formalized in terms of Hilbert space and Hamiltonian operators (and I'll come

[4] Hempel articulated an alternative view of models that doesn't quite fit either the wider or narrower senses above but rather took them to be, effectively, theories of limited scope (Lutz 2012, pp. 93–5; see also Mormann 2012).

back to this in later chapters), is perhaps better thought of as a kind of overarching theoretical framework in terms of which theories of specific kinds of systems could be elaborated (Wallace 2020). Thus, as has been emphasized, in order to get the theory to do any useful explanatory or predictive work, one has to specify the Hamiltonian for the particular system one is interested in, just as when deploying Newton's laws of mechanics, one typically needs to state the relevant force function, as in the case of the harmonic oscillator (as Cartwight puts it, physicists effectively take the Hamiltonian they need 'off the shelf'; Cartwright 1983). But these kinds of moves can be accommodated by the Syntactic Approach more or less straightforwardly—such models can be thought of as 'mini-theories', in a sense.

Likewise, when it comes to the other great theory of the day, namely general relativity, what we have is a general set of equations for which different solutions can be supplied. Einstein's solution yielded a closed, stable universe that is dense with matter. Again during the early development of the Syntactic Approach only one other solution, or model, was deemed acceptable, namely de Sitter's, in which the universe was also stable and closed but empty (see Gale 2017). Friedman and LeMaitre also published alternative models, in which the geometry of the universe changed through time, but these were dismissed (only to be revived following Hubble's observations confirming the expansion of the universe). In this case one might argue that such models should not be regarded as 'mini-theories' but insofar as the subject matter is cosmology, they should be treated as attempts to construct a theory of this, the actual world, with Einstein's equations regarded as providing a very general theoretical framework for a number of possible worlds.

In either case, then, one can understand how the role of such models was seen as restricted:

> It is important to realise that the discovery of a model has no more than an aesthetic or didactic or at best a heuristic value, but it is not at all essential for a successful application of a physical theory. (Carnap 1939, p. 68)[5]

Subsequently, of course, models came to be accorded not only a much more significant role, but a different structure and status from theories.[6] It might be speculated that this was due in part to developments within science itself, as with the establishment of the above overarching theoretical frameworks and in the post-revolutionary environments of quantum physics and space-time theory, scientists set to work applying these frameworks to particular systems. Still, it could also be pointed out that modelling—understood in a more epistemically robust

[5] Although again, as Lutz notes (2012, p. 97), Carnap did acknowledge that visualizable models could turn out to be accurate.

[6] Not everyone shared this dismissive attitude. The so-called 'modellists', including Hesse who famously emphasized the importance of analogy in science, insisted that models could yield understanding; see da Costa and French 2003, ch. 3.

sense than as mere heuristics—had long had such a role, before the quantum and relativistic revolutions (recall Duhem's dismissal of the 'English' penchant for modelling, for example). More significant—and reflecting the fact that this accordance of a larger role sits at the level of the philosophy of science—is the increased attention that philosophers of science gave to models through the late 1950s and '60s. Some of this might be due to the recognition that sciences other than physics were also deserving of philosophical attention—biology in particular, where, (in)famously there is a dearth of the kinds of laws that would feature in the relevant axioms under the scheme outlined above. But there was also a kind of feedback cycle as proponents of an alternative to the Syntactic Approach (to be discussed below) promulgated a characterization that placed models—formally construed—at its heart and also justified itself by drawing on examples of epistemically significant model construction in science. That in turn encouraged further light to be shed on such construction and development, even by those who were not adherents of this approach.

Whatever the history, it would be a brave soul to dismiss models as having merely aesthetic or heuristic value these days. Indeed, models are now taken to possess a kind of 'functional autonomy' (Morgan and Morrison 1993) in the sense that in such cases it is *them* and not the theory that have become the loci of scientific interest and practice. *That* claim not only seems plausible but is supported by numerous case studies and in general, one can accept that, first of all, the construction of models is a more complex matter than simply reinterpretation, involving a plurality of theoretical connections as well as non-theoretical elements; and, secondly, that having been constructed, the role of such models goes beyond the heuristic, or pedagogical, or even that of mere aids to understanding.

However, put in those terms, any apparent challenge to the Syntactic Approach that is represented by such recent developments seems much diminished. One could, for example, make an analogous move to the one that underpins the discovery-justification distinction and insist that however models are constructed—and it can be acknowledged that that may happen in all sorts of weird and wonderful ways—they can still be construed in terms of a partial interpretation of an underlying formal calculus. Indeed, the blurring of the theory-model distinction in terms of formal structure would mesh nicely with the blurring in terms of their role in practice.

The alternative would be to acknowledge that that practice involves at least two features—theories and models—that are separate foci of scientific activity, separate sources of scientific knowledge, and that have separate structures. One could then appeal to this in order to respond to the second question above and suggest that theories and models are ontologically different. That then raises its own difficulties: what is the nature of the relationship between these ontologically distinct entities, since they are clearly related (in some way) in scientific practice? If they are the same kind of thing then we can straightforwardly grasp how the inter-relationships between them that we find in practice—whether these are characterized as deductive or multi-aspected—are mirrored, in some sense, by the relevant ontological relationships between them as entities. But if they are ontically distinct, then the relations between them, also understood ontically, is a much more problematic kettle of fish and we have to say that what appears relatively straightforward in practice—again, whether the inter-relationship is deductive or more complex, it can at least be *described* straightforwardly—is not so at the ontological level. Indeed, one may wonder if talking of theories and models as having distinct structures even makes sense, given that we can apparently represent them—deductively or otherwise—in more or less the same terms.

However, careful consideration is required to determine whether there is anything in the above construal to prevent such a view embracing those further roles that models play that have been identified. Ultimately (spoiler alert!), I think that what is important is the relevant practice, rather than trying to individuate or establish appropriate identity conditions for either theories or models. But of course, it was concerns with capturing practice that ultimately led many philosophers of science to abandon the Syntactic Approach and adopt what is frequently acknowledged to be the currently dominant or hegemonic view, namely the 'Semantic' alternative. Before we discuss that, let us consider this presentation of theories as logico-linguistic entities in a little more detail.

The Impracticality of Formalization

On this approach, then, a theory (and hence, depending on one's view, a model also) *is* such a logico-linguistic entity. Specifically, as already indicated, it is an axiomatic calculus, closed under consequence within a particular logical framework and given a partial observational interpretation via a set of correspondence rules.[7] It is worth noting that, at the very least, a distinction should be made between this view of theories and the broader context in which it was elaborated—namely that of logical positivism (see for example Mormann 2007 and the blog post and subsequent comments in http://itisonlyatheory.blogspot.co.uk/2009/01/what-was-wrong-with-syntactic-view-of.html). Although few if any would subscribe to the core features of the latter these days, many would still cleave to some form of the former, even if some of the major concerns with the logical positivist enterprise—such as its perceived lack of contact with scientific practice—also stick to the Syntactic Approach, if only via a kind of philosophical guilt by association.

[7] Despite the name, it should not be inferred that there is nothing *non*-syntactic that constitutes a theory (Burgos 2007, p. 157). As Lutz (2012, p. 80) also records, Suppe noted some time ago that according to this view a semantic interpretation is included in the characterization of a theory (Suppe 1974, section IIE).

Here it might be useful to make a further distinction between 'strong' and 'weak' versions of this approach (Hendry and Psillos 2007). The strong form is as given above and identifies theories with a formal linguistic entity. The weak form, on the other hand, takes theories to be collections of statements that are interpreted 'literally', rather than via correspondence rules or the like.

Typically, with regard to the former, the formalism selected is first-order predicate logic with identity and the set of consequences of the axioms are taken to be closed under this logic. Other formal frameworks might be advocated, but the basic idea would be the same. A central worry is then whether this 'strong' form can accommodate the kinds of theories we come across in science (Hendry and Psillos ibid., p. 132). Thus, although it is possible to formalize certain simple theories in mathematics in terms of first-order predicate logic with identity, when we turn to more complex theories, such as we find in geometry, for example, the task becomes 'awkward and unduly laborious' (Suppes 1957, p. 248). Moving to even more complex theories, such as quantum mechanics, classical thermodynamics, or, say, quantitative versions of learning theory, where results from number theory and the theory of functions are appealed to, the task moves from the laborious to the 'utterly impractical' (ibid., p. 249; see also Suppes 1967, p. 58).[8]

Some have argued that this is an unfair characterization. According to Lutz, for example, it is the approach's critics who lumbered it with first-order predicate logic, whereas various exponents accepted alternative formalisms (Lutz 2012). And as he notes, such formalisms can capture (much of) the kinds of mathematics we find in science. Thus, as he puts it, the claim made by advocates of the alternative 'Semantic Approach'—that philosophers of science should turn to *mathematics* rather than meta-mathematics as the framework within which to conceive of scientific theories (a claim we shall return to below)—makes sense as a criticism of the Syntactic view only if it is first-order logic that is tagged as 'meta-mathematics' (ibid., p. 88).

A further related criticism is that, as just indicated, this formalization is going to have to include not only the relevant axioms for those parts of the theory that can be hived off as 'physical' but also those that are mathematical. So, wherever numbers are involved, we will have to give the full set of axioms of number theory and so on (Suppes 2002, p. 207). Even if we deploy some higher-order logic capable of capturing such axioms, we will have to present them all every time we characterize or identify the relevant physical theory. According to Muller (2011, p. 109) this exhaustive formal labour is mandatory within this approach.

[8] As noted in a previous work (French and Ladyman 1999), one can clearly detect an ambiguity in critical discussions of this approach (as pointed out by, for example, Collier 1992, p. 290) as to whether it should be considered to be 'strictly wrong' or 'just practically unworkable'. In that earlier piece I argued for the former but now I incline more towards the latter.

Now, you might wonder who or what is doing the mandating here! Of course, we might insist that unless the axiomatization is exhaustive, the characterization of the theory will be incomplete at best, or the identity conditions will be lacking, at worst. But the solution is obvious (Lutz 2012, pp. 89–90): we distinguish within the formalization a 'basic' calculus which axiomatizes all the mathematical features, and a 'specific' calculus added to the former, which covers the specific physical features of the theory (Carnap 1939, section 16). Since the former will be the same across various different theories (obviously there may be cases where we extend or otherwise change it to accommodate the use of new forms of mathematics within science), it is not necessary to mention it each time. When a theory is formalized and hence identified via a partially interpreted logico-linguistic calculus, the 'basic' part of that calculus can be left tacitly presupposed (Lutz 2012, p. 91).

Nevertheless, it is for the aforementioned reason of impracticality, at least partly, that we don't generally come across this kind of formalization in scientific practice itself. Of course, one can find examples of what are called 'axioms' in textbooks of quantum mechanics, for example, but they don't look at all like the kinds of things one is presented with in logic courses, nor do they seem to play the same kind of role—something that has been emphasized in the shift away from the Syntactic Approach to the alternative (van Fraassen 1980, p. 65). Now, again, the advocate of the former can appeal to the claim that what she is really interested in here is *rational reconstruction* (again see Lutz 2012, pp. 99–110) and in *that* context, it is really neither here nor there whether such formal axioms are actually found in practice.

Nevertheless, the above point of tractability still has some bite and here a comparison with the nominalist stance to mathematics is illuminating.[9] According to this, mathematical objects and structures either do not exist at all, or, at least, not as abstract objects (in which case, some concrete substitute might be acceptable; Bueno 2014) and we can reformulate our best scientific theories without them. Colyvan famously argued that there is 'no easy road' to this, in the sense that the only way to adopt such a stance is to actually roll up your sleeves and 'nominalize' each and every one of our theories—a task that is so daunting as to suggest to many a kind of in-principle impossibility. Likewise, one can propose a similar claim that there is 'no easy road' when it comes to this view of theories, in the sense that the lack of rational reconstructions of, for example, quantum mechanics and general relativity in terms of some higher-order logic strongly suggests that the default view should be that it is in principle impossible. To put it another way: without a positive argument provided by the advocate of this view that our best scientific theories can indeed be formalized this way, the practical difficulties

[9] I am grateful to one of the readers for this suggestion.

support a sceptical line. Thus, if the aims of such formalization are to both delineate the theory and to present its logical structure (leaving aside the question of whether theories may be said to have a *logical* structure), then there is little point to holding such aims if they are unachievable.

Furthermore, this focus on formalization within a given framework led to accusations that it further shifted attention away from practice by generating issues, concerns, and problems that had to do with that *framework* itself, rather than as arising out of that practice. This is the flip side of the expressive power of such a framework, of course. Thus, just to take one example that was at the centre of considerable debate 'back in the day', Craig's theorem (Craig 1953) states that 'every theory that admits a recursively enumerable set of axioms (in a first-order language) can be recursively axiomatised'. What this means is that if the primitive predicates of our formal language are divided into two classes, say, then the sub-theory consisting of all those theorems expressible in the vocabulary of one of those classes is itself recursively axiomatizable. In effect this then drains a given theory of theoretical content, since if the division of vocabulary is between the 'theoretical' and the 'observational' (although it does not have to be—it could be between the mathematical and the non-mathematical; see Ketland unpublished), then the sub-theory consisting of the theorems that are expressible in the observational theory is itself a recursively axiomatizable theory and since that sub-theory obviously contains all the empirical predictions of the original theory, we can then dispense with the original theory and its theoretical terms when considering its predictive success (for a useful discussion from the time, see Putnam 1965). And if we take predictive success as the principal aim of science, then we have a mechanism for dispensing with the theoretical terms that were so bothersome to logical empiricists and antirealists in general.

Of course, what we lose by reaxiomatizing away the theoretical terms is the expressive power they themselves provide within the theory. As a result the Craigian substitute for the original may be more complex but given the well-known difficulties in pinning down what is meant by simplicity in this context and why simpler theories should be preferred, this sort of objection can be rebuffed. Likewise, concerns that the Craigian pod-theory fails to explain the phenomenon via a small number of core principles (Field 1980, p. 8; Ketland, unpublished) can also be dismissed, on the grounds that this presumes not only a certain account of explanation but indeed, that explanation is, or should be, one of the aims of science. More significantly, perhaps, it has been argued that this manoeuvre, and the distinction between a theoretical and an observational vocabulary that underpins it, is little more than a logico-linguistic device that does not adequately capture a theory's empirical import.

So, consider Newton's theory of gravitation, taken to imply that there is something, namely absolute space, that neither has a position nor a volume. Or take quantum mechanics, on the Copenhagen interpretation, which states that particles sometimes have a position and sometimes do not (these examples are taken from van Fraassen 1976). In both these cases I have managed to assert certain consequences without using theoretical terms and so I can be taken to be making those assertions within the Craigian sub-theory. Yet I am still talking about unobservable things. In general, as long as those unobservable things are describable in observable terms, the sub-theory will assert their existence if the full theory does. Thus, the Craigian sub-theory is 'not a description of the observable part of the world of [the theory]; rather it is a hobbled and hamstrung version of [the theory's] description of everything' (ibid., p. 629). Some better account of the 'empirical import' of a theory is needed.

It is for this sort of reason, among others, that the Syntactic Approach and the philosophical views associated with it came to be seen as leading the philosophy of science into the (multiple) arms of logic and the philosophy of language and taking the discipline a 'mille milles de toute habitation scientifique, [leaving us] isolated in our own abstract dreams' (van Fraassen 1989, p. 225).[10]

Correspondence Rules and the Individuation of Theories

This distinction between the theoretical and observational vocabularies then leads to a further concern with this 'strong' form of the approach, which has to do with the issue with which I began this book, namely what theories are 'about' and how we capture that aboutness, formally speaking. So, we recall that having established the theoretical-observable distinction, we then 'bridge' the gap between statements on either side via correspondence rules that give meaning to the former set of claims and, crucially, assign the theory an interpretation (if only partial).[11] What this suggests, then, is that the interpretation of the theory, and therefore our understanding of what it is a theory *of*, is only established once the relevant correspondence rules are laid down and the bridge is built, as it were. But then that means that what the theory is *about* is only established at the point that it is applied (Hendry and Psillos 2007, p. 131). Thus, it would seem that the theory has been divorced from its content: 'what a theory is a theory *of* need not be a feature of the theory, conceived by itself; rather it is tacked onto it at the point of application' (ibid.).

On this view, the correspondence rules are not part of what individuates a theory. Now, the obvious response is that that implies that a theory is *just* a logico-linguistic calculus and that can't be what the adherents of the Syntactic Approach intended! The alternative is to internalize such rules and incorporate

[10] But see, again, Lutz 2012, pp. 87–8.

[11] For a detailed account of these rules and the differences in how they were regarded, see Suppe 1977, pp. 16–27; for more recent discussions, see Mormann 2012 and Lutz 2017.

them as part of what makes the theory the theory that it is, but that then leads to a further problem. Thus, according to Suppe, who was very much concerned with how we individuate theories, correspondence rules are indeed 'individuating proper parts of theories' (1989, p. 4). But then, any addition to or change in these rules must yield a new theory. So, if we extract a new prediction from the theory and express that via a new correspondence rule relating the theoretical principles to a new experimental or generally phenomenal situation or if we simply change a given experimental procedure—by the introduction of new measurement techniques, for example—or make a change in experimental design, then we have to accept that strictly we are dealing with a new theory.[12] Again, this does not mesh with scientific practice, where we find one and the same theory being taken to have multiple applications, offering multiple predictions, and so on. Indeed, on this view, what we have understood for years to be a given theory splinters into multiple theories, each one specific to a certain predictive or application context.

Thus, we have a dilemma: either the correspondence rules are internal to and hence individuating parts of the theory, or they are not and are external to it. In the former case, we have the problem that a change in experimental technique implies a change in the relevant correspondence rules which then implies a different theory, a consequence that may run counter to scientific practice. In the latter we have the issue, identified by Hendry and Psillos, of what the theory is about being decided only at the 'point of application', with the consequence that theories seem to become 'free floating' entities (Hendry and Psillos 2007, p. 132). Now, with regard to the second horn, we might respond with the exclamation 'how could it be otherwise?!' After all, we know that the same formal features—as presented within a particular theoretical context by the relevant mathematics—can be applied to different phenomena. Consider the classic (in all senses) case of the diffusion equation:

$$\frac{\partial \phi(\mathrm{r,t})}{\partial t} = \nabla.\big(D(r).\nabla \phi(\mathrm{r,t})\big)$$

where $\phi(\mathrm{r}, t)$ is the density of the material at location r and time t and $D(\phi, \mathrm{r})$ is the collective diffusion coefficient (see http://en.wikipedia.org/wiki/Diffusion_equation). It turns out that this also describes the variation of temperature in a material, with $\phi(\mathrm{r}, t)$ as the temperature of the material at location r and time t and $D(\phi, \mathrm{r})$ is now the thermal diffusivity, and hence is known as 'the heat equation'. Furthermore, it can also be used to represent certain kinds of financial transactions in 'options' markets within the context of the Black-Scholes-Merton

[12] This may seem absurdly strict but then, as Suppe notes, if the correspondence rules individuate the theory, then it follows that different rules strictly imply a new theory! If you're bothered by this, Schaffner offers a more nuanced view as we'll see shortly.

model (https://en.wikipedia.org/wiki/Black–Scholes_model). Clearly what this particular piece of mathematics comes to be 'about' cannot be decided purely 'internally' but only when the relevant terms are interpreted (and given the usefulness of this particular equation, we surely would not have it otherwise!).[13] But according to the Syntactic Approach, that interpretation (necessarily partial) is primarily established via a connection with the empirical phenomena, that is by means of the correspondence rules. Of course, one could always seek an alternative route to interpretation and meaning but that would be to deviate—I'll come back to this shortly.

So an advocate of this approach might be left wondering just how sharp this particular horn of the dilemma is—certainly sharpening it in the obvious ways runs the risk of question-begging. Furthermore, it hinges on taking 'the' theory to be just the formalism, as presented by the mathematics, so that we can talk about what 'the theory' is about being determined by 'its' application in the first place.[14] But that is undermined by the example just sketched: the theory of diffusion cannot just be *that* particular piece of mathematics, since it is multiply applicable to quite different kinds of phenomena. And proponents of the Syntactic Approach are quite clear that what a theory *is*, is the conjunction of theoretical principles (laws etc.) *and* correspondence rules.

That then leaves the first horn: a change in the correspondence rules implies a change in theory. One might try to insulate the view against such a consequence by suggesting that we should distinguish between different kinds of correspondence rules (cf. Schaffner 1969)—those that individuate and those that do not. So, in the context of a more or less sharp distinction between the contexts of discovery and of justification (in terms of which the notion of 'rational reconstruction' was developed of course), one might distinguish those 'initial' or 'heuristic' correspondence rules that give the relevant theoretical terms at least an initial meaning and interpretation. Some of these rules will effectively be inherited from previous theories that play a role in the heuristic process of theory construction. Thus, consider that classic (in not all of the senses above!) example of Bohr's theory of the atom. This was chosen by Nagel as an illustration of how correspondence rules work in establishing the relevant connections between such terms as 'electron', 'orbit', 'jumps', etc. and observable elements such as the line in the radiation spectrum of a particular element (Nagel 1961, pp. 94–5; also cited in Schaffner 1969, pp. 285–6). In establishing the connection, the correspondence

[13] And of course, as a reader pointed out, not only is mathematical structure multiply realizable in this way but there are many cases in the history of science where we take some theory or model to be about something and further theoretical or empirical investigations suggest that it is about something else. We'll come back to this.

[14] And there is an associated worry articulated by Hesse: how can observation statements feature in the confirmation or falsification of a theory if the latter can only be individuated via the connection with such statements (Hesse 1965, p. 16)? Her solution is to urge a shift in focus to models and analogies but here I just want to emphasize again the underlying assumption regarding theory individuation.

rule effectively encapsulates relations previously established within Maxwell's electromagnetic theory:

> On the basis of the electromagnetic theory of light, a line in the spectrum of an element is associated with an electromagnetic wave whose length can be calculated, in accordance with the assumptions of the theory, from experimental data on the position of the spectral line. On the other hand, the Bohr theory associates the wave length of a light ray emitted by an atom with the jump of an electron from one of its permissible orbits to another such orbit. In consequence, the theoretical notion of an electron jump is linked to the experimental notion of a spectral line. (Nagel 1961, p. 95)[15]

Thus, as Schaffner notes (1969, p. 286), what warrants the correspondences are theories that are effectively 'borrowed' (in his words), to be used in the heuristic process of construction. Of course, such correspondences may also function as routes for explanation that confer some initial plausibility upon the theory, although having been used in its construction they cannot be appealed to as part of any further confirmation. But the more important issue, in this context at least, has to do with their role as individuating parts of Bohr's theory. Note that this does not mean that earlier theories themselves constitute such individuating parts—as Schaffner emphasizes, *theories* such as Maxwell's, and also *models* such as Rutherford's, or '*hypotheses*' such as Einstein's with regard to the quantum (thereby illustrating the different kinds of 'elements' that might be so included), are not incorporated into the postulates of Bohr's theory; rather, as already mentioned, they underpin and warrant the relevant correspondence rules.[16] As individuating parts such rules delineate the theory and allow us to appropriately talk of 'the theory' being confirmed or falsified (and of 'it' making predictions; cf. Suppe 1977, p. 27) via subsequent connections established within the domain of justification.

Of course, insofar as the discovery-justification distinction becomes blurred, so does that between the construction of a theory and its empirical confirmation/falsification and consequently the delineation of the theory might be regarded as 'fuzzy' in certain respects. This move of distinguishing 'individuating' correspondence rules from those that are not also goes against the strict

[15] Schaffner analyses these kinds of connections in terms of 'causal sequences' relating the action of unobservable entities with observable events.

[16] It remains an interesting issue whether Bohr's theory, to pursue this particular example, can still be considered inconsistent in this case. As Vickers notes, whether or not it is so regarded generally depends on whether classical electrodynamics is considered to be 'part' of the theory or not (Vickers 2013; 2014). The conflicting stances on this issue lead him (in part) to propose theory eliminativism as a way out of the morass, and I will return to this in Chapter 7; but here I just want to note that a proponent of the syntactic approach, keen to cleave to classical logic, might point out that we can hope to avoid inconsistency as long as we are careful about what is warranted and to what extent.

understanding of the claim that the interpretation of theories is always partial because the stock of correspondence rules can always be added to, allowing for theory development. However, we can maintain the spirit of this claim by insisting, again, that it surely cannot be the case that any such correspondence rule constitutes an individuating part of the theory—consider a 'trivial' or minor experiment, for example, developed to establish the veracity of a previous claim using slightly different equipment. Again, we can adhere to the above distinction and insist that certain such rules will not just represent new applications of the theory but constitute a significant new development or extension, forcing us to admit that the interpretation of terms has changed and we have a new theory (cf. Shaffner 1969, pp. 289–90). Where one draws the line is obviously going to be difficult, but perhaps that is as it should be and certainly it should not be philosophers of science doing the drawing on the basis of some formal account of what theories are. Although some might argue for a more normative stance on this issue, many of us would be happy to defer to scientists themselves in such matters and allow for the distinction between constitutive and non-constitutive correspondence rules to be a matter of scientific judgement.

On this sort of modified form of the Syntactic Approach, what a theory is 'about' is established during theory construction, via correspondence rules that acquire their warrant from predecessor and background theories. This perhaps draws the sting from the above criticism of the 'strong' form of the approach. Nevertheless, the concern remains that this kind of formalization takes us 'a thousand miles' from scientific practice. So, we might still feel we have reason to shift to the 'weak' form of this approach, according to which, we recall, theories are conceived of as collections of statements that are interpreted 'literally', rather than via correspondence rules or the like (Hendry and Psillos 2007, p. 134). Although the relationships between such statements might be captured in formal terms, by means of some canonical logical framework, whether this can be done or not has no bearing on the status of the collection *as a scientific theory*.

However this seems to possess a number of disadvantages and none of the advantages of the strong form. First of all, it may not be clear what one is to take as the appropriate collection to begin with. Scientists differ in how they express a given theory, both in their scholarly publications and even more so in their more general or publicly accessible pronouncements. How are we to individuate a theory on this basis? Difficulties thus arise as to what we are to take as the relevant sentences of the theory concerned, perhaps, as Suppes noted, 'the main reason for this being that the notion of a sentence of the theory is not well defined when the theory is not given in standard formalization' (Suppes 1967, p. 58).

Furthermore, this sort of approach in general runs immediately into the well-known concern that a change in language then implies a change in theory, since the relevant collection of statements would be different (Suppe 1974, pp. 204–5). This concern runs along the following lines: a theory (whether Newtonian mechanics or general relativity) should be regarded as the same whether it is expressed in English or Portuguese; hence theories cannot be linguistic entities in the sense suggested by the 'weak' version of the Syntactic Approach. Such a concern then motivates a move to ways of individuating theories that are language independent.

Now, this might not seem such a powerful motivation. We can surely establish a translation manual for English and Portuguese such that the relevant terms of Newtonian mechanics or general relativity can be inter-translated. One might then identify 'the' theory with the equivalence class of such sets of sentences.[17] However, there is an obvious worry about the accuracy of such translations and the possibility of the indeterminacy of translation looms over this picture.[18] I'll come back to this issue.

Alternatively, one might insist that such statements merely stand for the relevant *propositions*, and that it is *these* that constitute the theory.

Thus, in answer to our central question: what a theory *is*, on this view, is a collection of propositions. Now, there is a huge amount to be said for and against propositions (for a useful introduction, see McGrath 2014).[19] But an obvious worry is that such a move effectively identifies a theory with a set of entities that are far removed from concrete practice and, as a consequence, for which issues of epistemic access arise.

Theories as Abstract Entities (First Pass)

There is a widely held view that propositions must be regarded as abstract, mind-independent entities (for a brief survey of recent versions of this view, the nuances of which I won't go into here, see Crawford 2005). The argument for such a view goes as follows (again, see McGrath 2014): consider the proposition that there are electrons, which we can denote by <there are electrons>. This proposition may be true, in the absence of any beings that entertain the requisite attitude—e.g. that of believing that there are electrons—or indeed that have the requisite mental states at all. In other words, it is possible for <there are electrons> to be true in the absence of any such mental states. But then (again following McGrath), if it is

[17] See Lutz 2014, p. 1484, for a more formal demonstration of how one might do this. Note however that this requires a further element—a 'pure structure' (see later)—in terms of which the theory can be identified and which ensures that the two translations are descriptions of 'the same' theory.

[18] This is the claim that there can be different translations of a language that are not mere stylistic variants of one another; see Hylton 2016, section 6.2.

[19] So, for example, I will follow McGrath in taking a proposition to be 'the sharable objects of the attitudes and the primary bearers of truth and falsity' (ibid.). That they are defined here as *objects* obviously suits my rhetorical purposes quite nicely!

possibly true in the absence of such mental states, then it possibly exists in the absence of such mental states. Hence it must be mind-independent.[20]

Furthermore, the proposition <there are electrons> is false, of course, in a world in which there are no electrons and, more generally, in worlds in which there are no concrete entities at all. But propositions, on this view, are the bearers of truth and falsity; hence, if the proposition is false in such worlds, it must exist in such worlds (to bear that truth-value). Hence propositions exist possibly and so must be abstract, actually.

The obvious concern, of course, is that which arises for all such abstract entities, namely the so-called Benacerraf problem (in its epistemic guise): if propositions are abstract and if abstract entities are understood not to stand in causal relations with concrete entities (such as ourselves) and if all knowledge is grounded in such causal relationships then how can we have knowledge of propositions? Now, there are a lot of 'ifs' in there, of course! But given that we do have knowledge of theories, an obvious tension arises with regard to the claim that theories *are* sets of propositions.

There are various ways of resolving this tension.[21] One could, for example, question the whole abstract-concrete distinction (see Rosen 2017) but then one still has to give some account as to how propositions can enter into causal relationships (via what properties?); or one could try some other account of knowledge, such as reliabilism, but then the problem still goes through (how can we have reliable knowledge of such entities . . . ?); or one could try to argue that propositions are not abstract in any sense that prevents them entering into the appropriate kinds of relations that would underpin knowledge claims. That last is what the views I am about to canvas suggest, in their different ways, and I mention them here because, as with the above claim that propositions are abstract, we shall be returning to them, or their variants, in an alternative context later.

Theories as Fictions (First Pass)

One such view suggests extending fictionalist accounts of other supposedly abstract entities, such as numbers, to propositions (see Balaguer 1998 and Yablo 2001 and also, more generally, Eklund 2011 and Woodridge 2006). So, the core idea behind fictionalism is to adopt the same stance towards such problematic entities as we do towards entities that feature in fictional scenarios, as described by novels, TV shows, or movies for example. (Here we see the first import of a

[20] Note we are not talking here of the mind independence of electrons but the mind independence of the proposition that asserts that there are electrons!

[21] Obviously the Quinean move of identifying propositions with sentence types in natural language is going to be of little help in the current context.

manoeuvre from philosophy of art!) Thus, consider the statement 'Buffy is a reluctant hero who wants to live a normal life' (taken from https://en.wikipedia.org/wiki/Buffy_the_Vampire_Slayer). This is not taken to be 'literally' true but only as true 'within the fiction' represented by the TV series (and subsequent comic) Buffy the Vampire Slayer. Furthermore, the entity designated by 'Buffy' does not actually exist, but only within the fiction. Thus, following Eklund (2011) we can distinguish a linguistic component to fictionalism from an ontological one: the former takes such statements to be not literally true, but as true only within the fiction and the latter that the entities referred to in such statements have the ontological status of fictional entities, or, more bluntly, do not exist. This can then be extended to mathematical entities, for example, so the statement that '3 is a prime number' is false, because there are no numbers (Balaguer 2015).

Nevertheless, such fictional entities can serve a useful purpose by functioning as *representational aids* and conveying information about the world, while not actually existing 'in' the world themselves. Thus, we might take Buffy as representing someone who appears insignificant and powerless but turns out to be a force to be reckoned with![22] Similarly, the supernatural creatures she encounters and, sometimes, fights, can be seen as metaphors for the anxieties of adolescence and the fiction as a whole tells us that such anxieties may be overcome, sometimes in surprising and unusual ways (see for example the selection of studies represented here: https://en.wikipedia.org/wiki/Buffy_studies#Works_in_print). There is considerably more to say, of course, and, I'll say some of it in later chapters. Here I shall focus on one well-known account of how fictions can function as conveyors of information, which has been explicitly taken up not only in fictionalist accounts of propositions but also, more recently, by philosophers of science with regard to scientific theories.

This is the 'fiction as pretence' view due to Walton (1990; 1993): the central idea is that by pretending that some entity is something else, we can convey certain information about the latter more easily or straightforwardly. Thus, to use Walton's own example—which I think plays an important motivational role in the extension of his approach to models in science—by pretending that Italy is a boot, I can convey to you the location of the town of Crotone. (Just to foreshadow what is to come: by pretending the atom is like a plum pudding, Thomson could more straightforwardly convey the relationship between the electrons and the overall positive charge; likewise Bohr, following Rutherford, pretended that the atom was like the solar system.) The attitude we adopt towards certain entities, on Walton's view then, is that of pretence or make-believe (again, we shall return to this attitude in Chapter 6 when we consider the ontological status of scientific models).

[22] Joss Whedon famously remarked that he wanted to subvert the Hollywood formula of 'the little blonde girl who goes into a dark alley and gets killed in every horror movie' and 'create someone who was a hero' (Billson 2005).

So, to take a much-used example, suppose some children playing in the woods pretend that some tree stumps are bears (Toon 2012, p. 34). In such a game, the tree stumps are *props*, to use Walton's terminology and the nature of the game is delineated by a kind of convention or *principle of generation* via the kids' agreement that stumps should be regarded as bears. Utterances such as 'Argh! There's a bear' or statements like 'That bear is smaller than the other one' are then made true, within the game, or pretence, by, for example, the location of a tree stump or its size relative to another one. Thus, '[t]ogether, props and principles of generation make propositions *fictional*' (ibid.).

Now, to forestall the obvious objection that mathematicians and philosophers of mathematics, or philosophers of science, or philosophers in general, are not like kids playing in the woods—or at least, not when they are talking about numbers or propositions—it is important to note that according to this account, for it to be true for certain entities referred to in a given discourse to be fictional, it is not required that those involved in that discourse be actively or explicitly engaged in the pretence or game; it is enough that they are participants, even if unknowingly so (Walton 1993, pp. 403–11).

Having said that, it is also important to be clear what the discourse is in this case and where it 'lies' in the following sense: there are the discourses of scientists, mathematicians and 'lay'-people and there are those of philosophers, particularly, in this context, philosophers of science. The former might be considered the 'object' level and the latter the 'meta'-level, to borrow a distinction from philosophy of language (and again I'll come back to this distinction later). When it comes to numbers, for example, mathematicians' discourse about them can be regarded as at the object level and we would not expect them to admit to being actively engaged in a pretence with regard to such entities (unless they were philosophically reflective, which of course some are). Philosophers' discourse, on the other, can be considered to take place at the meta-level and here we would expect those who adopt a fictionalist philosophy to acknowledge that they are explicitly engaged in a game when talking of numbers. Thus, as far as a philosopher of mathematics is concerned when a mathematician asserts a claim about numbers that appears to commit her to belief in their existence, she is actually only 'simulating' such belief (Yablo 2001). Simulation at the meta-level, however, would be taken as bad faith—here we need to be engaged.

Likewise, when it comes to propositions, there is an object-level of discourse entered into by some laypeople, at least, but also significantly by psychologists and others, who construct, develop, and defend theories of empirical psychology according to which ascriptions of belief are used to explain particular kinds of behaviour (Balaguer 1998, p. 810). Thus, consider the sentence 'Steven believes that trades unions are a force for good', which is invoked to explain why Steven joins a trades union, becomes a union representative, and so on. Such sentences—Balaguer calls them 'that-sentences'—not only seem to express facts about Steven's belief states but appear to be indispensable to characterizing those facts (ibid.). Now, consider the Quinean indispensability argument, much discussed in the philosophy of mathematics: we should be committed to the existence of those entities that play an indispensable role in our theories; certain mathematical entities play such an indispensable role; therefore we should be committed to the existence of those mathematical entities. Extend this to propositions and we may conclude that we ought to believe in the existence of those entities, namely propositions, which play an appropriate role in the explanation of particular kinds of behaviour.[23]

However, this conclusion can be resisted by noting first the lack of causal efficacy of propositions, mentioned above. So, the above statement expresses that Steven stands in a relation of belief to the proposition that trades unions are a force for good. However, since this proposition is not causally relevant to Steven's belief (in the sense that qua proposition it does not cause or produce that belief), then if that statement is true, it is so in virtue of facts about Steven and the proposition that are 'entirely independent of one another' (ibid., p. 814). Given that, such statements and the empirical theories in which they figure—such as those to do with belief psychology—can be accommodated by any form of scientific realism which takes the nominalistic content of our best scientific theories to be true, or approximately so, while that which refers to such entities such as propositions to be false.

Why, then, invoke them? Well, consider the position that takes mathematical entities to play only an 'indexing' (Melia 2000) or representational role (Saatsi 2011), rather than an explanatory one.[24] Likewise, when it comes to the statement 'Steven believes that trades unions are a force for good', we might say that we characterize it in terms of propositions because this provides an easy and straightforward way of expressing Steven's state of belief:

> [M]ore generally, the point is that just as the empirical structure of temperature states can be represented by the mathematical structure of the real number line, so too, the empirical structure of belief states can be represented by the logico-linguistic structure of propositions. (Balaguer 1998, p. 819)

The bottom line then is that at the object level of scientific discourse, at which belief psychology and other theories operate, we should regard propositions as *representational aids*; or, more specifically, within the fictionalist framework, as representational aids within a pretence or game of fiction, much like that played

[23] One might have various objections to the extension of this argument but let's just put those to one side for now!

[24] For an alternative stance, see Baker and Colyvan (2011).

by the children in the woods.[25] The suggestion, then, is that this account can be extended to the meta-level at which philosophers and in particular philosophers of science consider propositions.

So, to flesh out this suggestion, we begin with a certain discourse and characterize the statements within that discourse as figurative, or more generally non-literal; an account is then given of such figurative statements in terms of pretences or games of make-believe (Eklund 2011). Then the suggestion would be that this same analysis can be applied at the meta-level *to the discourse of philosophers of science*. Thus, when philosophers of science, within the Syntactic Approach, characterize theories logico-linguistically as sets of propositions, they are actually engaged in a pretence or game of make-believe in which those *propositions act as representational aids in support of the aims of philosophers of science.*[26] Thus, suppose one of those aims is to present or demonstrate the logical relationship between different elements of a theory, then in the context of a rational reconstruction say, one might present those elements first as sentences in English, such as, for example: 'Heisenberg's uncertainty principle' and 'the non-commutativity of pairs of self-adjoint operators representing observables', with the claim being that the former follows from the latter. Now, given the worry mentioned above about translations into other languages, the proponent of the syntactic approach will insist that these are just standing for, or are presentations of, the associated propositions. Demonstration of the relationship between the two and discussion of that relationship would then appear to commit our friendly neighbourhood advocate to propositions qua abstract objects. But according to the fictionalist, these propositions are just props in a pretence, or more generally, representational aids which allow us to convey certain facts about the inter-relationships between different elements of the theory, in this case quantum mechanics.

However, there is an immediate worry here, expressed in the question: facts about *what*, again? I said that the propositions help to convey certain facts *about* the theory which invites the further question, what I am taking *the* theory to be and the possibility of circularity arises if I take it to be a set of propositions! We can perhaps assuage the worry and break the circle by recalling the core nominalistic claim in play here: ultimately all there is are the physical entities of the 'empirical' world, with no abstract entities, whether numbers, propositions, or whatever. Those entities, their properties, etc. stand in certain relations[27] and these relations are represented at the object level of the discourse of science itself by the relevant statements and claims, some of which may be mathematical (suit-

[25] As Yablo notes (2001), such aids or props can also function as that which is represented. So, propositions function as representational aids at the object level of psychology, say, but can then be represented as entities by other representational aids, or further propositions even, at the meta-level of philosophical reflection on the claims and statements of psychology.

[26] I am leery of talking of *the* aim of the philosophy of science, for obvious reasons.

[27] Or, if one is of a structuralist persuasion, are constituted by certain relations.

ably interpreted of course), presented in 'the' theory as written down and spoken of in papers, presentations, etc. This theory is then itself represented at the meta-level at which the discourse of philosophers of science takes place in terms of sets of propositions. Within the fictionalist framework, these enable the philosopher of science to express and convey facts 'about' the theory where the latter now has to be understood as whatever is written down and spoken of at the object level of scientific discourse—that is at the level of scientific practice. Ultimately, then, there are only the entities and relations (or just the latter) in the physical world itself, and multiple levels of discourse and presentation/representation of and about those facts. And the fictionalist of course takes those discourses all to be games of make-believe.

I'll return to elaborate further on this sort of picture later in the book but there are two more things to note. The first is that it had better not be the case that claims like 'quantum mechanics is empirically adequate' come out as fictional—what fictionalism is supposed to be is an antirealist view about *theories*, qua entities, not a radical form of antirealism about the physical entities which the theory is supposed to be about. Again, this sort of concern can be assuaged by noting that the relevant proposition is just a convenient aid for understanding scientific practice—granted this will mean cashing out 'empirical adequacy' in terms of that practice—and this will of course have all the relevant features (experimental etc.) that justify such a claim. Things are different when it comes to the claim 'quantum mechanics is true' since in this case philosophers of science will disagree, perhaps vehemently, as to whether such a claim is justifiable or not. Certainly constructive empiricists, for example (see van Fraassen 1980; 2002), will argue that there is nothing in scientific practice itself that supports it and that it can only be asserted on the basis of a further inferential move—typically encapsulated in the infamous no miracles argument (see Chakravartty 2017)—that is itself hugely contentious. Here the empiricist would urge a shift in attitude towards the relevant propositions, from belief that they are true, to acceptance that they are empirically adequate. I will come back to this concern.

Secondly, note the role of practice in the above account. The sets of statements with which the advocate of the Syntactic Approach identifies a theory at the meta-level of the philosophy of science are manifestations or, better perhaps, presentations of the associated propositions which in turn, according to the fictionalist, are merely representational aids for conveying the relevant facts about theories *at the object level of scientific practice*. We will return to the role of scientific practice in this context in subsequent chapters.

Now, this is all well and good—adopting a fictionalist line seems to assuage the concern that by identifying theories with propositions, we are required to accept abstract entities. But of course, fictionalism itself is not without its issues and problems.

An obvious objection but one that is equally easy to dispose of is that when philosophers talk about theories within the Syntactic Approach, they are not explicitly engaging in a pretence or an imaginative game. However, as already noted above, it is not a requirement of this approach that the commitment to a pretence be explicit or deliberate. The point is that on further reflection our commitment as philosophers of science to propositions is revealed to be in some sense weaker than we thought (Yablo 2001). The reflective move might be compared to that urged by constructive empiricism where instead of believing theories/propositions to be true, we should accept them as empirically adequate.

A more worrisome objection is that fictionalism about propositions entails that all our belief ascriptions are strictly false (insofar as such ascriptions can be rendered as 'belief that *p*' where *p* is some proposition). As McGrath (2014) notes, that seems a bitter pill to swallow. It can be sweetened somewhat by, first, accepting that when it comes to communication, the falsehood of our belief ascriptions is neither here nor there. Think of numbers, again, for example (Yablo 2001): the Platonist believes they exist, the nominalist doesn't but both can agree that numbers are to be conceived as an omega sequence generated from 0 by the successive application of +1 (ibid.). Likewise, a non-fictionalist advocate of the Syntactic Approach and her fictionalist comrade in arms can both agree that the theory of quantum mechanics has certain features but one insists that the propositions with which it is identified are 'really there' in some sense, as abstract entities, say, while the other insists that they are not. That difference in insistence is no obstacle to communication.

Of course, both the Platonist about numbers and the realist about propositions may further insist that this is only true as long as that communication is not about the *existence* of numbers or propositions respectively. As soon as we touch on that topic, disagreement roars back into the frame. One way of responding might be to note that when it comes to numbers, we were actually talking about agreement in communication at the object level of mathematics itself and the disagreement over existence only arises at the meta-level of philosophical reflection. Obviously the presence of philosophically reflective mathematicians muddies the waters somewhat but one might still sieve their pronouncements and insist on that distinction. Likewise, when it comes to propositions, one might maintain that in practice we are all able to communicate our beliefs to one another more or less successfully without stumbling over the question whether propositions actually exist or not. Again, that question only becomes an obstacle at the level of philosophical discourse.

However, when it comes to *theories*, that is precisely the level we are operating at and the discourse we are interested in. It matters little to the current set of issues that scientists themselves are able to communicate with one another about various theories, their myriad qualities and defects, etc. What we are concerned with is how to understand the Syntactic Approach and whether its apparent commitment to abstract entities can be massaged away. At this meta-level the pill of false belief ascriptions is not sweetened one iota by pointing out that scientists can still talk to one another.

A second sweetener might be appealed to in the form of *figuralism*, a variant of fictionalism that notes that '[w]e are people apt on occasion to speak figuratively' (Yablo 2001, p. 85). Indeed, it may be argued that it is literal speech that is the exception (ibid.) and that a great deal of what fictionalists are trying to explain or massage away is already present and dealt with in figurative speech. And again, the terms deployed in such speech serve as representational aids meant to convey certain features or properties of concrete facts. We do not need to take such aids as referential themselves.

Still, qualms remain. One might agree, against the Platonist, that talk of numbers should be understood figuratively and that numbers themselves, indeed all of mathematics, should be understood as representational aids (albeit of vast and impressive power). Yet to say this of propositions means that not just one corner or niche within philosophy—namely the philosophy of mathematics—has to be given a figuralist gloss but huge swathes, if not the entirety, of philosophy have to be so understood—indeed, any aspect of it that involves or touches on belief ascriptions.

Now, one might counter-respond: such is the cost of avoiding commitment to abstract entities. All revolutions come at a price and in this particular case, that price is the realization that when we talk about theories we are actually engaging in a kind of game. Perhaps it is that word 'actually' that sticks in the throat, for what the fictionalist is saying to us is that we might think we are talking literally about propositions and hence theories but in fact (in some sense of fact)[28] we are not—we are and were all along only talking figuratively. Still, one can see how this might be easier for the old school logical empiricist to swallow, since she is already committed to a non-literal understanding of the object level talk of scientists—e.g. when they talk of (putative) unobservable entities, this is to be understood or translated away as talk of (long) strings of observable experimental outcomes. But as already noted, one of the things we have learned in the philosophy of science is that such parsing away is problematic, not least because it sets the philosophy of science at significant remove from the practice of science itself. And that insistence that scientists' claims do not mean what they appear to mean and must be translated into other (famously, empiricist) terms is partly to blame for the downfall of logical empiricism and positivism more generally. In setting herself up as the natural successor to the logical empiricists, the constructive empiricist rejects

[28] One such sense might be in terms of the relevant psychological or cognitive processes. Thus, Stanley has argued that it is not the case that the mechanisms involved in understanding the use of a discourse of which fictionalism is true (e.g. regarding propositions) are the same as those involved in games of pretence (Stanley 2001). Autistic people may have no problems with mathematical or modal discourse, nor with philosophical discourse in general, but may have problems with make-believe.

such insistence and maintains that scientists' talk *should* be taken literally.[29] Thus, we might also maintain that our talk as philosophers of science about the practice of scientists should also be taken literally and not have a fictionalist or figuralist gloss slapped over it.

Eliminativism (First Pass)

Can we achieve the same end as the fictionalist—namely avoidance of abstract entities—but without the cost of having to impose a non-literal understanding of scientists' own talk? Let's consider again the example of mathematical entities: an alternative to fictionalism about numbers, say, is structuralism, which holds, broadly speaking, that such mathematical entities are, at best, merely *placeholders* in the relevant mathematical structure (for an excellent survey, see Reck and Price 2000). Of course, this in itself does not avoid the problem of abstractness, since structures themselves, and their inter-related places, may be regarded as abstract and instantiated in various systems. The obvious move then is to nominalize structuralism and insist that there are just the relevant (concrete) systems, which are structured.

Now, there is a lot more to be said about mathematical structuralism (see for example Horsten 2014) but we can immediately see the difficulty in extending structuralism to propositions: we just don't have robust and easily identifiable structures in terms of which propositions can be regarded as placeholders.[30] Nevertheless we might take a leaf out of the structuralists' book and press for a kind of *eliminativism* in this context. Consider: structuralism has been extended into the physical domain with the claim that all that there *is*, in the physical world, is structure. Moderate structural realists maintain—like their counterparts in the philosophy of mathematics—that objects are mere placeholders in this structure and are contextually individuated by it. Radical or eliminativist structural realists, on the other hand, argue that objects can be eliminated altogether, leaving only the relevant structure as the focus of our ontological commitments (French 2014). Importing such a position into the current context would still have to face the issue of what would be the relevant structure but as we shall see, we can still appropriate the associated metaphysical manoeuvres that underpin eliminativism, at least when it comes to theories.[31]

[29] Although this is at the price of introducing the distinction between acceptance and belief, with the concomitant line drawn between empirical adequacy and truth.

[30] Although for an attempt to extend structuralism in this sense to language, see Meier, forthcoming.

[31] Obviously in this case the focus of our ontological commitments cannot be any relevant structure so this form of eliminativism will be even more radical than that presented by structural realism. What we will have, of course, is the relevant *discourse* and again, as we shall see, that will play a crucial role in allowing for eliminativism in this context.

There is a sense in which we have been here and done that: Quine famously rejected the existence of propositions on the grounds that we could not identify their identity conditions or come up with an appropriate standard of individuation in the sense of 'some standard of when to speak of propositions as identical and when as distinct' (Quine 1960, p. 200). His argument that no such standard could be given was that such a standard had to be articulated in terms of the synonymy relation for sentences, which in turn had to be understood in terms of dispositions to verbal behaviour but that move falls foul of his thesis of indeterminacy of radical translation (see, again, Hylton 2016):

> The totality of dispositions to speech behavior is compatible with alternative systems of sentence-to-sentence translations so unlike one another that translations of a standing sentence under two such systems can even differ in truth value.
> (Quine 1960, p. 207)

Although I am going to resist going further down that particular rabbit hole—and we have already noted that Quine's account of propositions will not help us here—he is of course not alone in voicing concerns about the identity conditions for propositions.

Thus, Woodridge, for example, notes that different linguistic, philosophical, and psychological practices pull us in different directions with regard to the nature of propositions and that although many competing accounts have been offered, none appears capable of accommodating all such practices in a fully satisfactory manner (Woodridge 2006, p. 345).[32] Indeed, as Bogdan (1986, p. 9) put it, philosophers have constructed such accounts with 'anarchic gusto'. Now, I'll come back to this issue shortly, but at this point I just want to note Woodridge's insistence that it is because there is a lack of philosophical consensus over the nature of propositions and a wide variety of theories about them, that we should conclude that *they have no determinate identity conditions*. As we'll see, Thomasson has suggested a similar move with regard to art objects and we can adapt and adopt such a manoeuvre with respect to scientific theories.

If you then agree with Quine's dictum of 'No entity without identity' (which is contentious of course),[33] then you should conclude that there are no such things

[32] One might be tempted to abandon propositions per se for something else entirely. Thus, Sperber (1975) suggests that the kinds of belief reports one finds in anthropological practice can be accommodated by replacing certain propositions with 'semi-propositional representations' that capture the vagueness and partiality inherent in such reports. An attempt at formalizing such a notion in terms of quasi-truth and in the context of a distinction between belief and acceptance can be found in da Costa and French 2003. Obviously one could envisage a version of the syntactic approach that incorporates this sort of approach as well.

[33] It is certainly so when it comes to (putative) physical objects in the context of quantum mechanics (see French and Krause 2006) but it is difficult to discern similar grounds for rejecting it when it comes to the likes of propositions and theories. After all, in the case of the former we have

as propositions (Woodridge 2006, pp. 345–6 and 348).[34] The issue now is to make sense of proposition *talk* in that case. Here the fictionalist raises her hand again. Within such a framework any indeterminacy in identity conditions generated by our practices can be made to disappear by virtue, for example, of insisting that the different objects and properties that the relevant expressions contribute to a proposition within the game of make-believe are themselves pretend entities (ibid., pp. 352–3).[35] Thus, on this account, we can accommodate cases such as

> Steven believes that the artist who painted *Catalan Peasant in the Light of the Moon* is brilliant but does not believe that the artist who painted *The Hunter (Catalan Landscape)* is brilliant.

and we do so by introducing different pretend objects picked out by the different expressions ('that the artist who painted…') and which function as constituents of different propositions. In effect the pretence removes any indeterminacy with regard to identity at the level of propositions by introducing a plethora of fictional entities in terms of which the relevant propositions can be distinguished.

Of course, these are all pretend objects, so the fictionalist would be sanguine about possible ontological inflation. Nevertheless, our games of make-believe have now become much more complex and involved and one might wonder whether we can't take the point about our practices indicating a fundamental indeterminacy in identity conditions and simply accept the resultant eliminativism as it is.

The question then is: how are scientific theories to be conceived on such a view? For Quine, the answer is: as presented and appropriately regimented within a language that by virtue of that regimentation goes beyond the 'natural' language of the scientist (Quine 1957, pp. 7–8). But as we've already seen, the same old problem comes back to bite us: if we identify a theory with a particular regimentation then Newtonian mechanics in regimented English will strictly not be the same theory as Newtonian mechanics in regimented Portuguese. An obvious response would be to insist that there is in this case, and will be in general, straightforward translations between these regimentations. So on that

experimental or causally mediated access to such objects of a kind that allows us to affirm their existence even if we grant their lack of such identity conditions; whereas on the case of the latter, we precisely have no such access and can only infer their putative existence to begin with on the basis of the very practices that pull us in different directions.

[34] One might be tempted to go down the pluralist route and accept different kinds of propositions but this runs the risk of undermining the standard connection between meaning and truth (Woodridge 2006, p. 348).

[35] This is to adopt, *within the pretence*, a Russellian account of propositions according to which belief reports are understood as relating the holder of the belief to a pretend complex of pretend objects, properties, and relations so that that-clauses involving different objects denote different propositions (Woodridge 2006, p. 352).

basis one could again identify the theory with an equivalence class of all such inter-translatable regimentations.

Of course, Quine himself would have problems with that response, given his thesis of the indeterminacy of translation (and so the rabbit hole yawns wide again! See Quine 1960, p. 27). Even granted that we should only read off any ontological implications once the translations have been appropriately regimented, insofar as that regimentation is still going to include interpretations of the relevant terms, those implications may differ dramatically. And of course, to go beyond such regimented translations and reify (what one claims is) the underlying theory might be regarded as just as problematic as reifying extra-linguistic meaning.[36] One might try to wriggle out of this by emphasizing that the thesis applies only to 'radical' translation, in the sense of that of the language of a hitherto untouched people but that seems a desperate move and open to the objection that we would still want to say that Newton mechanics in Kepler186f-ian is the same theory as our old friend.[37]

Direct Representation

However, there is a further concern regarding this emphasis on the linguistic explication of theories—that some regard as the final nail in its coffin—which is that it encourages the thought that 'a good theory, viewed as a collection of statements, directly represents the world in that the world (or a certain domain) directly satisfies the theory (i.e. makes it true or empirically adequate or what have you)' (Hendry and Psillos 2007, p. 135; see also Chakravartty 2001). Such a thought is of course hopelessly naïve for at least two reasons: first, because between the theory and the world (or some domain thereof) there will typically exist a whole range and variety of mediating steps, some involving the kinds of models already touched on; and secondly, in order to bring the theory into contact with the world, it will typically need to be adjusted, reconfigured, or generally beaten into shape in one way or another. The first reason has been expressed in terms of the need for articulating the nature and role of models of the data, models of the experimental set-up, models of the phenomena, and so forth, all the way up to models of the theory, and all of which 'mediate' between 'the' theory and the world (perhaps in different ways). The second has to do with the myriad

[36] I owe this point to Robbie Williams, who is not of course to be held responsible for what I've done with it.

[37] More reasonably perhaps, one might strive to resist the indeterminancy by rejecting Quine's underlying behaviouristic tendencies and insist that it is not the case that 'All the objective data [the translator] has to go on are the forces that he sees impinging on the native's surfaces and the observable behavior, vocal and otherwise, of the native' (Quine 1960, p. 28; so for a survey of the relevant issues, see Gaudet 2006).

forms of idealization, simplifications, and approximations, etc. that have to be introduced to make the theory more tractable, computationally speaking, easier to 'handle' in some sense, or just generally more amenable to being brought into contact with the phenomena.

Now, a proponent of the Syntactic Approach might simply respond by insisting that only the already naïve would be encouraged by the linguistic explication to adopt such a naïve thought! You don't have to be an advocate of the alternative to recognize, on the basis of either the history of science or close observation of actual scientific practice, that theories do not directly represent the world. Indeed, it is a mainstay of the 'deductive-nomological' account of the relationship between theories and phenomena—according to which a statement describing the relevant phenomenon is deduced from statements describing the relevant laws or hypotheses of the theory—that 'auxiliary hypotheses' must be appended to the laws or hypotheses of the theory in order to obtain something that applies to the specific situation in hand. These auxiliary hypotheses may include further laws, including laws pertaining to the instruments used in observing or measuring the phenomenon, as well as statements about the particular circumstances regarding the latter, and about the devices used, etc. Furthermore, it is a well-known issue with regard to theory confirmation or falsification that empirical support or the lack thereof can be diverted by such hypotheses away from the theoretical claim at issue (this is the so-called Duhem-Quine problem, of course). Thus, our proponent can once again chant 'Anything you can do...!' and simply replicate in linguistic terms the mediation between theory and the world mentioned above.

Likewise, she can accommodate idealization and approximation by introducing the appropriate statements into the set from which statements describing the phenomena are deduced, with the help of the above auxiliary hypotheses. Indeed, much of the initial discussion of the role of idealization in science took place in the context of the Syntactic Approach and there doesn't seem anything in principle to prevent such an accommodation. In particular, if one accepts that models as 'mini-theories' can be generated by reinterpreting the relevant terms in appropriate ways—as in the example of the billiard ball model of the ideal gas—then the Syntactic Approach doesn't appear to be facing any in-principle obstacle here.

Still, there is the further point that as well as acting as mediators, models have a certain autonomy that sets of sentences may not be able to capture (http://itisonlyatheory.blogspot.co.uk/2009/01/what-was-wrong-with-syntactic-view-of.html). However, one must be careful not to over-hype this idea, not least because such apparent autonomy will be in tension with any mediating role: on the one hand, if the model or whatever is autonomous from the relevant theory in the sense that there is no relationship, deductive or otherwise, between the model and that theory then it is difficult to see how the model can play any role in confirmation, falsification, or indeed 'mediate' at all; on the other, if all that is meant is functional autonomy in the sense that the model or whatever becomes the focus of epistemic attention (and we'll come back to this shortly), then it is hard to see why a (sub-)set of sentences couldn't perform such a role.

Conclusion: Where Do We Stand?

I began by outlining the so-called Syntactic Approach which takes a theory to be a set of logico-linguistic propositions, closed under classical logic. I then followed Hendry and Psillos in distinguishing 'strong' and 'weak' versions: according to the former, theories are to be identified with formal entities, whereas according to the latter, they are taken to be collections of statements, understood literally. The strong form runs into issues to do with the formalization involved and specifically whether this can capture the relevant theoretical practices. The weak version, on the other hand, finds itself re-expressed in terms of propositions and the issue of the ontological status of theories is then pushed back a step. One way of avoiding further commitment to abstract entities in this case is to adopt fictionalism, although that comes at a cost (but then, doesn't everything?). Alternatively one could avoid paying the cost by taking an eliminativist stance, which in turn requires some fancy footwork when it comes to our talk 'about' theories.

My aim here was not so much to argue against the Syntactic Approach, not least because it seems to have the resources to respond to the more well-known objections. Rather, I wanted to rehearse certain moves that we'll come across again in later chapters and the point is, however you initially characterize scientific theories, whether as sets of propositions or, as we'll see next, families of models, certain ontological commitments need to be faced up to. In the next chapter I will consider the alternative Semantic, or model theoretic, approach and as we shall see, fictionalist accounts have been advocated in this context as well. After spelling out the details, I shall again suggest that one can retain the benefits of such accounts while avoiding their costs by adopting a form of eliminativism towards theories.

2

Theories as Families of Models

Introduction

Although the Syntactic Approach may have the resources to resist the principal criticisms that have been levelled against it, many philosophers of science have now edged away from that view, at least as it is typically characterized. The precise reasons for this shift can be articulated in the context of the history of philosophy of science (see Suppe 1989). However, one principal cause was that just as many felt that the logical positivists did not pay sufficient attention to actual scientific practice (even though that has now been revealed to be something of a caricature), so they saw the Syntactic Approach as emphasizing the kind of formal analyses that just don't feature in that practice. Certainly, as already noted, within that practice itself we typically do not find axiomatizations of the kind mentioned in the previous chapter: again, what are called 'axioms' in a textbook on quantum mechanics, say, just do not look anything like the kind of thing we expect from our logic courses, nor do they play the same role (van Fraassen 1980, p. 65). And to denigrate this point and the previous criticisms in general as merely concerned with issues of *practicality* is to precisely miss the point: although the disparity between scientific practice and what is typically captured syntactically, 'will not affect philosophical points which hinge only on what is possible "in principle", it may certainly affect the real possibility of understanding and clarification' (van Fraassen 1989, p. 211). As is now well known, it is sentiments such as this which motivated the move to the alternative 'Semantic' or 'model theoretic' approach,[1] at the core of which is the claim that, 'theories are not collections of propositions or statements, but rather are extra-linguistic entities which may be described or characterized by a number of different linguistic formulations' (Suppe 1974, p. 221). The 'are' here is to be understood as that of identity—theories are to be *identified* with extra-linguistic entities (and we'll come back to the issue of what those

[1] Identifying 'fellow travellers' in this approach is itself an interesting issue. Take Kuhn for example, who, on the occasion of his handing over the PSA's presidential baton to van Fraassen, talked of a 'shared model-theoretic element' in his and van Fraassen's work (Kuhn 1993; and as he also noted, various features of his framework were subsequently given a formalized treatment in the work of the (set-theoretical) structuralists such as Sneed, Stegmuller, and others; see Schmidt 2014). Here we seem to have a nice reflexive example of the phenomenon Kuhn himself described (in his addendum to Kuhn 1978, 'Revisiting Planck'), where a scientist misremembers his own history to highlight aspects of his work that have acquired greater significance from today's perspective (a phenomenon that meshes with our discussion in Chapter 8).

There Are No Such Things as Theories. Steven French, Oxford University Press (2020). © Steven French.
DOI: 10.1093/oso/9780198848158.001.0001

entities are supposed to be).[2] Thomson-Jones offers a similar characterization through what he explicitly calls the 'Identification' claim: '*I*. A scientific theory is a collection of models', where such models are taken to be 'objects of a certain sort' (Thomson-Jones 2006, p. 529).[3] This, I would argue, is how the approach is typically framed—that is, via the *identification* of theories with models—and in these terms it has been claimed that, '[w]ithin a few short decades, the Semantic Approach has established itself as the new orthodoxy' (Halvorson 2012).[4]

Now, just as with the Syntactic Approach, there is some debate as how to delineate this approach or what positions might be included within it. Typically the above core claim is cashed out in terms of taking theories to be classes of set-theoretical models and characterized by what their linguistic formulations refer to when the latter are interpreted semantically, in terms of these models. It is in this sense that theories may then be understood as extra-linguistic and in turn it may then be claimed that according to this approach theories *are* nothing more than families of such mathematical models. In particular, what the Syntactic Approach formalized as the 'axioms' of the theory is now understood as serving to pick out the relevant models (by virtue of the fact that the axioms are true in those models). On the contrary, in order to present a theory on the Semantic Approach, the relevant class of models is defined directly.

Furthermore, these mathematical models, and hence scientific theories according to this view, are to be characterized (and here I am again deliberately trying to choose my words carefully) in terms of *structures*, of some kind, to be presented formally, in an appropriate fashion (Ladyman and French 1999, p. 105). Now, as is well known, the general notion of 'structure' has undergone extensive development in modern mathematics, culminating (for many) in the (essentially syntactic)[5] treatment of Bourbaki (1968).[6] Here one begins with a finite number of 'principal base sets' which effectively act as building blocks from which one can construct, by taking the Cartesian product, or forming the power set, for example, 'auxiliary' sets. This combination, of base sets, auxiliary sets, and a construction scheme, characterizes a 'species of structure' which can then be defined via an axiom yielding the relation satisfied by the specific construction used and the relevant sets. This relation is said to be 'transportable' in the sense that its

[2] 'The semantic conception is a view about the objects that are scientific theories' (Suárez 2011, p. 432).

[3] Interestingly, he also offers a 'less ambitious' methodological recommendation: '*M*. A scientific theory is best thought of as a collection of models'.

[4] Cf. also Frigg (2006, p. 51) and LeBihan (2012) who also refer to the Semantic Approach as the orthodox view of theories and mdoels. Here they are echoing Suppe, who made a similar claim some years ago: 'The Semantic Conception of Theories today probably is the philosophical analysis of the nature of theories most widely held among philosophers of science' (Suppe 1989, p. 3).

[5] But also see recent attempts to reconcile the syntactic and set-theoretic aspects in the context of the philosophy of science: Andreas 2014; Schurz 2014.

[6] For a useful survey of the notion of structure in mathematics, see Mac Lane 1996.

definition is not dependent on the specific natures of the construction or the sets themselves (see, for example, Corry 1992).[7]

As da Costa and Chuaqui (1988) noted, Bourbaki's idea of species of structures can be identified with Suppes's set-theoretical predicates, thus providing the bridge to standard (mathematical) model theory. Via such a bridge, Bourbaki's treatment had a significant impact in particular on the 'structuralist' school in philosophy of science, associated with Balzer, Moulines, Sneed, and Stegmüller, which explicitly attempts a rational reconstruction of scientific theories using Bourbaki's idea to achieve 'a kind of formalization different from that already employed by scientific theories themselves' (Schmidt 2014).[8] Note that if we take Bourbaki's notion as sitting at the foundations of mathematics then this formalization still meets Suppes's dictum that philosophers of science should use mathematics not meta-mathematics but by virtue of rationally reconstructing theories in set-theoretic terms, the structuralist formalization will clearly differ from what scientists do themselves. Here, then, we see a nicely articulated expression of the distinction between how scientists *present* their theories and how philosophers of science should *represent* them.

Of course, the issue then is what are the merits or otherwise of such a (meta-level) representation and many have balked at what is, for them, the overly formal nature of the structuralists' rational reconstruction.[9] Here I shall simply note (see da Costa and French 2003) that in these terms, to axiomatize a theory is to define a set-theoretical predicate and the structures which satisfy this predicate are to be regarded as the models of the theory (I'll come back shortly to the issue whether these formal structures should be so regarded).[10] Such structures (of first-order) can be characterized as follows:

$$\mathfrak{A} = \langle A, R_j, f_j, a_k \rangle_{\substack{i \in I \\ j \in J \\ k \in K}}$$

[7] Corry adopts a somewhat jaundiced stance towards the Bourbaki treatment, pointing out that for all the insistence on rigour, much of the development of this notion in the relevant part of the Bourbaki corpus proceeds in a less than entirely formal and actually rather ad hoc manner and, furthermore, it, and the associated concepts, are used only in a limited 'and certainly not highly illuminating or unifying' manner throughout the rest of the treatise (Corry 1992, p. 324).

[8] A useful survey and updated analysis of the structuralist school from A to Z can be found in Andreas and Zenker 2014.

[9] I suspect this is why they have not received the attention they and their supporters insist they deserve from Anglo-American philosophers of science (see the comment in Schmidt 2014). My own personal balking moment occurred when a member of the structuralist school attempted to accommodate the social aspects of science by stating 'Let S be the set of scientists . . .'. More seriously, one can adopt a pluralist or take-it-or-leave-it approach to such formalization, deploying it where one thinks it might be useful or illuminating and relying on less formal approaches—such as indicated here and developed in da Costa and French 2003—where one's aims do not call for such pitiless formal detail.

[10] 'Models are structures, usually mathematical structures' (van Fraassen 1993, p. 28).

where A is a non-empty set, R_i, $_{i \in I}$, is a family of relations, f_j, $_{j \in J}$, is a family of functions and a_k, $_{k \in K}$, is a family of distinguished individuals of A.[11]

Linguistic Independence

Insofar as this is claimed to be an account of theories that presents them as independent of any particular linguistic formulation, it may seem that it simply sidesteps the problems that I sketched in the previous chapter. However, a tension arises between this much vaunted claim and the characterization of theories in model-theoretic terms, one that runs through recent discussions of the adequacy of the Semantic Approach.

The tension lies in the fact that a '*model-theoretic* model', as usually understood, is a structure plus—crucially—an *interpretation* of a formal language in terms of that structure (that is, a map from the symbols of the syntax to elements of the structure; see for example Lutz 2014, pp. 1476–9). Now, this is enough to define an *L-interpretation*, U, for the language L, but such an interpretation is a model for a set of sentences Σ within L only if we add to it a *valuation* such that under this valuation, all the sentences of Σ are satisfied in U (see for example Bell and Machover 1977). Given *this* definition, it seems clear that the supposed (and much celebrated) linguistic independence of models cannot be true.

The point is well put as follows:

> to consider them [the models] in this setting [model theory], a particular formalization of the language…is presupposed.…if we do not have a formalized theory, how could we think of models in any reasonable sense, logical sense? Indeed, what really are these entities, the models?
>
> …from a logical point of view, it is not possible just to consider the models, and leave away the axioms—for the models are the models of these axioms!
>
> (Bueno, private email)

How to relieve this tension? One option is to try to draw back from the above apparently straightforward identification between structures and the models of the theory and take the definition of a structure to be part of the compound entity that is a model, but not all of it. Thus, when van Fraassen writes,

[11] By defining a 'simple pragmatic structure' in these terms, a form of 'pragmatic' or 'partial' truth can then be constructed, in which the simple pragmatic structures perform an analogous role to the set-theoretic structures in Tarski's theory (Mikenberg, da Costa, and Chuaqui 1986). For further consideration, see da Costa and French 2003.

> The impact of Suppes's innovation [switching to models] is lost if models are defined, as in many standard logic texts, to be partially linguistic entities, each yoked to a particular syntax. In my terminology here the models are mathematical structures, called models of a given theory only by virtue of belonging to the class defined to be the models of that theory
>
> (van Fraassen 1989, p. 366, n. 4 of ch. 8; cf. van Fraassen 1990, p. 483, n. 2 of ch. 1)

he should be interpreted as talking about *structures* by those who understand model in the sense of the 'standard logic texts'. It is then possible to make use of the formal machinery of isomorphism etc. to describe the relations theories have to each other and to the phenomena and so on. Hence, it can be argued, the tension dissipates: we can accept that the use of the term 'model' in the Semantic Approach is at least intelligible, and is not so far from its use in model theory, but we can read it to mean 'structure' if we are troubled by it (French and Ladyman 1999).

However, even if a structure is described independently of any sentences that are true in this structure, the point can be pushed that there is a further dependence on language, in that it still involves a mapping from a *vocabulary* to sets. Thus, Lutz has argued that although the formal characterization given above—that is,

$$\mathfrak{U} = \langle A, R_j, f_j, a_k \rangle_{\substack{i \in I \\ j \in J \\ k \in K}}$$

—in terms of an *indexed set I*, is language independent in the sense that any appropriate vocabulary can be used with the structure, the indexing effectively reintroduces a vocabulary by the back door (2014, p. 1480). Such a structure contains an indexed set of extensions of the relations $\{R_i,\}_i \in_I$ and the mapping from the I to the set of extensions is indistinguishable from an interpretation with a vocabulary given by I.[12]

Instead Lutz proposes a definition of 'pure structure' in which mappings obtain between arbitrary indexed sets and extensions (2014, p. 1481). That the indexed sets are not fixed is obtained by identifying any two mappings that differ only in their index sets. Here no specific vocabulary is introduced and we thus obtain a formal representation of language independence,[13] 'because the simultaneous

[12] One could insist that I is not a vocabulary, at least not in the sense that renders the structure language dependent, but Lutz insists this would mean appealing to factors beyond the formalism.

[13] The cost is that notions such as embedding must be redefined but this seems a small price to pay (Lutz 2014, pp. 1481–4).

switch of all structures to pure structures allows the choice between an infinity of vocabularies, and thus the freedom from any specific one' (2014, p. 1484).[14]

Unfortunately however this freedom may be illusory as Lutz's scheme replaces the index set by an order relation, that does the same work as the former, and so it seems we're back where we started.[15] Perhaps, then, we cannot escape this form of language dependence either. Indeed, Hudetz has argued if we allow higher-order logics in this context, then models that can be construed as set-theoretic structures can equally well be construed as model-theoretic (Tarksian) structures of such a logic (Hudetz 2017). In other words, for any set-theoretic structure there will be a 'naturally corresponding' (ibid.) higher-order logic structure as its model-theoretic counterpart.

This represents the latest iteration in a long-running debate: at the very beginning of what is now portrayed as the Semantic hegemony, it was argued that if the class of models in terms of which a given theory is characterized is 'elementary' in the sense that it contains the models of a set of sentences formalized in first-order (classical) logic, then the completeness theorem ensures the equivalence of the Semantic Approach, thus presented, and its syntactic rival (Friedman 1982, p. 276; Worrall 1984).[16] An obvious response is to deny that the class of models need be elementary and further to insist that in the case of models 'of' or characterizing scientific theories, it typically will not be insofar as such theories involve the real number continuum (van Fraassen 1985a, pp. 301–2; 1989, p. 211). Given that such a characterization will lead to non-standard models (according to the Lowenheim-Skolem theorem) we will not be able to characterize the theory in terms of a first-order axiomatic system.[17]

From the perspective of the Semantic Approach we can always exclude such non-standard models from consideration by stipulating the use of an intended model but that manoeuvre (of effectively pointing to or describing the model one intends) is obviously not available to the proponent of the Syntactic Approach

[14] Lutz then goes on to argue that Syntactic Approaches can also achieve this independence from particular vocabularies by characterizing theories in terms of equivalence classes of sentences (2014, pp. 1484ff.).

[15] I am grateful to one of the readers for pointing this out and for urging me to clarify this discussion more generally. As Hudetz (2017) also argues, Lutz's approach is limited by virtue of being confined to first-order structures, whereas scientific practice makes use of more complicated set-theoretic structures.

[16] Friedman phrases the sense of the class of models being elementary thus: 'it contains precisely the models of some first-order theory *T*' (1982), raising the issue as to what we're supposed to be taking as 'the' theory.

[17] Or, at least, not categorically, in the sense of confining that characterization to models that are isomorphic to one another. See also Suppes 1967, p. 58) who writes that when considering theories of physics that involve real numbers, as well as those of pure mathematics, 'it is very much simpler to assert things about models of the theory rather than to talk directly and explicitly about the sentences of the theory, perhaps the main reason for this being that the notion of a sentence of the theory is not well defined when the theory is not given in standard formalization.'

(van Fraassen 1989, p. 211; see also Suppe 2000). Nevertheless, she has a range of options at her disposal (see for example Lutz 2014, pp. 1478–9), the most obvious being to shift to a higher-order logic (in which the Löwenhein-Skolem theorem does not hold) as already mentioned.

More interestingly, from my point of view, such an advocate can ask what it is, precisely, that we are trying to do with these characterizations: if it is to do with answering certain questions that can be phrased within the language of our first-order framework—and one would expect that any question that we as philosophers of science would be interested in could be so phrased—then any answer that could be given in terms of the Semantic Approach, could also be given within the framework of the alternative. Thus, consider the question whether a theory should be regarded as empirically adequate or not. As is very well known, van Fraassen explicated this in terms of the formal notion of one structure being 'embedded' in another and claimed that this notion could not be captured syntactically. However, analogous syntactic notions—e.g. of 'implantability'—*can* be defined (see Turney 1990; also Lutz 2014 pp. 1481–4) on the basis of which one can obtain syntactic equivalents of the notions of empirical adequacy and empirical equivalence.[18] And in general, the proponent of the Syntactic Approach can respond to the claim that 'descriptions of structure in terms of satisfaction of sentences is much less informative than direct mathematical description' (van Fraassen 1989, p. 212) by insisting that formally, at least, anything the Semantic Approach can do, its Syntactic rival can do also (see the discussion in Lutz 2014, pp. 1476–8 and also Hudetz 2017). Now, that's not necessarily a problem for me, since I want to argue that neither should be taken as providing the means by which theories might be said to be constituted, or the conditions in terms of which they could be identified. Indeed, Hudetz's result sheds a helpful light on a recent further debate regarding that very issue.

Capturing the Identity of Theories

The exchange begins with the claim that the Semantic Approach 'makes incorrect pronouncements about the identity of theories' (Halvorson 2012, p. 184), since it identifies theories that are distinct and distinguishes theories that are identical.

Thus, consider the question: given two classes of models, M and M', under what conditions would an adherent of the Semantic Approach say that they characterize the same theory? One, perhaps obvious, answer would be: when they are

[18] Nevertheless, and in the context of the point I'm about to make, Turney agrees that the Semantic, or model-theoretic, account is an improvement over the Syntactic approach, on the grounds that it is better able to represent the relationship between scientific theories and observations.

isomorphic (Suppe 2001, p. 526). Now, there are different ways of spelling out what it is for two *classes* of models to be isomorphic. One is to treat the classes as sets themselves and to then suggest that M and M' are isomorphic if there is a bijection from one to the other, just as with sets in general.[19] However, as Halvorson shows, this is too permissive in equating theories that we would not otherwise take to be the same. So, he constructs, within a language that has a countable infinity of propositional constants p_i, toy theories T and T' in first-order propositional logic, where T is just the theory whose only consequences are tautologies and T' is formed by adding to the language a new propositional constant q, so that T' is given by the infinite set of axioms $\{q|—p_i\}$. If M is the set of models or truth-valuations for T and M' for T' then it is straightforward to show that M and M' will have the same cardinality, so obviously a bijection will exist between them and hence under the above criterion T and T' must be deemed to be equivalent when intuitively they are clearly not. And just in case one were to protest that when it comes to the Semantic Approach we are primarily concerned with predicate languages, Halvorson constructs a further example that makes the same point (2012, pp. 192–3).

He then argues that the Semantic Approach makes the opposite mistake of distinguishing theories that we should regard as equivalent. So, returning to the issue of what it would be to establish an isomorphism between the classes of models M and M', the strongest way of doing so would be to say that they are the same theory when they are identical. But now it seems that examples can be constructed of what we would naturally take to be the same theory, which can be given alternative axiomatizations, yielding different models and hence inequivalence on this criterion. Thus, Halvorson gives the example of autosets and groups: the former is a set, S, with a transitive action on itself, formulated in a language with a single binary function symbol o, with a theory given by three straightforward axioms and for which a model would be $<S, o>$; the latter is formulated in a language also with a binary function symbol o, but also a unary function symbol i and a constant e, and the relevant theory T' covers the usual group-theoretic axioms with the associated model being $<G, o, i, e>$. In terms of the aforementioned criterion, T and T' should be regarded as distinct, since $M \neq M'$, clearly. However, one can show that the theory of autosets is definitially equivalent to the theory of groups, since each autoset carries definable group-theoretic structure (2012, p. 194).

Even if one were to liberalize the relevant criterion, by accepting that if two theories are equivalent then each model in M should be isomorphic to one in M'

[19] We recall that a bijection is a one-to-one function between the elements of two sets, where every element of one set is paired with exactly one element of the other, and every element of the latter is paired with exactly one element of the first.

(which would be the obvious way to articulate this notion) one can still find apparent counter-examples (2012, pp. 194–6). Furthermore, 'the semantic view also gives wrong answers about when one theory is a subtheory of another and about when one theory is reducible to another' (2012, p. 197). The conclusion drawn is that the Semantic Approach is plagued by various ills (2012, p. 203) and the diagnosis is that where it went wrong was in asserting that theories are *nothing more* than collections of models. If instead of regarding M and M' as akin to 'bare sets', we take into account that they possess topological structure[20] then one may be able to distinguish them in a way that matches the distinguishability of the relevant theories. This suggests that the Semantic Approach can be rehabilitated by taking it to assert that a theory is a structured set, or topological space, of models (Halvorson 2012, p. 205).

Setting this last, intriguing point to one side, how might the adherent of the Semantic Approach respond to this claim of inadequacy?

An obvious move is to question the criteria for theory equivalence indicated above and propose an alternative. Thus, Glymour argues that the notion of isomorphism that Halvorson articulates in his final characterization of theory equivalence above, is 'purely structural' in the sense that it maps any nth order relation in one model to any other nth order relation in another, so the relations come to be regarded as mere placeholders (Glymour 2013, p. 291). Although Halvorson's motivation for adopting such a general notion lies in his understanding of the aforementioned lifting of the yoke of language as rendering the Semantic Approach language-free in some sense, this is not the standard notion of model theory. The latter requires the relevant structures to be elementarily equivalent, so the given relations are understood to be interpretations of a predicate in the relevant first-order language. Seen through the lens of *that* notion, Halvorson's apparent counter-examples fail.

Of course, Glymour insists, that means giving up the 'untenable' claims that theories are 'language free', according to this approach (Glymour 2013, p. 292). Nevertheless, the content of a theory can still be cashed out in terms of the relevant class of models and with this standard notion of isomorphism to hand, one can give a 'purely' model theoretic explication of the equivalence of theories in different languages, via the notions of common definitional extensions and expansions (Glymour 2013, p. 292):

> A definitional extension of a language L by a new predicate R is a formula $R(x) \leftrightarrow \Phi(x)$ in the language $L + R$ of L extended with R, where x is a vector of variables and no other variables occur free in the formula, and $\Phi(x)$ is a

[20] That is we define a topology on M 'by saying that a sequence (m_i) of models converges to a model m, just in case for each sentence p, the truth value $m_i(p)$ converges to the truth value $m(p)$' (Halvorson 2012, p. 204).

well-formed formula of L. Two theories have a common definitional extension if each has a definitional extension such that the two extended theories (now in the same language) are logically equivalent.

A definitional expansion by relation R of a model M of a theory T with language L_T enriches, (or in other terminology, expands) M with R to form a structure for the language L_{T+R} satisfying a definitional extension of L_T by R. A definitional expansion by R of the class of models of T expands each model of T by R forming an expanded class of models such that there is a definition of R (in terms of L_T) in L_{T+R} that is satisfied by all models in the expanded class.

Two theories with disjoint non-logical vocabularies are then said to be formally equivalent if and only if they have a common definitional extension or if their classes of models have a common definitional expansion. With this understanding of theory equivalence, Halvorson's examples lose their bite (Glymour 2013, pp. 293–6).

This captures the global aspect of theory equivalence that Halvorson identifies, in the sense that the it applies to the classes of models in terms of which the theories are presented, but at the cost, it seems, of admitting a certain language dependence—e.g. via the axiomatization required for interpretive isomorphism (Glymour 2013, p. 297). As far as Halvorson (2013) is concerned doing so leads to the collapse of the Semantic Approach into the Syntactic alternative and the 'rhetoric' of the likes of the proponents of the former cannot be maintained.[21] Furthermore, he insists, the common content of theories cannot be captured solely via the relevant class of models, since, as he argued previously, different formulations of the same theory may have different classes of models.

Of course, such claims of collapse depend on a clear idea of what the *collapsandum* is! As van Fraassen notes (2014, p. 281), historically speaking what the adherents of the Semantic Approach thought they were rejecting was the view that insisted that just as the whole of mathematics could be grounded in or reduced to formal logic, so the same could be achieved for natural science, via a unified axiomatic edifice. What these early adherents then urged was a shift away from *logic* as the appropriate device for characterizing theories and science in general, as discussed in Chapter 1, and towards models and modelling. One might try to argue that the tool chosen for that first conjunct is inadequate for characterizing

[21] Halvorson has a further axe that he insists on grinding, namely that since recent work on structural realism derives its plausibility from the Semantic Approach, if the latter collapses into the alternative, 'then structural realism is no more plausible today than it was in the time of Carnap and Hempel' (2013, p. 476; see also p. 478 and 2012, p. 186 and pp. 196–7). In a sense this is quite a bizarre claim, since some versions of structural realism (e.g. Worrall's so-called 'epistemic' form) *are* articulated within the framework of the Syntactic Approach to theories. The plausibility of this realist stance turns on much more than such a characterization. Indeed, a variety of such possible ways of characterizing theories can be presented, even though reasons can be found for preferring the Semantic Approach in this context (French 2014).

the second. Certainly in those early, heady days of the Semantic revolution, many took their cue from Suppes, who famously argued (as we have already noted) 'that the meaning of the concept of model [in the sense of Tarski] is the same in mathematics and the empirical sciences. The difference to be found in these disciplines is to be found in their use of the concept' (Suppes 1961, p. 165).

Granted, that just as Newtonian mechanics continues to be actively developed within physics (see for example: https://arstechnica.com/science/2014/08/the-never-ending-conundrums-of-classical-physics/) so the Syntactic Approach has been further pursued by the likes of Hutteman, Lutz, and others, to the point where differences that previously seemed stark, no longer appear so. Nevertheless, it remains the case that adopting model theoretic tools within the philosophy of science does not require one to eschew language altogether. As already noted elsewhere (da Costa and French 2003), if one follows Suppes's set theoretic line, then one is of course committed to acknowledging the role of the set-theoretic predicates which the relevant structures satisfy.[22] van Fraassen, of course, is not a follower of Suppes (indeed he sees the latter as offering only a 'precursor' to the Semantic Approach; van Fraassen 2014, p. 280) but while acknowledging that some of the rhetoric may have been misleading, he continues to insist that the aim was never to banish language from our theorizing about science but to 'put it in its proper place' (ibid., p. 281). The issue thus resolves down to where that place is!

One can imagine a kind of spectrum of positions on this, from the first proponents of the Syntactic Approach at one end (although even this placement might be disputed, as we have seen), to the likes of Suppe, Suppes, and van Fraassen at the other, with some delicate history of the philosophy of science forensic work required to situate them in some historically appropriate order. Somewhere up from the Syntactic Approach we might find Lutz and Halvorson, who could be crudely characterized as advocating a form of this view, whereas we can place Glymour, for example, more squarely in the middle, as he writes,

> Like it or not, on the semantic view, language and logical syntax are indispensable tools for the presentation of theoretical content, and that being so, there is no reason why proposed syntactical equivalence relations corresponding to model-theoretical equivalence relations should not be considered and no reason why the fact that relations in different models are co-denoted by the same expression in a language should not be taken into account. (2013, p. 289)

Setting aside the indispensability issue (and again, van Fraassen would insist that language and logical syntax are indispensable only in a trivial sense), here I

[22] Halvorson fails to note this acknowledgement in da Costa and French (2003) and the surrounding discussion of the role of language in this context, despite citing the work as extolling the Bourbaki approach.

want to focus on the idea that logical syntax on the one hand, and classes of models, on the other, are merely *tools* for the presentation of theoretical content and of theories more generally. As such, one may have broadly pragmatic reasons for preferring one set of tools over another (and this may cut both ways of course; see Hudetz 2017) but this is a crucially different stance from taking theories to *be* logico-lingustic or model-theoretic in some sense.

This difference is clearly played out in the debate with Halvorson who repeatedly insists that what is at issue here is the *identity* of theories (indeed, the whole thrust of his paper is encapsulated in section 4, which is entitled 'Identity Crisis for Theories'). Thus, he writes, '[a]ccording to the semantic view, a theory *is* [my emphasis] a class of models' (Halvorson 2012, p. 190; later on he talks of the semantic view 'reducing' theories to sets of models; ibid., p. 192) and his aim is explicit, namely that by seeing how his proposals for defining an isomorphism fail, 'it will become clear that it is impossible to formulate good identity criteria for theories when they are considered as classes of models' (ibid. p. 190; see also p. 201). The discussion throughout is framed in terms of individuating theories and an obvious move to make would be to reject such a framework as inappropriate.

Indeed, framing the debate over the viability of the Semantic Approach in this way leads to the possibility of question begging over what counts as 'the same' theory to begin with. Thus, to demonstrate that the Semantic Approach identifies theories that should be regarded as distinct, Halvorson's strategy is to syntactically formulate a couple of theories, show that they are inequivalent by the standard criterion of definitional equivalence and then point out that the relevant sets of models are isomorphic and hence the theories must be counted as the same.

As Glymour notes, however, the question of what 'is' the theory is precisely what is in dispute, so to maintain that a theory 'is' its syntactic formulation in terms of which it can be show to be inequivalent to another which the Semantic Approach renders as equivalent is precisely to beg the question against the latter view (Glymour 2013, p. 287). Glymour himself thinks this objection doesn't go through because of the aforementioned role of language: to present a theory as a class of relational structures is to describe that class in some language (ibid.). And to dismiss the role of language as 'trivial' in this context is to miss the point. Again as Hudetz notes, there is a sense in which the families of structures in the Semantic Apporach encode syntactic information that can become a 'linguistic yoke', to use van Fraassen's metaphor (Hudetz 2017). However, the traffic goes both ways: we can take whatever criterion of equivalence that is applicable within the Syntactic Approach and carry it over to be applied to the structures of the Semantic Approach (ibid.). Thus, the former retains no distinctive advantages in this respect at least.

Throwing the Toys out of the Pram

Note the implication of this inter-translatability: neither the advocate of the Semantic Approach, nor her Syntactic rival can now claim that the purported advantage that one has over the other in terms of providing criteria for theory equivalence or distinguishability can help resolve the issue of what a theory *is*.

It is also worth noting that the examples of 'theories' presented in the exchange between Halvorson and Glymour are either 'toy' logic cases or taken from mathematics where, in both cases, clearly articulated formulations can be given in terms of which equivalence, or not, can be explicitly demonstrated via some standard technical device and then contrasted with the relevant relationship obtained via the appropriate such device at the level of classes of relational structures.[23] In such cases, and leaving aside the above issue of question begging, the identity criteria of the theories can be made clear, at one level or another. But this is not the case when it comes to examples of *scientific* theories. Should Newton's theory of mechanics or Maxwell's theory of electrodynamics or Einstein's general relativity be identified with certain syntactic formulations? To do so would clearly beg the question against the Semantic Approach and in these cases we don't have the clearly articulated formulations that Halvorson presents. Instead we have... what? Well, that's a good question actually and one that we shall return to but for now let's say that we have some equations, interpreted of course, written down in various texts, in various languages, sometimes 'expressed' or presented in quite different ways. We could, of course, attempt to construct a syntactic formulation of any or all of these theories, along the lines of the Syntactic Approach but again, to insist that *that* formulation *is* the theory and that in such terms the Semantic Approach misidentifies it is precisely to beg the question. (And equally, the proponent of the rival view can say the same if we were to articulate the criteria of theory identification in model-theoretic terms!)

The point is that we don't have the nice clear and clean examples that the above debate focuses on. What we have is something a lot more complex and a lot messier, in the context of which the articulation of criteria of theory identification is a much less straightforward and much more contentious business. I shall be pursuing this particular line throughout the book but the core point I want to make here is, we, as philosophers of science then have to decide on what basis we are going to select those features of this messy collection of statements and diagrams, supposed axioms that don't look like anything you were taught in logic class, equations and claims, that we then take to be 'the' theory in question. One way forward—drawn from the recent developments of the Semantic Approach and

[23] The same is true of Hudetz (2017).

thus to foreshadow the discussion of subsequent chapters—is to focus on the *representational role* of these scientific models.

Thus, van Fraassen, while also dragging the debate with Halvorson into the context in which the Semantic Approach was originally proposed, namely that of scientific theories, writes that when a scientist presents a theory 'she provides a class of models for the representation of those phenomena' (van Fraassen 2014, p. 277). Of course, that immediately raises the further issue of determining the characteristics of representation by which we may pick out scientific representations from the melee that is scientific practice in any field. That will be the topic of the next chapter but, just to steal that and subsequent chapters' thunder, van Fraassen draws the time-honoured comparison with representation in *art*: 'we properly speak of a model of combustion or of the San Francisco Bay [cf. Weisberg 2013] in the way we speak of a painting of fire or of the Giaconda' (van Fraassen 2014, p. 277).

Given that scientific models are, primarily, representations, in what sense may they also be mathematical structures? The answer is straightforward: 'A model is a mathematical structure in the same sense that the *Mona Lisa* is a painted piece of wood' (ibid.). In other words, both the representational content of the painting and the actual painted piece of wood are what make the *Mona Lisa* the artefact that it is, and similarly, there is more to a model, as a scientific artefact, than the relational structure in terms of which we can define embeddability, isomorphism, and so on.[24] In particular, if we restrict our considerations to the former, and take a model to be a structure plus an interpretation which maps expressions in some language to elements of that structure, so that sentences may come out true under such an interpretation, we stand to overlook the representational aspect that is so crucial in the scientific context.

Indeed, Thomson-Jones (2006) argues that that not only should we keep these two roles—the truth-making and the representational—distinct, we should drop the former from our characterization of the Semantic Approach entirely.[25] His principal motivation for this view is that,

> When it comes to showing the naturalness and plausibility with which theories in the empirical sciences can be viewed as collections of models...it is quite

[24] Lutz (2015) argues that on van Fraassen's own account of representation, there may not be anything 'more', since according to this account representation is cashed out in terms of the embeddability of data models, which is a formal relation preserved under isomorphism (see also Hudetz 2017). Whatever the merits of this argument, the account of representation I shall present in the next chapter differs from van Fraassen's and the issue of what 'more' should be added over and above the formal structure will be central to the discussion.

[25] Again, he frames the debate in terms of *identifying* scientific theories with objects of a certain sort, namely models and distinguishes two broad versions of the Semantic Approach: the stronger which takes a scientific theory to be a collection of models and a weaker form that takes it as 'best thought of' as such a collection (Thomson-Jones 2006, p. 529).

unclear that the models in question are, as constituents of those theories, functioning as truth-making structures in any substantial way. (ibid., p. 530)

Even if we eschew the kinds of toy examples that Halvorson favours, and consider Suppes's presentation of Newtonian mechanics via the appropriate set-theoretical predicate, a model taken from the collection picked out by this predicate is a truth-making structure for the relevant statements only in the 'thin' sense that it provides a domain of discourse for the quantifiers featuring in these sentences. And this is because the latter are not, of course, uninterpreted and requiring interpretation in the way that a Tarski-type model provides an interpretation for some sentence of a first-order language. On the contrary, they are already interpreted sentences of 'mathematical English. So, the model picked out by the set-theoretic predicate is not a 'serious interpreter' of these sentences but only a 'description fitter' (ibid., p. 531).

What the set-theoretic predicate provides us with, then, is 'a perfectly good tool for picking out a collection of mathematical models' (ibid., p. 532). And the representational character of the latter is precisely what the advocates of the Semantic Approach need to focus on if they are to maintain the aversion to all the linguistic issues besetting the Syntactic view and stay close to scientific practice (ibid., pp. 533–4). Indeed, Thomson-Jones argues, shifting this focus yields a much more flexible form of Semantic Approach since the many kinds of mathematical structures and concomitant different ways they can serve representational ends, puts a 'rich palette at our disposal' when it comes to understanding scientific practice (ibid., p. 534).

I would maintain that this 'palette' of resources is made even richer by drawing on further technical developments such as the partial structures framework and the associated notion of partial isomorphisms, presented in previous work (see for example da Costa and French 2003; or more recently Bueno and French 2018). Here we tweak the core characterization of a model within the Semantic Approach, as given above, by taking the family of relations, R_i, $i \in I$, to be partial in the following sense:

> A partial relation R_i over A is a relation that is not necessarily defined for all n-tuples of elements of D [see da Costa and French 1990, p. 255]. Each partial relation R can be viewed as an ordered triple $<R_1, R_2, R_3>$, where R_1, R_2, and R_3 are mutually disjoint sets, with $R_1 \cup R_2 \cup R_3 = D^n$, and such that: R_1 is the set of n-tuples that (we take to) belong to R; R_2 is the set of n-tuples that (we take) do not belong to R, and R_3 is the set of n-tuples for which it is not defined whether they belong or not to R.[26]

[26] To avoid a possible confusion between R_1, R_2, and R_3 and particular occurrences of a partial relation R_i, I will always refer to the former as R_1-, R_2-, and R_3-components of the partial relation R_i.

> The set-theoretic construct $A = \langle D, R_i \rangle_{i \in I}$, is then said to be a 'partial structure' and partial isomorphisms can then be formally defined between such structures.
> (Bueno 1997; French and Ladyman 1999)

Using this formalism, we can represent the hierarchy of models—what Suppes (1962) called models of data, of instrumentation, of experiment, as well as various kinds of 'intermediate' models—that take us from the phenomena to the theoretical level (Bueno 1997); the partial relations are extended as one goes up the hierarchy, in the sense that at each level, partial relations which were not defined at a lower level come to be defined, with their elements belonging to either R_1 or R_2. It also provides an appropriate characterization of both scientific theories and models, particularly with regard to their open-ended nature and both the way in which they can be further developed and also 'draw' on further features from both other scientific theories and mathematics (da Costa and French 2003; Bueno and French 2018). In particular, this framework can capture the way in which idealizations feature in such models—such as that of the simple pendulum, for example—contrary to the claims of Pincock (2005) and Weisberg (2013). It can also handle so-called 'caricature' models such as Schelling's famous model of racial segregation (French 2017), as well as 'scale' models, taxonomic models, and physical models more generally (French 2010). That last claim may seem odd but remember: I am not claiming that any scientific model, including 'theoretical' ones, *is* a set-theoretic structure; rather the point is that these structures can be used to characterize or represent, at the meta-level, the relevant features of practice, including physical models such as Crick and Watson's steel-plate-and-wire model of DNA and species models held in natural history museum collections (Ladyman and French 1999).

Furthermore, with partial isomorphisms understood as holding both 'horizontally' between theories, and 'vertically', between data models and theoretical models, these resources can also capture the multitude of relationships that constitute scientific progammes (Bueno 1997 and 2000; da Costa and French 2003). Further extended again to include partial homomorphisms, it can also accommodate the relationships between such theories and the mathematics in which they are 'framed' (Bueno, French, and Ladyman 2002; see also Bueno and French 2012); in particular, this approach can capture the 'surplus structure' of mathematics, which plays an important heuristic role in scientific developments (Redhead 1975; Bueno and French 2018). As Thomson-Jones said, a 'rich palette' indeed![27]

As we'll see in the next chapter, this framework nicely accommodates the representational role of models and in that context van Fraassen observes that we

[27] Further interesting work on models, including their role in exploring the modal features of theories, can be found in Gelfert (2016) and Massimi (2018).

could, perversely perhaps, apply the Syntactic view to the philosophy of art and rationally reconstruct the *Mona Lisa* in terms of a mapping from certain natural language expressions to features of the painted piece of wood, such that certain statements made in artbooks, say, come out true under that interpretation. However, he suggests, this is just as un-illuminating when it comes to artistic practice as adopting the above stance towards scientific representations (van Fraassen 2014, p. 278). In both cases, it is more natural to point to the painting or the scientific representation and say 'that *is* the Mona Lisa/Newtonian mechanics (respectively)'. (However, as we'll see, things are not quite that simple.)

Given that a scientific model is a representation, 'it does not follow that the identity of a theory can be defined in terms of the corresponding set of mathematical structures without reference to their representational function' (ibid.).[28] And if we focus only on such structures while ignoring the representational function then of course we will identify putative theories that are distinct—but we always knew that, as the well-known examples of the equations describing gas diffusion and temperature distributions over time demonstrate (ibid., p. 279). It is only by appreciating their distinct representational functions that we can see that they are not the same, even if the relevant mathematical structures are.[29]

Conclusion: Where Do We Stand (Again)?

The issue of how we might appropriately capture those representational functions is what I shall turn to in the next chapter. Before I do so, let's just pause to consider where things stand: in the last chapter, we looked at the so-called 'Syntactic Approach' that takes theories and models to be sets of statements or propositions. That of course pushes the answer to our core question—what *are* theories?—back a step and we looked at two possible answers: fictions and abstract entities. In this chapter we've looked at the currently dominant alternative—the 'Semantic Approach'—that attempts to break free of the linguistic chains of its syntactic predecessor by taking theories and scientific models to be families of set-theoretic models. However, we've seen, first of all, that the grip of language cannot be

[28] Cf. Halvorson who suggests, 'the semantic view was not wrong to treat theories as collections of models; rather, it was wrong to treat theories as nothing more than collections of models' (2012, p. 204). But the point is that it is Halvorson's conception that is wrong. As French and Saatsi noted some years previously, '[i]t seems to be a popular misconception of the semantic view that it says *nothing* but the following about theories: theories *are* (with 'is' of identity) just structures (models)' (2006, p. 552).

[29] Hudetz (2017) argues that this misses the point, since Halvorson's concerns are not so much about representation as about how to present the *formalism* of a theory: 'The question is which kind of entities models should be taken to be when we reconstruct the formalism of a theory in philosophy of science (e.g. in order to provide a rigorous account of what it is to empirically and ontologically interpret a formalism or in order to define intertheoretic relations).' Note again the emphasis on what kind of *entities* models should be taken to be—as I've already indicated, I shall argue that *they should not be taken to be any kind of entity at all!*

thrown off so easily and, second, that our core question also underlies recent critiques of this approach. Again, of course, the answer to that question is just pushed back a step and what we haven't done is look at the fictionalist and 'abstractionist' options in this new context. We will, however, examine both in some detail in subsequent chapters.

Of course, that the Semantic Approach cannot serve as a multi-purpose tool for all issues in the philosophy of science has long been recognized. Consider: how can the realist avail herself of standard accounts of truth and reference if she eschews language in the above sense?[30] One response is to draw an 'internal/external' distinction that goes back to Suppes (see da Costa and French 2003, pp. 29–33): from an 'external' perspective we can characterize theories in set-theoretic terms, deploying appropriate formal tools to capture their openness, their inter-relations, their empirical adequacy and so on; but from an 'internal' perspective we can also characterize them via propositions and thereby, if we are of a realist inclination, apply the usual frameworks of truth and reference. Isn't this just having one's cake and eating it?! Not if one resists the insistence on reifying theories as either one or the other—that is, as either set-theoretic structures or propositions—to begin with (ibid.).[31]

But that is to jump ahead; let's now consider how the representational function of theories might best be characterized. As we'll see, this encourages a comparison with artworks that will help us get a better grip on our central question.

[30] Not a problem for the likes of van Fraassen of course!

[31] But that is not to say that one or the other characterization might not have some pragmatic edge over the other: '[w]e may allow that in their descriptive or representational role, models offer significant advantages of both practice and principle over formalized languages' (da Costa and French 2003, p. 31).

3

Theories as Representations

Introduction

How, then, should we go about capturing this representational function of theories and models (for useful surveys, see Boesch 2015 and Suárez 2010)? An obvious starting point is to make an explicit comparison with those artworks that can be regarded as representational, such as drawings and paintings. Well-known examples of the latter have been invoked to both support certain views of scientific representation and undermine others (see van Fraassen and Stigman 1993; van Fraassen 2008 and also Suárez 2009).[1] Even more interestingly, some of these moves have been constructed on the back of certain devices and manoeuvres imported from the philosophy of art (for a recent such example, see Frigg and Nguyen 2017a). Although I am broadly sympathetic to such appropriation—not least because we have to start somewhere in getting to grips with the concept of representation—I think we should be cautious about the extent to which specific examples from the world of art can be taken to bear on issues having to do with models in science, at least insofar as such examples are supposed to have meta-level evidentiary value. Furthermore, although van Fraassen's move of drawing attention to the *representational function* of theories and models undermines the critical concern regarding their identity, this comparison with artworks can further muddy the waters when it comes to their *ontological status*.

Let's begin by considering this importation of some pre-established analysis of representation from the philosophy of art, language, cognitive science, or wherever. Which kind of analysis one favours may well depend on how we conceive of—dare I say, *represent*—theories and models in the first place. If we think of them in terms of an axiomatized set of logico-linguistic statements, or some collection of literally understood statements, as sketched in Chapter 1, then we might be naturally drawn to accounts of linguistic representation in which notions of *denotation*, for example, feature prominently (cf. Elgin 1983 for example). If, on the other hand, we conceive of theories and models in non-linguistic terms, as in the Semantic Approach, then we might look to analyses of representation based on notions of *resemblance* and *similarity*, not least because one obvious way of formally capturing such notions is via *isomorphisms* and *partial isomorphisms*

[1] And here we recall how van Fraassen invoked the example of *The Mona Lisa* in his defence of the Semantic Approach.

There Are No Such Things as Theories. Steven French, Oxford University Press (2020). © Steven French.
DOI: 10.1093/oso/9780198848158.001.0001

holding between the theory or model and the target. These analyses have generated diverse criticisms, but here I want to focus on those that are either motivated by examples drawn from the art world or deploy devices taken from the philosophy of art. And one lesson that I want to draw is that due consideration needs to be paid to the differences between works of art and scientific theories and models when it comes to evaluating such formal accounts.

From Art to Science

The *locus classicus* when it comes to representation in the philosophy of art is Goodman's *Languages of Art* (1976), where he begins by dismissing what he calls the 'most naïve view of representation' (Goodman 1976, p. 3), namely that it involves, at its core, *resemblance*. This cannot be the case, he maintains, because first, anything resembles anything else in one respect or another and hence we need some appropriate framework in order to pick out the relevant resemblances—but then all the emphasis needs to be on that framework. Secondly, resemblance, or similarity, is symmetric in the sense that if *a* is understood to resemble or be similar to *b*, then we would also take *b* to resemble or be similar to *a*, but representation is not symmetric in this way: Freud's painting *Benefits Supervisor Sleeping* (https://en.wikipedia.org/wiki/Benefits_Supervisor_Sleeping) can be said to represent Sue Tilley, but most of us would balk, intuitively, at claiming that Ms Tilley represents the painting. Likewise, as Suárez says, 'a glass of water [for example] is similar to itself, and similar to any other glass that is similar to it' (Suárez 2003, p. 233), but, if 'an equation represents a phenomenon, the phenomenon cannot be said to stand for the equation' (ibid., p. 232), and we would not claim that the phenomenon represented itself.

Given the first point, then, Goodman argues that pictorial representation must be taken to be relative to the relevant conceptual framework or interpretive system and just as with perception, we must reject the 'myth of the innocent eye'—that is, we cannot take a painting to 'innocently' represent in the absence of the appropriate conceptual scheme. In place of resemblance, he sets *denotation* at the core of representation (Goodman 1976, p. 5), in the sense that pictorial depictions *denote* their subjects, akin to linguistic predicates. Thus, just as the name 'Sue Tilley' denotes the woman who is the subject of Freud's painting, so the painting also denotes her.[2] However, one difference between names and paintings is that the latter, unlike the former, have to be understood within a system that can be regarded as 'dense' in the following sense: any two marks in a painting, no matter how small the difference between them, can be understood, from within

[2] Depictions of generic subjects have multiple denotaitons and those with fictional subjects have an empty or null denotation. I'll come back to cases of 'impossible' or contradictory subjects shortly.

the appropriate system, as representing different features of the subject. Conversely, any two such features of the subject, no matter how small the differences between them, can be represented—again, within the appropriate system—by different marks on the painting.[3] This density then affords a certain 'open-ended' nature to be attributed to depictions, in the sense that any mark on the canvas of the painting, for example, may turn out to be significant for our understanding of what the painting represents and hence there is always the possibility of more being discovered about it.[4]

The Open-Ended Nature of Artworks

So for Goodman, a painting could be considered to be a set of points on a planar surface, with subsets of points, constituting regions, assigned a particular hue, brightness, and saturation. Given any two paintings that have regions of colour that differ, no matter how slightly, there will always be a third that is more similar to each than they are to each other. Likewise, given any two distinct 'extensions' of a painting, in the sense, say, of what it is taken to represent, then no matter how similar those extensions are to one another, we can always find a third that is more similar to each of the two than they are to each other. Thus, '[i]n order to figure out whether a given scene is within the extension of a picture, one needs an indefinitely precise characterization of that scene in terms of angles, sizes, distances, colours, and so on.' (Kulvicki 2006, p. 20).

Given this feature of semantic density, and the consequent requirement of indefinitely precise characterization, what a painting is 'about' may be open-ended in the sense that for any scene which is claimed to be the extension of the painting, there is always another—characterized slightly differently in terms of angles, sizes, colours, and so on—that could be put forward as an alternative. And as a result, we can always find new and different interpretations of a given painting. Of course, in many cases, the relevant differences may be so slight that any difference in interpretation constructed on that basis will be dismissed as not a genuine difference at all but one can easily imagine other examples where the differences are significant.

Thus, for Goodman, representation is always conventional, with apparently realistic depictions only taken as such from habit or familiarity. Such dependence on convention is clearly evident with the example of perspective: as is well known, attempts to develop systems of perspective go back to at least the fifth century BC

[3] This is clearly drawn from mathematics, where a continuum of points is said to be 'dense' if, for any two points, there is (at least) a further point between them.

[4] Although of course, given that even the finest grained painting is not actually dense in the mathematical sense, that possibility might be regarded as not limitless.

in ancient Greece and Chinese artists used 'oblique perspective' since the first or second centuries AD, but different conventions were often deployed. However, it was in the fifteenth century that the rules of geometrical perspective were established by the likes of Brunelleschi[5] and Lorenzetti (rules which were of course subsequently played with and flouted). And it is only by having such a prior conceptual framework that we can determine which resemblances between a painting, say, and its subject count as relevant.

There are a number of well-known criticisms of Goodman's account (see for example Kulvicki 2006, pp. 24–7)[6] but the issue I want to focus on is to what extent, or even whether, it can be exported into the philosophy of science. Note that the considerations sketched in previous chapters have a bearing on this issue, given that for Goodman pictorial representation is symbolic, with pictures understood as akin to linguistic predicates. One might not be surprised if an advocate of the Syntactic Approach to theories were to embrace Goodman's view, while a proponent of the Semantic Approach rejects it as unable to capture the non-linguistic nature of theories qua representations.

First of all, semantic density may be sufficient for this 'open-endedness' with regard to what the painting is about, but it is arguably not necessary. Consider: some years ago, the Van Gogh Museum in Amsterdam authorized the production and sale of a limited number of expensive and, supposedly, exact reproductions of certain paintings, using a three-dimensional printing technique that even reproduced the thickness of van Gogh's brushstrokes (as well as the imprints of stamps and stickers attached when the paintings were loaned; see http://www.startribune.com/van-gogh-museum-issues-35-000-replicas-of-artist-s-work/239769841/). Are such paintings *about* the same thing as the original? The temptation to respond 'of course!' might be tempered by the appreciation that the person 'loading up' the three-dimensional printer has never seen 'that thing' (for example, the wheatfield under thunderclouds). Indeed, an artist may copy an artwork with the specific intention that the copy not be about the same thing. Sturtevant made an exact copy of Jasper Johns's 'Target with Four Faces' in order to raise questions about the nature of art, of authorship and the authority of the artist (http://www.bbc.com/culture/story/20141112-great-artists-steal; both artworks can be found in the Museum of Modern Art, New York). Given such cases, it would seem, semantic density cannot be necessary when it comes to what art is about—to determine that some appeal must be made to the relevant context, including the artist's intentions.[7]

[5] Possibly helped by his mathematician friend Toscanelli.

[6] Thus, as Kulvicki notes (2006, pp. 20, 39–41), pictures are, for Goodman, essentially analogue representations and an obvious counter-example would be digital representations, such as one finds in much of today's media (from newspapers to iPads!). In response Kulvicki suggests replacing syntactic and semantic density with notions of sensitivity and semantic richness.

[7] Again I am grateful to one of the readers for pressing me on this point.

Can scientific theories be regarded as syntactically 'dense'? It's not clear whether this makes any sense. Of course, if a theory is characterized in terms of a particular set of logico-linguistic axioms and their consequences, or some set of statements in general, as suggested in Chapter 1, then even if the number of such consequences is infinite, it is not the case that between any two members of this set there will be a third, at least in principle! On the other hand, if theories are identified with families of models, where these are taken to be set-theoretic, then insofar as any such model can be numerically indexed, we can understand such families to be syntactically dense in a trivial sense. But of course, just because we can avail ourselves of the real number continuum in indexing set-theoretic models does not mean that in *practice*—that is, as these models are used to characterize 'actual' scientific theories—we have the requisite sense of density in play. Indeed the question whether we do or not just seems a bit silly.

And the reason it seems so becomes apparent when we shift to the question whether theories are 'semantically dense': insofar as the 'extensions' of theories are systems in 'the real world' or, at the very least, observable phenomena, the answer is 'yes' (or at least as much as pictures are, ignoring the obvious issues of whether reality is fundamentally granular or not). Again, physical properties can be represented mathematically and hence can be 'indexed' on a continuum—at least so far as classical physics is concerned. But—and here's the rub—is it then the case that in order to determine what the extension of a theory is, we need an 'indefinitely precise characterization', as Goodman insists we do when it comes to a painting? Leaving aside issues regarding the range of possible interpretations when it comes to either paintings or theories, scientists seem able to achieve such a determination without such a precise characterization.

Consider again Einstein's theory of special relativity: this has as its extension a whole range of phenomena that is classified under 'relativistic'. Now, as I suggested in Chapter 1, there are clearly issues as to what this theory might be taken to be about. Einstein himself originally took it to be about clocks and rods, whereas Minkowski and subsequent commentators, including Einstein, interpreted it as about space-time. In large part, of course, this had to do with philosophical orientation, with antirealists of an empiricist bent opting for the former and realists for the latter. In the current context of the realism-antirealism debate, an advocate of the most prominent form of antirealism, namely constructive empiricism might say that the theory might ostensibly be about space-time but we can't actually know whether it actually represents that entity or not. Here someone might be tempted to draw comparisons with the way artworks can take on new meanings over time and certainly we can find examples of changing stances towards certain features of theories: the infamous cosmological constant in the general theory of relativity was originally introduced by Einstein so that his theory could appropriately represent a static universe but when Hubble provided evidence for an expanding universe, he dropped it. The subsequent discovery that the expansion of the universe is accelerating led cosmologists to give the constant a positive value and it currently features in the standard 'λ-CDM' model of the universe (see https://en.wikipedia.org/wiki/Lambda-CDM_model).[8]

However, insofar as we can make this comparison between the changing understandings of theories and artworks, it doesn't seem to have much to do with 'semantic density'. Indeed, perhaps all I need to note is that when it comes to science, the 'density' that might be associated with the relevant properties at the level of the data acquired in the relevant observations and experiments is effectively washed out or coarse-grained away via the construction of appropriate models of the phenomena. And it is to *these* that theories may be said to apply (see again Bogen and Woodward 1988; for discussion in the model-theoretic context, see da Costa and French 2003, ch. 4). For this reason, Goodmanian conventionalism can be resisted, to some extent anyway.

Nevertheless, as sketched above, the density of representations is taken to support or 'enhance', at least in part, the open-ended nature of our investigations of what it is they are taken to represent (Walton 1990, p. 349; also mentioned in Giovannelli 2010). The idea is that just as it seems there is 'always more to be learned by examining things more closely or more carefully' (Walton 1990, p. 307) when it comes to the real world, so we can continue to discover 'more or less indefinitely' (ibid.) new aspects of a painting, for example, unlike (it would seem) a work of literature, which isn't 'dense' in this sense.[9] Of course there are limits (ibid.) to this process, but what enables us to go further than first impressions, say—as when we first look at Turner's *Rail, Steam and Speed—The Great Western Railway* and then notice that the blur or blob in front of the on-coming train is in fact a hare (see https://thebeautyoftransport.wordpress.com/2013/05/08/so-fast-its-just-a-blur-rain-steam-and-speed-by-j-m-w-turner/)—is that between any two representational marks (broadly delineated perhaps) on the canvas, there is a third, which is then open to representational denotation, in the sense of what it is meant to denote (in this case a hare) and interpretation (in the sense of the hare standing for that which once was the fastest thing in the country (!), or nature in general, about to be ground under the iron wheels of technology...).

The open-endedness of theories is also a significant feature of science, at least on most views.[10] The idea is that there may always be new phenomena within a given theory's domain that comes to be discovered and to which it may then be

[8] λ stands for the constant and 'CDM' for 'cold dark matter'. Again, I must thank one of the readers for pressing me on these issues and providing this example.

[9] One might dispute whether it is not the case that there is 'always more to be learned by examining things more closely or more carefully' when it comes to literature but the point is that if true, that's not because the text is 'dense' in the required sense.

[10] Some might argue for a view of theories as 'closed' but I think it is better to understand this in terms of their domain being clearly delineated so that within that domain, the theories themselves may be considered to be open for further development; consider the case of classical physics in that regard.

applied.[11] Theories are progressive in a manner that involves their extension as they bring more and more phenomena within their purview, including in some cases phenomena that one may not have expected them to explain. Such an important feature of theories has not gone unnoticed of course and both the Syntactic and Semantic Approaches can accommodate it: the former does so via its notion of a partial interpretation of the relevant theoretical principles through correspondence rules, with the possibility of further such rules, or bridge principles, being laid down as new phenomena are discovered which requires connection to and explanation by the theoretical principles; the latter has the capacity to handle such open-endedness more explicitly, via the incorporation of 'partial relations' which indicate those features to which we are currently unclear whether the theory applies or not. The point is, of course, that the philosophy of science has various resources for accommodating the open-ended nature of scientific research without resorting to notions of *density*, exported from mathematics or wherever.

We might have further qualms regarding the relevance of Goodman's scheme for representation in science. Let's consider again his two core critical points: first, we need some appropriate framework in order to pick out the relevant resemblances to begin with and secondly, that this resemblance will be symmetric and hence cannot underpin representation.

There is clearly a bit of a tangled relationship between the philosophy of art and the philosophy of science when it comes to Goodman's insistence that there is no 'innocent eye' and hence some conceptual scheme is required to enable us to establish which resemblances are significant. We might recall Hanson's famous dictum 'There's more to seeing that meets the eyeball' (Hanson 1958), often invoked in the context of discussions about the theory-ladenness of observation. Like Goodman, Hanson claimed that the nature of our perceptual experience depends on the concepts that we can deploy and argued that this was also the case in science, a line that was also pursued by Kuhn, of course (1962). Thus, famously, Hanson argued that Tycho Brahe and Kepler 'saw' different things when they perceived the sunrise because their astronomical theories were different (the former advocating a geo-heliocentric system in which the sun orbited the earth and the other five planets orbited the sun; the latter, of course, defended the Copernican heliocentric system, within a geometric framework that yielded elliptical orbits).[12] However, any relativism inherent in such a view and the associated claim that there can be no theory independent observation has generally been abandoned by most philosophers of science, not least because of the now

[11] And of course, if the theory cannot be applied to these new phenomena then the latter are regarded as anomalies and possibly indicative of the shores of a new domain, to be explained by and accounted for by entirely new theories.

[12] And not just Brahe and Kepler but also Simplicius and Galileo, Hooke and Newton, Priestly and Lavoisier, Soddy and Einstein, De Broglie and Born, Heisenberg and Bohm...(Hanson 1958, p. 13).

widespread recognition of the extent of the 'checks and balances' involved in observation and also of the multiple models employed in the move from data to 'phenomena'. In addition, we can find numerous cases from the history of science and scientific practice in general that indicate that despite radically differing theoretical frameworks, scientists will still agree on the relevant evidence, or on the impact of new evidence and so on (Kelly 2014).

Furthermore, there is something a little odd about the claim that because any *x* can resemble some *y* in numerous different ways, we cannot use resemblance as the basis of representation in either art or science. After all, it is not typically the case that in either domain we are presented with either an artwork or theory and then have to figure out what it is that the artwork or theory represents. Of course, that may be the case in certain historical or archaeological work (see Currie 2017), but generally we know what it is that the artist or scientist *intends* to represent, whether it is a group of young people desporting themselves in a park, the horrors of war, the phenomenon of superconductivity, or the structure of space-time, where that knowledge is obtained either directly from the artist or scientist concerned or from the historical record. Furthermore, and crucially, that intention is cashed out through the construction of the given artwork or scientific theory, and even though the methodology of that construction will be different in certain ways (but also similar in others),[13] we can confidently say that a core part of that methodology will involve grounds for focusing on those resemblances that are taken to be significant.

Consider the numerous sketches and drafts that many artists produce prior to the finished painting (e.g. Constable's sketches for *The Haywain* or Turner's of the sea, of the countryside, the sky, and so on) that reveal their focus on certain features of the scene, say, such as the looming clouds or the reflected light on the water. In the case of art, what drives that focus is typically only the intentions of the artist, where those intentions may be multifaceted and complex, of course—so Turner's intentions in *Rail, Steam and Speed* were not to give an accurate representation of a 'Fire Fly' class locomotive crossing the Maidenhead Railway Bridge (the configuration of the bridge is distorted, the proportions of the rail engine are not quite right, even for broad gauge...), nor are they just to represent the speed of the train; rather he intended (or so it is claimed) to represent the rushing of technological progress, with the train leaving the boat in the river behind and crushing the unfortunate hare beneath its iron wheels! When it comes to science, of course, the scientist's intentions may come to naught, or at the very least be over-ridden by other considerations, not least those deriving from sometimes dramatic experimental results (and indeed, the same may be said of the artist's intentions as well).

[13] So, just as one can identify certain heuristic devices used in science, one can similarly find certain moves used by artists in the development of their work.

A nice example of this is provided by the case of superconductivity (see French and Ladyman 1997; Bueno, French, and Ladyman 2012): for many years after its discovery it was characterized, qua phenomenon, as having to do with zero electrical resistance. But attempts to construct appropriate and empirically adequate representations on that basis failed. With the empirical discovery of the expulsion of magnetic fields from a superconducting material as it passed below the critical temperature—known as the 'Meissner effect'—that characterization was abandoned and a new and much more adequate representation was constructed in terms of the phenomenon known as diamagnetism (a quantum effect whereby an applied magnetic field induces an opposing one). We can view this episode in terms of a shift in focus from one set of (inadequate) resemblances to another: prior to the discovery of the Meissner effect the resemblance of the putative theoretical representations was with zero resistance, but afterwards it shifted to zero internal magnetic field. The important point is that it was experiment that determined which resemblances were the important ones and this was then incorporated into the construction of the relevant theoretical models. In other words, we do not simply model a phenomenon, of course—we model it *as* something (for further consideration of the notion of 'representation-as', see, again, van Fraassen and Stigman 1993; or Frigg and Nguyen 2017a).

There is more to say of course on these issues but here I just want to emphasize that resemblance as (in part) the basis for representation cannot be dismissed simply by noting that anything resembles anything else in some respect or other. In both art and, especially, science there is a context, heuristic in both cases and also experimental in that of the latter, which drives the focus on certain resemblances at the expense of others.

Likewise, Goodman's second point—that resemblance is symmetric but representation is not—should not be taken to conclusively rule out accounts based on the former. Thus, it is not always the case that an artwork represents a scene, or person say, and not vice versa. Consider the sub-genre of cosplay that involves dressing and self-painting to look like a character from a work of art (as in this wonderful example of van Gogh's self-portrait: http://www.kartonista.com/art/vincent-van-gogh-cosplay.html). Of course, in these cases one could maintain that the cos-player is representing the artwork which represents the scene or person concerned, so we have nested representation rather than some sort of inversion of the relationship. Less flippantly, the nature of representation seems to require a conceptual distinction between the representation and that which is represented and this is typically cashed out in terms of an asymmetry between the two. But this in and of itself does not rule out resemblance-based accounts: one could insist, for example, that resemblance, itself represented formally in some manner, is necessary but not sufficient for representation, with further requirements imposed that generate the requisite asymmetry. One such might have to do with the relevant intentions of the artist or the user. Alternatively one might explore ways of extending the formal representation of resemblance itself to incorporate some form of asymmetry.

I will come back to these issues when we consider a particular way of capturing the notion of resemblance or similarity within the Semantic Approach. Let me for the moment continue with the emphasis on Goodman by considering a classic approach to representation in science that draws on his account. Along the way I want to raise a number of further questions concerning this comparison between art and science and the way representation can be characterized in both.

Scientific Representation and the DDI Account

Consider the classic model of the simple pendulum. We set the bob swinging (not too wildly of course) and we observe that for small amplitudes of swing, the time it takes to complete one swing (the period) is independent of the amplitude and dependent only on the length of the line (the well-known formula is,

$$T = 2\pi\sqrt{\frac{l}{g}}$$

with T = period, l = length of line, g = acceleration due to gravity). Now, as is well-known certain idealizations have to be made to obtain this expression, such as taking the bob to be point-like, so the air resistance can be ignored, taking there to be no friction at the end where the line is held fixed, taking the line itself to be massless, and so on. Hence, it has been claimed (Hughes 1997) the model itself cannot be regarded as physical and must be taken to be an *abstract object.* I'll come back to this move to the abstract realm in Chapters 4 and 5 but the important point here is that conceived as such the *model pendulum* has no length, nor is there any time in which it completes an oscillation and therefore it is 'similar to material pendulums in no obvious sense' (ibid., p. 330). Again, then, similarity does not seem to offer an appropriate means of capturing the representational relationship.

Now, this claim that relationships of similarity cannot hold between objects of different categories ('abstract' and 'physical' in this case) is clearly a very strong one to make. Indeed, it seems unclear how such relationships might be blocked on the grounds of a difference in the ontological nature of the entities concerned—think of numbers, which are surely abstract if anything is, and the claim that there are structural similarities between the number system and sets of physical objects. And of course, Giere has famously argued both that models are abstract objects and that the relationship between such models and real systems is best characterized in terms of similarity (1988; again, we'll come back to this claim).

In support, we could argue that ideal objects—objects that involve any idealization, however slight—exist in other possible worlds (see for example Nowak 1995). The similarity relationship between a model, regarded as such an object, and a 'real' system would then be reduced to a relationship—some 'counterpart' relation perhaps—holding between some possible world and the actual one. Alternatively, we could take idealized objects to be fictional and hence the models themselves should be considered as fictions—a view we touched on in Chapter 1 and which I shall return to in Chapter 6—and that just as we can assert that Sherlock Holmes bears a certain resemblance, or is similar in certain respects, to Joseph Bell,[14] so we can say likewise with regard to the model of the simple pendulum. Or, finally, we could adopt an ontologically less extravagant view of idealization, such as the following:

> idealization consists in specifying in advance the kinds of predicates expected to occur in claims being made in a given context about objects of a given kind, rather than in referring to some fictitious, idealized objects. (Grobler 1995, p. 42)

In particular, describing a pendulum in terms of a point-like bob and a massless string does not amount to substituting for it by some abstract object existing in another possible world; rather the description is simply an indication of the relative irrelevance or lower significance of certain properties in that theoretical context (ibid.).

This last option meshes nicely with the Semantic Approach to models as set-theoretical structures. Here, in constructing such a structure certain relations which hold in the real system will be represented by corresponding relations holding between elements of the set, but others will not (see French and Ladyman 1998). Of course, if we take the model *to be* that set-theoretic structure then insofar as that structure is itself an abstract object and as such does not actually possess a length, say, it might be insisted that talk of similarity remains inappropriate. However, my overall argument in this book is that we should not reify models and theories in this, or any other, way and that the Semantic Approach should be seen as a meta-representational device useful for various purposes within the philosophy of science (I'll return to this in Chapter 9). So the concern evaporates and I will take it that from the perspective of that Approach, what the material and ideal pendulums have in common are *aspects of the relevant structures.*[15]

[14] A medical lecturer at the University of Edinburgh, generally regarded as a pioneer of forensic science.

[15] Interestingly, it has been suggested that the material pendulum participates in the 'form' of the ideal one but although this has been proposed as '[m]ore illuminating, albeit less fashionable' than the abstract objects account (Hughes 1997, p. 330), it has not been developed any further.

Neverthless, if you're bothered by such Goodmanian concerns, you might be attracted to the so-called 'DDI' account of representation, where DDI stands for 'denotation, demonstration, and interpretation'. As it says on the tin, at the core of this account is the notion of *denotation* taken directly from Goodman.[16] A model then 'stands for' or refers to a physical system, and elements of the model *denote* elements of the subject. Furthermore, scientific representations possess an 'internal dynamic' which allows us to make predictions of both everyday and—crucially—novel facts. Through this internal dynamic we *demonstrate* the relevant results. And a given model may provide the internal dynamic of another so that we get a hierarchy of models, and of representations, in which a model at one level of the hierarchy is the subject of and represented by a model at a higher level. Finally, the aforementioned results that are demonstrated within the model have to be *interpreted* in terms of its subject. It is only after such interpretation that we can see 'whether theoretical conclusions correspond to the phenomena, and hence whether the theory is empirically adequate' (Hughes 1997, p. 333).

Of course, we need to know *which* features of the model, or more generally, the representation, correspond to which features of the target system and that is achieved via a 'dictionary' or 'key'. A more recent account that also draws on Goodman (and Elgin) emphasizes this aspect of the representational relationship: according to the 'DEKI' account (Frigg and Nguyen 2017a and 2017b), representation involves *denotation* (understood broadly), *exemplification*, in the sense that only certain properties are selected (those that are said to be exemplified by the representation), a *key* that specifies how those exemplified properties are associated with those of the target system (cf. the term 'code' in Shech 2015), and *imputation*, in the sense that the representation imputes at least one of those properties to the target system.

This framework then nicely accounts for the role of symbols in art, as in the example of Frans Pourbus the Younger's painting of Anne of Austria (Frigg and Nguyen 2017b; see https://commons.wikimedia.org/wiki/Category:Female_portraits_by_Frans_Pourbus_(II)#/media/File:Anne_of_Austria_mourning_her_father_Philip_III_of_Spain_in_1621_by_Frans_Pourbus_the_younger.jpg). Pourbus was a renowned portrait artist, noted for his depiction of the clothes and jewellery of his subjects. But in this case, Princess Anne is portrayed in simple, dark robes, mourning the death of her father. In her left hand she holds the paw of a small dog, sat on a pedestal of some kind. According to the DEKI account, the painting denotes a princess-with-dog but although it is also a representation-of

[16] As one of the readers of an earlier draft noted, if 'to denote' just means 'to refer' then denotation is clearly necessary but not sufficient for representation. Moreover, on many accounts of semantic content, reference is determined by meaning/content, so it is the latter that should be taken as the core of representation, since it is both necessary and sufficient (see Shech 2015). And meaning/content is determined in part by similarity or resemblance (and in part by intention or stipulation or convention).

the princess, it is not a representation-of the dog since the dog does not denote anything; rather,

> Under the conventions used at the time, the dog was a symbol for fidelity, and so the painting should be read as coming with a key associating a dog with fidelity (much in the same way in which litmus paper comes with key associating the colour red with acidity). The painting then imputes the thus keyed-up property to the princess and represents her as faithful. (Frigg and Nguyen 2017b, p. 54)[17]

Likewise, the Phillips-Newlyn machine (one of the two examples of which we have here at the University of Leeds: https://www.youtube.com/watch?v=nfDfcPywjHg) represents the national economy (in Keynesian terms) by means of a system of pipes and reservoirs with water flowing through. Originally designed to be a teaching aid, it turned out to be a surpisingly accurate simulator of the economy. Here, of course, the interpretation (and associated key) maps hydraulic properties onto economic ones (ibid., pp. 50–1). And the DEKI account can be extended to non-concrete models as well, such as Newton's model of the sun-earth system as consisting of two perfect and homogenous spheres interacting gravitationally (ibid., pp. 54–5). One can understand the underlying two-body system as an 'imagined-object', which in the Newtonian model is a solar-system-representation and which can then be interpreted in the same way as material objects are.[18] Interestingly, that imagined-object can be retained but the interpretation changed to give a different representation: so, with one sphere interpreted as a proton and the other as an electron, we get a hydrogen-atom-representation in the context of the Bohr model of the atom (ibid., p. 55).

We'll come back to that particular model shortly but in the case of the solar-system-representation we need to say more about the nature of the gravitational interaction—for example, by incorporating certain structural properties, as related by Newton's law of universal gravitation. That's no problem for the DEKI account, since such properties can 'be imputed onto their target systems in virtue of hypothesizing that there is some structure-preserving mapping that holds between the two' (ibid., p. 50). Hence this account is compatible with the partial structures version of the Semantic Approach as outlined in Chapter 2 (ibid.).

However, as sophisticated and flexible as the DEKI account is, there are two concerns one might have about it. The first is quite general and centres around this idea that representation involves supplying a 'key' via which certain elements of that representation—the relevant exemplified properties—can be understood,

[17] She also features in Dumas's novel, *The Three Musketeers*, and in the BBC's loose adaptation *The Musketeers*, where the imputation of fidelity acquires a certain ironic aspect!

[18] The role of imagination in science has recently come to be a topic of significant discussion and we'll touch on it again when we consider models as fictions (see for example Stuart 2017; Salis and Frigg forthcoming).

in terms of the target system. Consider again the cases of the portrait of Princess Anne, on the artistic side, and the Phillips-Newlyn machine, on the scientific. When it comes to the former, the representational role of the little dog is cashed out in symbolic terms: it represents the 'keyed-up property' of fidelity that is then imputed to the princess. In this case, of course, the property is not one that can be straightforwardly visually represented, like the colour of her eyes, for example, or the shape of her face. Take another example: the 'Arnolfini portrait' by Jan van Eyck (https://www.nationalgallery.org.uk/paintings/jan-van-eyck-the-arnolfini-portrait), which has been the subject of multiple interpretations (to get the flavour of some of them, see https://en.wikipedia.org/wiki/Arnolfini_Portrait). Here too we find a small dog in the foreground, standing between the couple, that is usually understood as symbolizing fidelity, but also, apparently, could mean lust! Again, these are properties imputed to the target (the couple, whoever they are and whatever their relationship, whether that of husband and wife, betrothed, or widower and ghost) which have to be keyed in this way because of the difficulties in representing them directly, within the constraints of that kind of painting.

This idea of a 'key' can be found in a number of discussions of representation: Bolinska (2013) refers to 'informativeness'; Contessa (2007) uses 'scope'; Giere (2004) talks of 'relevance' in this context; and more explicitly, perhaps, Shech (2015) refers to the 'representational code', as mentioned above.[19] And it does seem an appropriate element to include when it comes to certain artworks at least—indeed, the burgeoning field of iconography in art hinges on decoding certain 'keyed-up' properties (see any number of Dutch 'vanitas' paintings for example; https://en.wikipedia.org/wiki/Dutch_Golden_Age_painting#/media/File:Pronkstillleven.jpg).

However, here again we should be careful in transposing such features to scientific representations—when it comes to the Phillips-Newlyn machine, for example, we don't find such symbols in play. In this case, the relevant hydraulic properties are mapped straightforwardly onto the economic ones via the appropriate *interpretation*. So, for example, water represents money, obviously, a large reservoir towards the top of the machine represents the treasury, other reservoirs represent the health service, education, and so on. Here we don't have 'invisible' properties that need to be keyed up in the manner of the dogs in the paintings above. So, consider, what is the basis of the painterly convention that dogs represent fidelity? Obviously it is grounded in the loyalty and devotion that dogs exhibit, especially house-trained ones and lapdogs as in the case of Pourbus's painting. (I'll leave to one side the issue of why they might be used to represent lust!)

Of course, in the case of the Phillips-Newlyn machine, we might argue that the metaphor of money 'flowing' through an economy is what underpins the

[19] Again I would like to thank one of the readers for reminding me of this and for helping me to sharpen the subsequent discussion.

representational value of water in this case and clearly there is a 'key' that enables us to read off the economic properties from the hydraulic ones but this is effectively folded into the interpretation. The point is, in the case of certain paintings we need to distinguish such a 'key' from the interpretation in general in order to highlight the role of such symbolic elements, whose meaning lies hidden within the work, but in science we don't, because these kinds of elements are typically absent. This is not to say that science doesn't involve symbols (!) nor that it doesn't use representations of 'invisible', that is, unobservable, properties. Obviously it does, on both counts.

Take the example of 'spin', in fundamental physics, which famously can be regarded as a non-classical property in that, unlike charge, say, it wasn't identified, posited, characterized, etc. prior to being reconceptualized within the framework of quantum theory.[20] This is standardly symbolized by s, where the value of s depends on the kind of particle (fermions taking on half-integral values, bosons integral and the relationship is captured by the 'spin-statistics theorem') and its component in a given direction—that is along a certain axis—can take on the values $2s+1$ (so a spin-½ particle can have the values +½ and −½, or spin 'up' and 'down'). Here we are, in effect, setting out the 'key' that enables us to impute an unobservable property, represented in the model or theory, to the system in question. In doing so we are also, of course, interpreting the relevant mathematics, and hence those associated features of the model or theory. But this is very different from the case of Princess Anne's dog—spin is represented directly within the model, not indirectly via some further convention-based and contextual understanding. An analogy might be if I introduced some further element into my model such as 'torque', say, and insisted that in this context, there was a well-established convention under which we should understand 'torque' as exemplifying spin, which is then imputed to the system under question.[21]

The point, then, is that in the case of scientific representations we do not need to draw a distinction between a 'key' and an interpretation.[22] To put it bluntly, in science there is no need for iconography! This marks one of the differences between such representations and artistic forms, and as in other cases, suggests a degree of caution when it comes to importing devices from the philosophy of art into the philosophy of science; we'll come back to this shortly in the context of another example.

It also underscores the point about denotation noted above and whether, as imported from Goodman into the current context via the DDI and DEKI accounts,

[20] Thus, although it was initially thought of as akin to rotation around some axis, this understanding was soon abandoned, not least because the value of the spin can be half-integral and there's no way of making an elementary particle 'spin' faster or slower!

[21] In the context of the emerging field of spintronics, there is something known as 'spin-transfer torque' (https://en.wikipedia.org/wiki/Spin-transfer_torque).

[22] I'd like to thank my MA class for discussions that helped to clarify this point.

it should be accepted as an appropriate element of scientific representations. Thus, in his defence of an 'inferentialist' approach to representation, Suárez highlights the role of denotation, or stipulation:

> Suppose that I stipulate that the paper upon which I am writing represents the sea, and the two pens that I use to write represent ships on the sea. This act of denotation allows us to correctly draw a few inferences about the ship-on-sea system on the basis of a consideration of the pens-on-paper system, such as, for instance, that the trajectories of ships may cross and that they may crash.
>
> (Suárez 2004, p. 772)

However, Shech takes issue with the claim that just by stipulating that the paper denotes the sea, and the two pens denote ships, we can then 'correctly' draw inferences about the ship-on-sea system. He asks,

> How, for instance, may we correctly infer by inspecting the paper and pens that 'the trajectories of ships may cross and that they may crash?' The answer is that we cannot because there is no representational code licensing any inferences from the vehicle of representation to the target. To see this, imagine that instead we stipulate that the paper denotes the color blue and the pens denote the numbers three and four. Are we then to infer 'on the bases of consideration of the pens-on-paper system' that the trajectories of the numbers three and four may cross or crash as they traverse on top of the color blue? (Shech 2015, p. 3469)

As he goes on to note, we don't infer that the sea has the same texture as writing paper or that ships are useful writing implements and the reason is that we are immersed in certain representational norms, encoded in, well, the 'code', or 'key' as Frigg and Nguyen would have it. But then it is that code/key or, rather, those norms that are doing all the work in the sense of telling us what inferences are appropriate to draw and which are not. Denotation, or stipulation, is merely a labelling procedure that in and of itself tells us nothing in that regard (ibid.).[23]

The concern, then, is that a focus on denotation, as incorporated into the 'DEKI' account via this notion of a 'key', downplays the role of interpretation and thereby blurs an important distinction between artistic and scientific representations. Granted that interpretation features in both domains, there is much greater transparency in the scientific when it comes to the iconographic elements. There is then no need to specifically invoke a 'key' in order to decode anything as the meaning is effectively given 'up front'.[24]

[23] Once again, a big tip o' the hat to one of the readers for encouraging me to think further about these issues.

[24] This is not to say that there aren't issues to do with meaning and interpretation in science! My point is that these issues do not entirely overlap with those in art.

The second concern emerges when we consider certain kinds of representation that are not so prominent in art but crop up again and again in the history of science. These include apparently inconsistent representations, such as Bohr's model of the atom and a closer look at that example again illuminates revealing differences between those accounts that emphasis denotation and structuralist approaches.

Denotation and Inconsistent Representations

Bohr's model can be described as a 'local' model insofar as it represents a clearly specified type of system (Hughes 1997, p. 330). The theory of quantum mechanics, on the other hand, can be characterized as a 'global' theory insofar as it deals with a heterogenous collection of physical systems, such as atoms, but also their constituents. In the latter case we can still say that we have representation on the DDI account since each individual system in this collection can be represented by a particular model defined in terms of the theory (ibid.). What a global theory defines, then, is not a particular model but a *class* of models and in the application of the theory to a particular system it is a 'local' member of this class that represents. Thus, '[t]here is…no significant difference between the representations supplied by local and global theories' (ibid., p. 331).

As it stands, this distinction between 'local' and 'global' theories also fits nicely within the Semantic Approach. But how can we accommodate the famous (purported) inconsistency of the Bohr model, where this (supposedly) arises from a combination of *classical principles*, such as the claim that accelerating charges radiate, and *quantum postulates*, to the effect that charges only radiate when they 'jump' between orbits? This raises the broader issue of whether and in what sense one can have inconsistent representations. I'll come back to this but for the moment let me just note the following.

The most extreme answer would be to say no, inconsistent representations are simply not possible. Now, of course, putative examples of such 'impossible' representations can also be found in art (think of Escher's work, for example, which we'll also come back to)[25] but more importantly, perhaps, insisting that Bohr's model did not represent *anything* would be a hard line to hoe. On the other hand, taking it to be both representational and inconsistent presents obvious problems. Within a Goodmanian approach, such as the DDI or DEKI accounts, one would have to appeal to some notion of *inconsistent denotation*, and then either swallow the dialetheic consequence of an ontology of inconsistent objects (Priest 2006) or, more plausibly, articulate a non-classical version of that notion (via paraconsistent logic, say) that accommodates the denotation of ordinary, consistent objects somehow.

[25] Gombrich wrote that Escher's work 'presents so many interesting comments on the puzzles of representation' (Poole 2015).

Alternatively, one could try to accommodate such cases within the framework offered by the Semantic Approach. Thus, putting things a little crudely, Bohr's model contained elements of both quantum and classical physics and if we were to focus on each element to the exclusion of the other, we might be tempted to say the theory represents a quantum or classical system respectively. However this would be to ignore what effectively 'binds' the theory into a whole and allows the two disparate elements to co-exist, as it were, and that is Bohr's central notion of a 'stationary state'.[26] It is here that the two contradictory elements come together: classical mechanics applies to the *dynamics* of the electron in the stationary state, while quantum theory comes in when the transition between such states is considered. However, it is important to note that it is not only in the discreteness of the stationary states that we have conflict between quantum and classical physics but also in the assertion that the ground state is *stable*, so that an electron in such a state will not radiate energy and spiral into the nucleus as determined by classical physics. *This*, it can be claimed, is the central inconsistency of the Bohr model and together with their discrete nature it is what makes the stationary states so peculiar.

Of course, as the Bohr model evolved and came to be supplanted, this peculiarity was eventually understood in terms of the new quantum mechanics and as a consequence, the formal inconsistency evaporated (or, better, came to be replaced by 'higher-order' or interpretational incongruences which the principle of complementarity was intended to resolve; see da Costa and French 2003, ch. 5). At the time the model was proposed, however, the notion of a stationary state was not understood at all, or at best, only *partially*, and characterizing the theory in a way that can accommodate the partial and conceptually 'blurred' nature of the stationary states would then allow for a certain internal 'looseness of fit' between the component elements of the theory. And this in turn, gives us an idea of how the theory can still be said to represent: what it presents to us is a model which possesses elements of classical and quantum physics but which has at its heart this poorly understood and conceptually indistinct notion of stationary state; and what it represents is, crudely, an atomic system or, less crudely, certain features of that system.

This 'looseness of fit' cannot be accommodated by a notion like denotation, at least not directly. Of course, one can understand denotation very simply as yielding a form of linguistic 'picking out' of the system, so that one can say the Bohr model 'denotes' the hydrogen atom, say. But as jobs go, that's pretty thin and all the work in articulating how the putative inconsistency can be accommodated, via looseness of fit or whatever, is still going to have to be done via some other means, which will have to track the relevant (structural) similarities.

[26] It is this fact, that the theory is held together as it were, that undermines attempts to accommodate the inconsistency by separating the classical and quantum elements and applying a non-adjunctive logic (for further discussion, again see da Costa and French 2003, ch. 5).

Finally, another possible way out from under this problem would be to deny that Bohr's model is 'really' (in some sense) inconsistent. Thus, Vickers has argued for a deflationary approach that eschews talk of whole theories 'being' inconsistent and instead considers, first, whether a relevant[27] set of putatively inconsistent propositions can be identified and secondly, whether the scientists involved were actually committed to these propositions (Vickers 2013). In the case of Bohr, he notes, first of all, that, granted what I so confidently asserted above,[28] commentators in fact differ as to where the inconsistency sits and thus how it should be characterized. More importantly, however, he argues that 'at least some of the propositions needed to reach genuine inconsistency are not historically relevant' (ibid., p. 46) and that in addition, they neither do any scientific work together as a group nor do they feature in the commitments of any individual scientist at a particular time. Thus, the claim that *Bohr's model* was inconsistent simply evaporates, since what is taken to be Bohr's model may vary, depending on the contexts, both historical and philosophical.

Underpinning Vickers's analysis is the idea—to which I am obviously sympathetic—that we should not simply assume that a given theory is 'out there', as it were, to be brought into the spotlight when certain issues need to be addressed. In the case of apparently inconsistent theories, we may select a particular set of propositions and that set will either be inconsistent or not.[29] The issue for us, as philosophers of science, is whether that selection can then be appropriately justified, and whether it tells us anything interesting about scientific practice (ibid., p. 108). As I shall discuss later, making that selection is in effect to *determine* what we take the theory to be; that is, by representing, or perhaps better, presenting to ourselves, as philosophers of science, theories using the Syntactic or the Semantic Approach, or whatever kind of account we choose, we are at the same time determining what the theory is qua representation of some physical system. The issue then is whether that particular meta-representation/object-level construction appropriately captures the features of scientific practice we are interested in and I would argue that in the case of apparent inconsistencies in science, the Semantic Approach does a pretty good job.

Representation and the Toolbox

Let me make my stance clear(ish): I think we should be cautious in constructing our (meta-level) framework for understanding representation in science on the basis of importing certain devices, such as denotation, used for understanding

[27] Relevant in an appropriate historical sense.

[28] For Vickers's analysis of the da Costa and French account sketched above, see 2013, p. 43ff.

[29] And so indeed, as one of the readers has also pointed out, Vickers's approach can also be accommodated by the Syntactic Approach but, again, that is to miss the point of the exercise.

representation in art. Instead I suggest that we begin with those meta-level frameworks that are currently used in the philosophy of science to characterize the structure of theories, inter-theory relations, and theory-data relationships.[30] Of the two most well-known, namely the 'Syntactic' and 'Semantic' Approaches, by virtue of making the relevant relations explicit in its meta-level formal characterization, I would argue that the latter offers a more straightforward 'handle' on such relations as are claimed to hold when it comes to representation. In my opinion, there are certain pragmatic advantages that accrue *at the level of the philosophy of science* from adopting the structuralist framework.

Nevertheless, I am more than happy to *appropriate* elements from approaches to representation in art, if I think they can be put to service in an account of representation in science. This is in the spirit of what I have called elsewhere the 'Viking approach' to metaphysics (and history; see French 2014) but which, perhaps less aggressively, has been renamed the toolbox approach (French and McKenzie 2012; 2015). So, bluntly, the idea is to look at aesthetics as providing a 'toolbox' of devices, moves, and approaches that philosophers of science may appropriate to develop their own discipline-specific frameworks. And having said *that*, it may be that certain of those 'tools' appear very similar to devices already deployed in the philosophy of science. Thus, as we'll shortly see, at least one aesthetician has articulated an account of representation in art that looks a lot like the approach to scientific representation I shall sketch below, following Bueno and French (2011). This may give some hope of providing a unitary account of representation in both domains. That hope aside, much of the discussion to follow will be concerned with illuminating the similarities and differences between the different approaches, not only when it comes to representation but also towards artworks and theories more generally.

Return to Resemblance

We recall how artistic and scientific representations come apart:[31] first of all, although it may be true with regard to the former that almost anything may stand for anything else (Goodman 1976, p. 5; consider, for example, Craig-Martin's *An Oak Tree*: http://en.wikipedia.org/wiki/An_Oak_Tree), this is not the case in science: not anything can serve as a scientific model of a physical system since if the appropriate relationships are not in place between the relevant properties, then the 'model' will not be deemed scientific to begin with (French 2002).

[30] With, of course, the caveat that I am going to read 'of theories' here in a certain, eliminativist way!

[31] This section is based on Bueno and French 2011 and 2018, ch. 3; see also Bueno and Colyvan 2011.

Thus, denotation should be replaced with '*immersion*', by which a mapping is established between the empirical set-up and the representational structure (Bueno and Colyvan 2011). This can then be explicated using the 'tool' of a partial isomorphism holding between the relevant structures (Bueno and French 2011; for the deployment of this framework in the context of the application of mathematics to science, see Bueno and French 2018).[32] This can be regarded as the 'informational' stage of modelling (see Chakravartty 2009; Shech 2016), since there is information transfer from the empirical set-up to the (mathematical) model. Consequences are then drawn using the representational structure, in the '*inferential*' stage, and it is here that we might describe the model as having a 'functional' role, whereby it is used in order to derive particular results (Chakravartty 2009). The final step is *interpretation*, taking the results of the derivation stage back to the target system via another mapping (which may be different from the first), which, again, can be formally described in terms of partial isomorphisms (or some other suitable mapping).[33] The focus is, once again, on the 'informational' aspect (Chakravartty 2009), given the reverse information transfer from the (mathematical) model, suitably interpreted, to the empirical set-up.

If we grant that an important role for models in science is to allow scientists to perform various forms of 'surrogative' reasoning (Swoyer 1991), then this approach accommodates this by allowing the 'surrogative' nature of such reasoning to effectively ride on the back of the relevant partial isomorphisms, since it is through these that we can capture the kinds of idealizations, abstractions, and, contentiously perhaps, inconsistencies that we find in the models that actually feature in scientific practice. It also captures the core features of the so-called *inferential* and *interpretational* accounts of representation (Contessa 2011). The former takes it as a necessary condition for something to be a representation precisely that a user is able to *perform* these surrogative inferences (see Suárez 2004; Suárez and Solé 2006). However, this seems to turn the relationship between representation and surrogative reasoning upside down, since it suggests that a model is a representation of a system by virtue of its user being able to perform surrogative inferences from the model to the system, whereas it would seem that it should be the reverse—we can perform such inferences *by virtue of the model being a representation* (Contessa 2011).

Alternatively, the 'interpretational' account emphasizes that for a model to count as a representation, the user must first *adopt an interpretation* of the model in terms of the system targeted. That interpretation will then, of course, provide the user with a set of rules via which she can translate features of the model into features of the system and thereby run the above kinds of inferences. As it stands,

[32] But other kinds of morphisms, such as partial homomorphism, can be—and often are—invoked (see e.g. Bueno, French, and Ladyman 2002; and Bueno and French 2011).

[33] In effect this folds together DEKI's 'keying up' and 'imputing' moves.

that would allow representation to come too cheaply, so further conditions must be satisfied for a given model to count as a *faithful* representation (Contessa 2007). These include that the model should be *similar*, in certain respects and to certain degrees, to its target system (Giere 2004; Teller 2001). And of course, this can be formally captured via a partial isomorphism being taken to hold between the model and the system (where the latter is taken to instantiate an appropriate structure, of course; see French and Ladyman 1999).[34] In this case, *complete* faithfulness would be satisfied by an isomorphism holding between the model and the system and *degrees of faithfulness* could then be accommodated through the framework of partial isomorphisms as outlined in Chapter 2 (and if one really wanted to, one could even adopt a particular metric for this; see da Costa and French 1990).[35]

The above three-stage framework of 'immersion, inference, and interpretation' thus explicitly accommodates both the interpretational and inferential aspects of modelling, in a way that does not privilege one over the other.[36] Here we see how, from the perspective of capturing the ways that scientific models represent, the advantages of adopting the Semantic Approach.[37]

The take-home message, then, is that first, the establishment of the relevant mappings needs to be at the forefront of any such account, since the inferences that are at the heart of scientific modelling ride on the back of these; but second, to base one's account of representation on such inferences alone, without accepting the underlying formal aspects, is to engage in a confusion (Chakravartty 2009; Shech 2016). Let's look again at a couple of the examples I've already mentioned.

[34] This 'epistemic' form of representation can be further distinguished from 'ontological' faithful representation that has to do with the extent to which the vehicle or model is 'on the road to the truth', with respect to the target (see Shech 2015). Typically one might expect these two senses of faithfulness to align but they come apart in the case of certain models of statistical mechanics, which represent phase transitions in infinite systems and hence are epsitemnically unfaithful, even if such models bring us closer to the truth (see also Bueno and French 2018). I am again grateful to a reader for bringing this distinction to my attention.

[35] Contessa himself worries that faithfulness is a 'gradable' notion, whereas morphism is not, in the sense that whether a given morphism holds or not is a 'yes–no' question. Granted that a partial isomorphism either holds or not (and it doesn't, the representation can be said to be completely unfaithful!), the partiality of partial isomorphisms *is* gradable. Indeed, Contessa goes on to suggest that 'faithful epistemic representation is a matter of structural similarity, and that the more structurally similar the vehicle and the target are (i.e. the stronger the morphism between the structures instantiated by the vehicle and the target is), the more faithful an epistemic representation of the target the vehicle is' (Contessa 2011, p. 130).

[36] Thus, as Chakravartty (2009) suggests, the distinction between 'informational' and 'functional' approaches really amounts to little more than that between means and ends.

[37] One of the readers has suggested that there is nothing to stop the Syntactic Approach from accommodating these interpretational-inferential-informational aspects—perhaps not, but then its up to an advocate of that approach to go ahead and do it!

Case Study 1: Superconductivity (Again)

We recall that in this case, a core experimental result (the 'Meissner effect'), effected a change in our understanding of this phenomenon: from one having to do with zero resistance, to something that was analogous to the essentially quantum property of diamagnetism.[38] Following this shift, and on the back of that analogy, the London brothers constructed a 'macroscopic' model that adequately represented the phenomenon without explaining the (ultimately quantum) mechanisms behind it. By taking the magnetic behaviour of a superconductor to be *similar* to that of a diamagnetic substance they were able to shift the characterization of superconductivity from a phenomenon involving a current that persists in the absence of an electric or magnetic field, to one in terms of the current understood as a kind of *diamagnetic volume current*, whose existence is necessarily dependent upon the presence of a magnetic field. Thus, the critical feature that was drawn upon at this stage of the construction of the representation was the relationship between the magnetic field and the current, which, within the Semantic Approach, can be captured by the R_1 components in the relevant partial structures.

However, those aspects of the diamagnetic structure that were relevant for the *microscopic* understanding were not carried over and these can be captured by the R_3 components. It was this further structure that proved to be of critical importance for the subsequent development of the 'microscopical' interpretation of superconductivity (see Bueno, French, and Ladyman 2012). In particular, by considering a superconductor as a single big diamagnetic atom, and investigating the form of the current that would be obtained on quantum mechanical grounds, an expression for this current could be obtained that had *exactly* the same form as the expression for superconducting current flow. In subsequent work the quantum mechanical account of diamagnetism was assumed right from the start, and riding on the back of the analogy, as it were, was used to give the outlines of an explanation of superconductivity (ibid.). Here again further structure was imported from the diamagnetic domain, yielding something close to an isomorphism between the relevant structures.[39] Thus, as developments proceeded, elements previously captured by the R_3 components fall under the R_1.

Here we see the way in which representations in science can be motivated and *shifted* not only by empirical considerations but through similarity based analogies with other domains. And these representations at the different stages of

[38] Historically of course this is a pretty crude summary; for a more nuanced discussion, see Potters 2019.

[39] Close but not quite an isomorphism, because of course there are still differences between a superconducting ring and a diamagnetic atom, so we are still invoking an analogy here.

theoretical development can be characterized by suitable partial morphisms holidng between partial structures within the Semantic Approach. First, structure was brought across from the domain of diamagnetism into the model of superconductivity, where the relationship is understood to be a formal one (in the sense that the models in question satisfied the same mathematical equation).[40] Second, once it was realized that the analogy goes beyond such formal aspects, further structure in the form of the relevant relations between electrons, was transferred across to yield a derivation of the core mathematical equation, now suitably interpreted at the microscopic level. At this point the phenomenon came to be characterized by its fundamental constituents.

Case Study 2: Bohr's Model of the Atom (Again)

As noted already, according to this model an electron in a hydrogen atom can exist in one or other of a discrete set of orbits or 'stationary states'.[41] Classical mechanics can then be applied to account for the dynamical equilibrium of the electron in one of these stationary states but not to transitions between states, where the relation between the amount of energy and frequency of radiation emitted is given by Planck's formula (Bohr 1981, p. 167). However, it is not only with regard to the discreteness of the states that we have an apparent conflict between quantum and classical physics but also when it comes to the assertion that the ground state is stable, so that an electron in such a state will not radiate energy and spiral into the nucleus as determined by classical physics.

As we've already seen, this example is particularly interesting in this context since it raises the issue of whether, and in what sense, one can have *inconsistent* representations in either science or art.[42] We'll consider examples of the latter shortly but let us look at Bohr's model in a little more detail: it contains elements of both quantum and classical physics and if we were to focus on each element to the exclusion of the other, we might be tempted to say the theory represents a quantum or classical system respectively. However, this would be to ignore what effectively 'binds' the theory into a whole and allows the two disparate elements to co-exist, as it were, and, as we noted above, that is Bohr's central notion of a 'stationary state'. It is here that the two contradictory elements come together:

[40] That the superconductor was taken to be *represented as* a kind of diamagnet is clear from statements by the Londons themselves (London and London 1935, p. 88).

[41] This section is drawn from Bueno and French 2011.

[42] Callendar and Cohen (2006) treat the claim here, that the partial structures account can accommodate inconsistent theories, as constituting a form of *reductio* of this account. However it is not the apparent examples of inconsistent models in the history of science that alone motivates it. Rather, having constructed the account for other reasons, it counts in its favour that as well as accommodating more straightforward cases of scientific representation, it can also handle apparently inconsistent theories.

classical mechanics applies to the *dynamics* of the electron in the stationary state, while quantum theory comes in when the transition between such states is considered. The central inconsistency of the Bohr model can then be located in the *stability* of the ground state and together with their discrete nature it is this that makes the stationary states so peculiar. Of course this is not to say that there is a corresponding inconsistent object, since I'm not taking the model to be true, and, indeed, neither did physicists at the time (again for the pitiless details, see da Costa and French 2003, ch. 5).

As the Bohr model evolved and was supplanted, this peculiarity was eventually understood in terms of the new quantum mechanics and as a consequence, the formal inconsistency evaporated—or, better, came to be replaced by interpretational incongruencies associated with wave-particle duality, for example, which the principle of complementarity was intended to resolve (ibid.). At the time the model was proposed, however, the notion of a stationary state was not understood at all, or at best, only partially, and if one were to represent Bohr's theory in terms of partial structures, the stationary states would have to be located among the R_3 components, as relationships which had not yet been established to hold or not for those particular elements of the domain, namely atoms. As quantum theory developed, this notion came to be better understood—in particular in terms of the eigenstate of the relevant Hamiltonian—and it can be thought of as shifting from the R_3 component in one partial structure (corresponding to Bohr's theory) to the R_1 component of another (corresponding to the new quantum mechanics of the mid-1920s). By 1926 the concept of a stationary state came to be formulated in a mathematically precise way, and it is this stationary state, more precisely understood, that was ultimately incorporated into the new quantum theory. Of course, something was lost along the way, most notably Bohr's notion of well-defined circular electronic orbits, which are components that were not mapped into the new quantum theory via any partial morphism.

We can perhaps think of the notion of a stationary state in Bohr's theory as doing 'double duty' in being understood in both classical and quantum terms (we'll come back to this idea of certain elements performing double duty when we consider some examples from art). Indeed, we can think of Bohr's skill here in terms of tinkering with the conventional understanding of stationary state, so as to preserve the 'illusion' of an entire theory or model that was in fact inconsistent. And when it comes to the crucial surrogative reasoning, a long as one derives results from only selected parts of the theory, the collapse into triviality that follows from inconsistency is avoided, but not if one takes the theory as a whole, as an entire object as it were. The way in which parts of the predecessor theories, such as Maxwell's, and the way in which parts of Bohr's theory can be used in obtaining the relevant derivations can both be captured in terms of partial isomorphisms.

What we place in the R_1-, R_2-, or R_3-components can only be decided in retrospect, of course, since we are characterizing—or, *representing*—the theories from the point of view of a philosophical understanding of science. Not surprisingly, different philosophical views may highlight different features of scientific practice, and disagreements about the importance of the chosen features may emerge. The point is that by characterizing the theories in the way indicated here, we can accommodate the partial and conceptually 'blurred' understanding of the stationary state in Bohr's theory that allows for a certain internal 'looseness of fit' between the component elements of that theory. And this, in turn, gives us an idea of how the theory can still be said to *represent*: what it represents are atoms, but it does so in terms of elements that incorporate aspects of classical and of quantum physics and has at its heart this poorly understood and conceptually indistinct notion of stationary state.

(Partial) Isomorphism in Artistic Representation

I have mentioned how we might import certain devices from aesthetics into the philosophy of science but we might ask whether the traffic might go the other way. In particular, we can ask, can this kind of formal account be extended to certain examples of representation in art? The answer would appear to be 'yes', with Budd's (1993) account of depiction standing as an artistic counterpart to the Semantic Approach.[43]

There are three components to this account. First of all, there is the focus on pictures and depiction, regarded as a distinct kind of representation:

> it is definitive of a picture that it represents what it depicts by depicting it, and depiction is a form of representation different from any other. (Budd 1993, p. 154)

What distinguishes pictorial representation from other kinds can be articulated initially and intuitively in terms of a picture *sharing* properties with its subject. This intuition can be captured more precisely by drawing on the crucial distinction between our 'visual world' and our 'visual field':

> My visual world at any time is the complete way the world is then represented to me by my visual experience. My visual field is a certain aspect of the way the world is represented to me by my visual experience. (ibid., p. 158)

The relationship between the two can be understood in terms of *abstraction*: my visual experience represents the world as a collection of objects 'spread out in three-dimensional space' and my visual field is what is left when we abstract the

[43] Another useful comparator is Hopkins's 'EROS'—'experienced resemblance in outline shape'—account (Hopkins 1998; 2005); again I am grateful to one of the readers for alerting me to this work.

apparent distance of these objects from me. Thus, for example, if my visual world contains a circular object that is tilted away from me (ibid.), then within my visual field this object will appear elliptical. However, one should not characterize this difference in terms of the visual world and visual field *containing* different objects, but rather in terms of different accounts of how my visual experience represents the world as being, one complete and one partial. Nor should we understand the visual field as some kind of two-dimensional entity which we perceive: 'a representation in my visual field is the manner in which the world is in some way visually represented to me in two of the three spatial dimensions' (ibid., p. 159). The claim that a painting represents by sharing properties with its subject can then be firmed up as follows:

> a picture looks like what it depicts only with respect to properties of the spectator's visual field, not those confined to his visual world. (ibid.)

The third and final element is the one that invites comparison with the Semantic Approach, namely the explicit introduction of *isomorphism* into this context, where this holds between the *structure* of the surface of the painting and the *structure* of the relevant visual field. So, when you look at a painting, what you see is the structure of the surface of the painting as being isomorphic with the structure of the visual field of the state of affairs that the painting depicts. This gives the following account of pictorial representation:

> First, your experience must involve a visual awareness of the presence before you of a marked surface. Secondly, you must see the structure of the surface as being isomorphic with the structure of the visual field representation of the picture's subject when seen from a certain point of view, namely, that from which it has been depicted. (ibid., p. 161)

Putting it in a nutshell, *representation consists in the perceived isomorphism of structure* (ibid., p. 162).

Budd illustrates his account with a variety of examples, some of which also bear interesting comparisons with representations in science.

First of all, consider that old chestnut, the Necker cube (Bud 1993, pp. 159–61; see https://en.wikipedia.org/wiki/Necker_cube), much loved by certain philosophers of science as an illustration of Hanson's point about the theory-ladenness of perception (see Kuhn 1970, p. 111). In this case, the visual field representations of the different orientations of the cube have the same two-dimensional structure. Hence your visual experience contains the same two components: 'a visual awareness of the lines as lying on the drawing surface and the same representation of the lines in your visual field' (Budd 1993, p. 160). What distinguishes your seeing the picture with one orientation as opposed to the other is your experience of what your visual field representation is isomorphic to: first, you see the structure of lines as being isomorphic with the visual field representation of the cube with one orientation and then as being isomorphic with the visual field representation of the cube with the other (ibid.).

Perhaps even more significantly, cases of 'abstract' art and 'inconsistent' representations (or representations that contain contradictory elements) can also be accommodated. This requires a little further refinement. First of all, in many cases the visual field representation of the picture surface may be 'strikingly unlike' that of the relevant state of affairs. The perceived dissimilarities can take two (obvious) forms: the visual field representation of the picture-surface can lack features possessed by the visual field representation of what is depicted, or the former can possess additional features that the latter does not. Of course, this absence or presence of features does not imply that the spectator sees *what is depicted* as lacking or possessing the relevant features. The absence of colour in a black and white drawing, for example, is not understood by the spectator as indicative of a lack of colour in what is being depicted but only as not indicating any colour; as Budd puts it, 'the object, as depicted, has an indefinite appearance in the dimension of colour' (ibid., p. 164). Referring to intentions once more, in this case, by eschewing colour, the artist's intention is simply 'to depict only the spatial structure of a state of affairs and the comparative brightness of its parts, perhaps, not the colours of its constituent objects' (ibid.). Likewise, a spectator viewing a painting or drawing in which stippling or cross-hatching is used to represent variations in tone does not read these features into the objects being depicted. In general, we typically—but not always—abstract from certain details of the picture-surface.

Consider, for example, Picasso's 'Head of a Girl': here the loops and spirals of Picasso's drawing correspond to the visual field representation of a girl's hair only when other features are abstracted away. In general, schematic depictions resemble the visual field representations of the corresponding objects 'only with respect to the structural features they depict' (ibid., p. 164). Thus, the perceived isomorphism refers to structure 'at a certain level of detail' (ibid., p. 165) and one can easily see how this account also accommodates Cubist works, for example. Indeed, Budd himself explicitly applies his approach to Braque's 'Pitcher and Violin' in which 'naturalistic' perspective is abandoned (see http://www.artchive.com/artchive/B/braque/v_pitchr.jpg.html). Here we can draw an obvious comparison with abstraction in the case of models. Furthermore, as noted in French (2003), this kind of account may be extended to more complex works such as Picasso's *Guernica*, although here, as in other cases, there may be more than one possible 'target' of the representation.

More intriguingly, this framework can also accommodate depictions of apparently 'impossible' or contradictory objects as exemplified in the works of Escher, for example. If we consider the famous case of the three-pronged tuning fork, or 'impossible trident', which has two rectangular bars at one end and three cylindrical bars at the other (Budd 1993, p. 166; see https://en.wikipedia.org/wiki/Impossible_trident), then what we have, in effect, are two incompatible isomorphisms:

> if we direct our gaze at one end of the configuration of lines we see the structure of the lines as being isomorphic with that of visual field representation of two rectangular bars, whereas if we look at the other end the lines look to be structurally isomorphic with the visual field representation of three cylindrical bars.
> (Budd 1993, p. 166)

However, we cannot impose on the lines a single, consistent interpretation and attempts to do so lead to well-known visual discomfort. Here we can draw another suggestive parallel between the artistic and scientific cases: if we focus on, or abstract out, the classical and quantum aspects of Bohr's model, then each can be taken to represent, respectively, classical and quantum objects. It is when these aspects are brought together and we wonder what the model as a whole represents that we experience a kind of cognitive tension akin to the visual discomfort we encounter when we look at one of Escher's drawings,[44] or the trident illusion. The tension arises precisely because in each case we are forced to flip between incompatible aspects of the representation.

Here's another example: Escher's famous waterfall drawing (see http://en.wikipedia.org/wiki/Waterfall_(M._C._Escher)). There we have two sets of mappings: one from the world to the surface of the drawing, and another from the surface of the drawing to the visual field. Focusing first on the relation between the world and the surface of the drawing, note that in so far as no existing physical waterfalls can possibly instantiate the arrangements depicted in the drawing, the latter carries more information than the former. In that sense, any mapping from the world to the surface of the drawing will lose some information. However, any waterfall in the world is a three-dimensional object, and Escher's waterfall drawing is only a two-dimensional construct. In that sense, some information is lost when the former is mapped into the latter. Of course, by using a suitable perspective, that information can be recaptured, thus producing the illusion of a three-dimensional scene on a two-dimensional surface. Thus, we don't have here a simple case of informational gain or loss. Informativeness is a context-dependent notion, and depending on the features we are attending to, certain mappings may entail information loss and others information gain. Given the *partiality* of information involved, capturing such mappings in terms of partial morphisms seems the appropriate way to do it (Bueno and French 2011).

[44] Escher's 'Ascending and Descending' was apparently inspired by L. Penrose and R. Penrose's paper 'Impossible Objects: A Special Type of Visual Illusion' (Penrose and Penrose 1958).

Just as the case of Bohr's model, Escher's drawing creates the illusion of an inconsistent waterfall because certain elements in the painting do 'double duty': for example, consider some of the pillars supporting the aqueduct in the drawing. At one end, they rest in the inner wall of the aqueduct, but at the other end they support the outer wall. Also, close examination reveals certain arches that vanish into thin air. Referring back to our earlier discussion, Escher clearly violates here the conventions of perspectival drawing. Part of his brilliance lay in discouraging us from examining closely these aspects of the work, thereby maintaining the illusion that it represents an inconsistent object.[45] Here, the cognitive dissonance arises when one attempts to apply these conventions to the whole drawing.

Of course, some care must be taken in drawing such parallels. In the Bohr case, the cognitive discomfort was eventually assuaged—or, perhaps, merely transposed to the level of interpretation—through the introduction of a formally consistent theory (namely the later quantum mechanics of Schrödinger and Heisenberg, as put on a sound mathematical footing by von Neumann; see Bueno and French 2018). In this case, that discomfort carried some heuristic force, leading to further theoretical developments (see da Costa and French 2003, ch. 5). However, no such developments can, or should, be expected in the case of Escher's work. Again, this reflects an obvious difference between scientific representations and artworks: in the case of the latter, part of the aim may be precisely to generate a measure of discomfort.[46]

Budd also invokes another example, Holbein's famous painting, 'The Ambassadors', but here complications arise for his account, as we shall see. Holbein portrays two apparently wealthy and well-educated individuals,[47] standing by a table festooned with various instruments, scientific and musical, indicating their education, their cultural upbringing, and, it is claimed, aspects of the then current political context (the iconography of this painting is still subject to considerable debate).[48] But most strikingly, as is well known, the foreground of

[45] Escher himself wrote 'If you want to express something impossible, you must keep to certain rules. The element of mystery to which you want to draw attention should be surrounded and veiled by a quite obvious, readily recognisable commonness' (Poole 2015).

[46] This does not preclude an exploratory role for such 'impossible' objects. Shech (2016, p. 327) uses the example of anyons—two-dimensional particles that obey a non-standard form of quantum statistics—to illustrate how the structure of the theory can be further illuminated by such entities. Insofar as the mathematical description of anyons can be regarded as part of the 'surplus structure' associated with quantum statistics (along with the representations of paraparticles), this can be accommodated by the three-stage framework involving partial structures, together with the heuristic fruitfulness that arises from the exploration of such structure (Bueno and French 2018). Escher's works and other examples of 'impossible' objects may also be regarded in a similar way (Shech 2016, pp. 327–8).

[47] One is supposedly Jean de Dinteville, French ambassador to England in 1533. The other is his friend, Georges de Selve, bishop of Lavaur, who acted as ambassador to the Holy Roman Emperor, the Venetian Republic, and the Holy See. However, there is still some dispute over their precise identification.

[48] Thus, there is a lute beneath the table, with a broken string, which is supposed to represent religious discord. On the DEKI account, such an interpretation can only be made on the basis of some 'key', as discussed above (Frigg and Nguyen 2017b).

the painting includes a depiction of a skull that is represented anamorphically, so that it appears as a skull only if looked at from one side of the picture. Whether Holbein's intention here was to emphasize the transience of life and our own mortality (see for example Kieran 2005, pp. 92–6), as in the case of the Dutch 'vanitas' paintings mentioned previously, or to create a striking effect dependent on the painting being hung in a stairwell, or whether he was just showing off his artistic skills, is unclear. But within this account, the depiction of the skull is taken to be isomorphic with the representation in the visual field of a skull in the precise sense that there exists a one-to-one mapping between the points that compose the two, although:

> you see it as being structurally isomorphic only when you see it from the side and accordingly, you see it as a depiction of a skull only from that unusual, oblique point of view, not when looking at it from straight on. (Budd 1993, p. 162)

This kind of perspectival anamorphosis developed in western art during the early Renaissance, although earlier examples have been claimed and more recent forms can be seen in certain chalked pavement art (see for example: http://www.julianbeever.net/index.php?option=com_phocagallery&view=category&id=2:3d-illusions&Itemid=7). The crucial point is that in order to understand what it is that the chalked image or the particular feature such as Holbein's skull represents, the observer needs to shift her perspective with respect to the painting and stand in a particular relationship with it (for an extensive discussion, with historical and modern examples, see Collins 1992 and also Di Lazzaro, Murra, and Vitelli 2019 which nicely focuses in particular on the mathematics involved).

Now, we just need to pause a moment at this point. In a sense, Budd is right—the depiction of the skull is isomorphic with the representation in the visual field of a skull, but only from the vantage of a specific point. It is only by shifting to this vantage point that the role of the anamorphism within the representation as a whole can become apparent. Doing so opens the possibility of further interpretations of the painting: we may, for example, interpret the skull as a 'vanitas' element, representing one of the three 'planes': the heavenly, the earthly, or realm of the living, and death.[49] Thus, '[s]ciences and arts, objects of luxury and glory, are measured against the grandeur of Death' (Bätschmann and Griener 1997, p. 184). Here we might again recall the model of superconductivity, where the Meissner effect forces a shift in perspective on the phenomenon and drives a new interpretation of it.

However, the difference is that in the case of *The Ambassadors*, the 'driver' behind the shift is incorporated within the representation itself (an obvious

[49] There are of course many interpretations of this painting—just Google 'interpreting The Ambassadors'!s

difference perhaps, given the role of experimental results in science). More interestingly, the skull encourages the viewer to become an 'eccentric' observer (Collins 1992; Di Lazzaro, Murra, and Vitelli 2019), in the sense of one that oscillates back and forth between the two perspectives, yielding an active transformation that allows the observer to play a role in constructing the representation in a way that is not so straightforwardly captured by talk of a one-to-one correspondence. Thus, there is a kind of visual indeterminacy within the painting, that is only resolved into a determinate image once the shift has been effected. Indeed, what we have here are two correspondences, with one embedded within the other and only available by undergoing a specific transformation. But although it is interesting to consider other ways in which indeterminacy is expressed within art, including literature for example, and how these might compare to indeterminacy in science, particularly quantum physics (see Darby, Pickup, and Robson 2017), it is not clear that there are any scientific models that present a similar set of embedded correspondences (but if anyone knows of any, please let me know!).

This demand that the observer play a more active role is also a feature of many other artworks, of course. However, the 'pay off' for adopting such a role in this particular case may seem rather meagre: a mere *memento mori*, the effect of which could have been equally as well or even more effectively achieved by placing a clear image of a skull under the table, or via any of the other means typically employed at the time (e.g. dancing skeletons and the like).[50] For this reason it has been suggested that there is more to Holbein's painting than is revealed by simply getting a grip on the symbolism of the various objects placed on the table (Kenaan 2002). Indeed, taking the painting to present a 'matrix of puzzles' (ibid., p. 63) begging to be resolved may in fact obscure (perhaps deliberately) what it is 'really' trying to convey, which is something hidden, or a secret that cannot be presented openly. Again, the anamorphic skull is crucial here:[51] granted that such techniques were not unusual by this time, it is significant that typically they were used as a kind of trick, or visual illusion, to 'hide' an erotic scene in some cases, or a political figure, to be satirized or secretly revered in others. Holbein does not deploy such a technique in any of his other works and here he does not use it in any of these typical ways; rather, it is incorporated within an apparently quite formal portrait but in a way that can't help but catch the viewer's attention.

[50] Here we recall that Holbein himself produced a famous series of woodcuts ('The Dance of Death') in which death, disguised in various ways, confronts various individuals as they engage in everyday activities, all of whom fall into his clutches, despite their protestations. The series is in fact one long *memento mori*.

[51] But not the only one—the background curtain also suggests obscuration and something hidden (Kenaan 2002).

The point is that this secret cannot by revealed by unravelling the symbolism of what is already 'in' in the standard viewer's visual field but only by adopting the perspective of the 'eccentric' observer:

> Thinking of the view, or of the visual field, opened by the painting, we can tentatively make a distinction between that which structures the specific visuality of this field and that which appears *in* the already structured visual field. This is a distinction between the structural conditions and the contents of the painting
> (Kenaan 2002, p. 68)

In other words, the anamorphism should not be seen as an emblematic object in the painting but rather as an 'element of a second order', or as a structuring element that helps lay down the conditions via which that which the painting secretly represents can be grasped.[52] The use of anamorphism to convey the idea that one must give up one's ordinary or conventional perspective is crucial to this endeavour and, furthermore, it tells us not to seek for a 'deeper' meaning but to see, from a fresh perspective, that which is given on the surface of the representation and is thus 'right before our eyes' (2002, p. 69).

Note, first, that trying to account for this in terms of a 'key', as in the DEKI approach outlined above, doesn't quite capture what is going on. Yes, the skull, as it stands, is a symbol—a *memento mori*, as we noted—and as such must be 'keyed' but as Kenaan suggests, this is to miss the real point, conveyed by the anamorphism, which is to shift perspective on the scene. And secondly, relatedly, what is important here is not that there is a one-to-one correspondence between the smeared patch of paint in our visual field and a skull, when that field undergoes a certain transformation, but that *we need to make such a transformation to see what the painting really represents*. The painting qua representation, includes certain conventional or standard elements, captured via the appropriate 'key'—the figures, posed in a certain way, the various objects, presented in ways that invite political or religious interpretations, almost as if to deliberately mislead us—but also crucially incorporates a further element that by its very presence and *what we have to do to see it as itself a representation*, undermines the likewise conventional or standard interpretation associated with the aforementioned elements (namely, that the painting is 'about' two wealthy, erudite and politically powerful figures living in a time of religious discord), and encourages us to see what the painting is 'really' about.[53]

[52] That anamorphic techniques may be used in encryption is mentioned in Di Lazzaro, Murra, and Vitelli 2019.

[53] OK, OK, I'll tell you—according to Kenaan, the secret that the painting is both hiding and revealing to us, once we drop our conventional perspective and shift our stance, is that our two ambassadors are a gay couple. He invites us to consider the way they are framed and presented as a couple, drawing a comparison with van Eyck's 'The Arnolfini Marriage', mentioned above. He also draws our attention to the respective stances of the two men, and their positioning, with de Dinteville on the left hand side, in what would be the man's place in a conventional marriage portrait, holding a phallic looking dagger

That is not to say that no deciphering at all of the painting is involved in figuring out what it is 'about', of course, but the iconographic move involved (literally!) is not straightforwardly accommodated by a 'key' or 'code'—nor should we expect it to be given the revelation of what the painting is actually about. If Kenaan is right (2002, n. 121; and I'm not saying he is), this wouldn't be the sort of thing that could be uncovered through the usual *memento mori* symbolism. More significantly for us, the deciphering itself is entirely different from the standard cases of skull = death, peeled lemon = the bitterness of life, and so on.[54]

Just as importantly, it is at the very least unclear whether we have anything like this phenomenon of the 'eccentric observer' in science, although we do have elements or features of models and theories that force us to shift our perspective and consequently, our understanding of the relevant phenomena. Thus, consider the equivalence between inertial mass and gravitational mass in Newtonian mechanics. The former is a measure of the resistance of the body to acceleration under the application of some force and as such it can be regarded as passive. The latter has both an active and passive aspect: it is active insofar as it generates the gravitational field and it is passive insofar as it reacts to a gravitational field (for a discussion of Newton's struggle to reconcile these different features within his metaphysics, see Harman 1982). Within experimental error, the two have been shown to be equivalent (to an accuracy of 2×10^{-13} using lasers and the reflector left on the Moon by Apollo 11, which is utterly cool of course; see https://en.wikipedia.org/wiki/Lunar_Laser_Ranging_experiment) and indeed, Einstein elevated this equivalence to the status of a general principle in the development of general relativity (Norton 1985).

Now, in his detailed discussion of heuristics in science, Post suggested that this equivalence could be conceived of as a kind of 'footprint' of general relativity within Newton's theory, in the sense that, seized upon by Einstein, it led him to the former. In other words, like Holbein's *Ambassadors*, Newton's theory contained within itself the element that forced a shift to an entirely new understanding of the world. But of course, unlike Holbein, Newton did not intend to force such a shift[55] and we only identify the footprint in the scientific case retrospectively, from the vantage point afforded by the successor theory.

So, it's not enough to simply note, when it comes to *The Ambassadors*, that the depiction of the skull is isomorphic with the representation in the visual field of a

near his crotch, while at the same level, de Selve demurely draws close his robe and adopts what was commonly portrayed as a feminine bodily posture. In this context, the *memento mori* acquires further significance: intimations of death were not uncommon in depictions of married couples, not least to signify not only the impact of mortality on one's relationship with another but also the mutual commitments exemplified by the vow 'til death us do part'. Of course, this interpretation, like so many, can be contested; what I'm interested in is the way the representation contains an element whose viewing via a transformation may encourage us to re-think what the representation is about.

[54] Once more I am grateful to one of the readers for pressing me on this.

[55] Of course, Holbein may not have intended a shift to the particular interpretation that his painting represents a gay couple but, assuming that he didn't decide to paint an anamorphic skull on a whim, he did intend for it to have some such effect.

skull from the vantage of a specific point; nor is it enough to advert to the role of some 'key' that captures the veritas element. Budd's account must be augmented with a consideration of how that shift to a specific vantage point potentially generates a shift in how the painting as a whole should be interpreted. Likewise, when it comes to scientific representations, such as Newton's theory of gravitation and Einstein's general theory of relativity, the Semantic Approach must be supplemented with some account of how the 'footprint' of the latter within the former was exploited by Einstein. That in turn will require serious consideration of the heuristic moves that were made, something we shall come back to in later chapters.[56]

It is worth noting that Budd's account is not restricted to isomorphisms relating to spatial structure but can also incorporate colour, texture, brightness, and so on: 'there is not only a perceived isomorphism of spatial relations but a correspondence of perceived colour: the spectator sees the picture not merely as a structural spatial isomorph but also as a chromatic icon' (Budd 1993, p. 167). Thus, this approach can be generalized and if we characterize both the surface of the painting and the relevant visual field in more formal structural terms, then the central idea that the visual field representation of the picture-surface can lack features possessed by the visual field representation of what is depicted, or the former can possess additional features that the latter does not, can be captured in terms of partial correspondences holding between these structures in a way that meshes nicely with the partial structures account.

There are of course other accounts of depiction that are available (see for example Lopes 1996). However, my interest does not only concern the examination of possible commonalities and differences between representation in art and in science but also the comparison of how representation is *presented* within the philosophy of art and the philosophy of science. Given the prevalence within the latter of isomorphism-based approaches, Budd's art-based formulation offers a useful bridge between the two fields[57] and illuminates suggestive parallels with the Semantic Approach by virtue of the formal similarities.[58]

[56] That some such supplementary consideration of heuristics is required here has long been noted (see da Costa and French 2003; Bueno and French 2018).

[57] However, as Sanchez-Dorado notes, Budd's concerns are different from mine, in that he was concerned to distinguish pictorial from non-pictorial representation, rather than trying to establish the conditions for representation (Sanchez-Dorado 2017, p. 16). Furthermore, as the above summary makes clear, Budd articulates a notion of 'experienced similarity' between a painting, say, and the relevant scene, whereas I am understood to have a more 'objective' notion in mind. Thus, 'the central role Budd concedes to the subject's intervening in the process of representing reconfigures the notion of isomorphism, to the point that it means something considerably different to what French is assuming it to mean in this view' (ibid., p. 17). These are fair points and indeed I am sympathetic to Sanchez-Dorado's overall concern that care must be taken when importing both examples and particular views from aesthetics to the philosophy of science or vice versa. Nevertheless, I would still maintain first, that Budd's approach gives the lie to those who insist that isomorphism-based accounts have no place in the philosophy of art and secondly, and more importantly, that it affords a useful bridge between the two fields, even if we grant that adjustments must be made on each side, as it were.

[58] Obviously, some features of Budd's account, such as those dealing with the visual field, appear to have a limited applicability to science. Likewise, the notions of models and data structures that are so important in a scientific context do not find a natural home in art.

More than this, one can identify certain 'meta-parallels' between the ways that representation is itself represented in the philosophies of art and of science. Indeed, Shech (2016) has argued that just as, following Chakravartty (2010), we can distinguish informational from functional accounts of representation in the philosophy of science, so we can make as similar distinction in the philosophy of art. Budd, of course, would be on the side of the former, whereas those who argue that pictorial representations facilitate certain cognitive functions, such as recognition (Lopes 1996; 2005) would align with the latter. Thus, it has been argued that we need more than isomorphism in such cases—we need some *knowledge* of what it is that the picture depicts (Downes 2009). Consider the following challenge: 'does visual field similarity in fact explain how we experience pictures as like their subjects, independent of knowledge of what they depict?' (Lopes 1996, p. 23; see also Lopes 2005). It has been suggested that this challenge must be faced by Budd's account, and hence by extension by partial isomorphism approaches in general (Downes 2009). Can it be met?

Let us get clear on the precise nature of this concern. The challenge was originally thrown down against Peacocke's account of picturing, according to which, for example, the shape of Salisbury Cathedral is presented by Constable's famous painting (http://en.wikipedia.org/wiki/Salisbury_Cathedral_from_the_Meadows) as 'a shape in the visual field that is experienced as similar to one which the cathedral itself might present' (Lopes 1996, p. 22; Downes 2009, p. 424). The representational relationship is then described in terms of something being appropriately related to an object 'only if it is presented in a region of the visual field which is experienced as similar in shape to a region in which the object could be presented when seen from some point of view' (Lopes 1996, p. 22). Here we see the similarities with Budd's account: when we 'experience' a painting we experience its visual field properties.

However, in order for this account to work, these visual field properties need to be associated with the relevant features of the painting. Obviously that association needs to be established on *independent grounds* in order for it to be adequate. Unfortunately, we have no such independent grounds for ascribing the same properties to the intermediary visual field and what it stands for. What would provide such grounds are the relevant rules of optical projection, such as Alberti's rule, which dictates that a picture should be seen as a transparent surface on which the outlines of objects seen through it may be inscribed (Lopes 1996, p. 24). With such rules in operation, the above relation holds (because in effect the definition of the visual field and the rule for drawing in perspective are identical). However, those pictures that are not produced in accordance with such rules will not satisfy it. Those that involve projections onto interposed planes that are not perpendicular to the line of sight between artist and subject—such as *The Ambassadors*, again—will satisfy the above relation only under special conditions; and those such as the trident/blivet illusion (and presumably Escher's drawings) are simply not explicable in these terms at all (ibid., p. 27).

Extended as a criticism of Budd's account, however, this seems a bit odd, given that the latter explicitly claims to be able to accommodate *The Ambassadors* and the impossible trident. And indeed, this account sidesteps the challenge by virtue of being much broader than Peacocke's approach, not least in that it allows for the visual field experience of the painting's surface to be strikingly *unlike* the relevant state of affairs in possessing features that the latter does not have or vice versa (see also Shech 2016, pp. 321–2).[59] More importantly, as already highlighted, when experiencing a painting, 'you must see the structure of the surface as being isomorphic with the structure of the visual field representation of the picture's subject *when seen from a certain point of view*' (Budd 1993, p. 161; our emphasis).[60] That 'point of view' will then embody the relevant rule that allows the viewer to appropriately interpret the painting—granted, in the case of both *The Ambassadors* and the blivet, this is not so straightforward, as we have seen in the case of the former, where the relevant rules were tacit and conventional—and provides the independent ground required for the above explanation. Perhaps then, the challenge can be met.

However, there is a more important issue here: how should we respond to the challenge when it comes to *scientific* representation? In that case what would provide the relevant 'independent grounds' that would underpin the ascription of the same properties to the visual intermediary and the system? Obviously we cannot expect anything like 'Alberti's rule' to operate here but when it comes to *observable* entities and systems, at least, there are corresponding rules—embodied in our understanding (partly, if not considerably, theoretical) of the relevant apparatus and instruments—that precisely underpin such an ascription.[61] In certain cases, such as that of the optical microscope, the relevant rules will be comparatively straightforward, based, in part at least, on our knowledge of how light behaves when refracted, reflected, diffracted, and so on.[62] As the level of instrumental complexity increases, with a corresponding increase in the theoretical component of our understanding of the instrument concerned, so the rules, if they can be called such, become correspondingly more complex.[63] Thus, in the case of images of systems at the nanoscale, for example, a theoretical image may be produced, corresponding to what is thought to be going on and that is used to guide the development of the experiment (see Bueno 2008). In such cases, the empirical image produced by the instrument—e.g. a scanning tunnelling microscope—may

[59] This is what suggests that Budd's account should more properly be viewed as involving partial isomorphism, of course!

[60] Likewise, Lopes emphasizes that depiction should not be understood as constrained by a rule that it represents things from a single viewpoint (Lopes 1996, p. 120). This also allows him to accommodate Cubist paintings, for example.

[61] Here again, I am drawing on Otàvio's work.

[62] Although even here, significant issues arise; see for example Kusch 2015.

[63] 'Rules' is used here as a catch-all term for a whole complex of theoretical, experimental, design principles, rules of thumb, and so forth.

be experienced as resembling this theoretical image and the relevant inference of similarity made between the former and the sample under examination:

> Partial mappings between the theoretical image and the empirical image can be explicitly identified, given that researchers have direct access to both sorts of images. Given the partial mappings, researchers then *infer* that there is a corresponding match between the empirical image and the sample.
>
> (Bueno 2008, p. 135)

In the case of representations of *unobservable* objects, the situation is different, of course. Here it is difficult to avoid entering the realism-antirealism debate as the relevant grounds for ascribing the same properties to the elements of the representation as are supposedly possessed by the system will have to do with further considerations, which I shall return to in Chapter 9. However, it should already be obvious that the concern that lies behind the above challenge can be met when it comes to scientific representation of such unobservable systems.

More generally, the point remains that it is because some aspect captured by informational accounts such as Budd's is satisfied that cognitive activities such as recognition or the drawing of certain inferences regarding representational targets are successful in the first place (Chakrvartty 2010, p. 203; Shech 2016, p. 321). Having said that, when it comes to Lopes's challenge above, I did not mean to suggest that visual fields and the associated rules of perspective etc. play any role with regard to representation in *science* in general. Granted, then, that certain formal frameworks are plastic enough to be applicable in both the artistic and scientific domains, we must still be careful when exporting examples, moves, and devices from one context to another. And the reason to take care, at least in part, has to do with issues concerning the nature of artworks and theories qua *objects*, as we'll now see.

Criticism: Resemblance Is Not Directional

We recall the concern that resemblance, or similarity, and hence its formal counterpart, isomorphism, are symmetric and reflexive but representation is not (see Goodman 1976). Representation, in short, has *directionality*.

Now, we need to be clear about the level at which we are operating when we deploy such formal notions as partial isomorphism: these are introduced *at the meta-level of the philosophy of science* to represent scientific theories and models at the 'object-level', as it were, of scientific practice. The former is, formally, an equivalence relation insofar as it relates two (partial) structures. However, the models and theories themselves are constructed, developed, presented, etc. within a particular scientific context. And if we consider that context as a whole, then other factors come into play that may be appealed to in order to effectively add 'directionality'. Elsewhere I have noted the role of heuristic factors in scientific practice and acknowledged that these cannot, of course, be formally considered as part of the *structure* representing, again at the meta-level, the theory or model.

It is the relations between these formal structures that are claimed to capture the way in which theories and models represents sytems in the world. So, the question now is whether these further factors should be considered as *constitutive* of those representation relations, by virtue of adding the requisite directionality. Note, however, that although it is true that all representation must involve such factors, it is not the case that these factors are always the same. Thus, if they were taken to be constitutive, we would obtain different representation relations depending on the context. Alternatively, we might take these factors not to be *constitutive* of the mechanism of representation (French 2003), at least not in the particularities, and thereby extract from these different contexts the underlying mechanism, which can then be formally represented via partial isomorphism.

So, consider again the role of *intentions* here: van Gogh, for example, intended his self-portrait to represent himself (see https://en.wikipedia.org/wiki/Portraits_of_Vincent_van_Gogh); there is obviously no such intention on the part of the painting. The importance of intentions in this context is often illustrated by comparing a face drawn in the sand by someone with what appears to be a face, etched in the sand by the wind and waves, say (see van Fraassen 1994). The former counts as a representation, whereas the latter does not—why? Because in the case of the former, it is usually claimed, there were the requisite intentions. But now compare some squiggles that look like a face, etched in the sand by the action of the wind and waves, and the equation '$E=mc^2$' (a consequence of the special theory of relativity), similarly etched (improbable perhaps but not impossible, given enough beaches). Here, I would argue, the crucial issue is whether the squiggles in the sand should be regarded as *an art object to begin with*. To decide this, one must appeal to the relevant intentions or its causal provenance more generally, which will include the relevant causal history, the role of the artist, and so forth. Indeed, it is confusion over such provenance that has led to art installations being thrown away as rubbish: one of the most famous such examples being Damien Hurst's ad hoc 'installation' at his gallery showing of the detritus of the artist's studio which was subsequently swept away and put into bin liners by the gallery's caretaker (see http://www.theguardian.com/uk/2001/oct/19/arts.highereducation1).

On these grounds, the squiggles would not be taken to be an artwork to begin with and hence not representational. In the case of the *equation*, on the other hand, even if we had observed someone inscribing it in the sand and therefore were assured that the relevant intentions were playing an appropriate role and that the relevant provenance was in place, we would not take it as *the* theory, or the relevant part thereof, because of the kinds of concerns already discussed in Chapters 1 and 2. Whatever the intentions of the person doing the etching, we would not say that those particular squiggles *are* the special theory of relativity (or relevant part thereof) that represents the relevant phenomena. Given that the theory is something other than the inscription in the sand, the issue of the intentions behind that specific inscription becomes less significant when considering its representational nature.

What this brings up, of course, is the core theme of this book: what *is* a theory? Now, as we shall consider in more detail later, a similar question can be asked about certain kinds of art—such as musical compositions for example, and as we shall see, some of the discussions here bear a similarity with those in the philosophy of science—but in the case of paintings, say, or sculpture, an answer is forthcoming comparatively easily. Putting things rather crudely for the moment, we could point to the massive rectangle of canvas on the wall of the Museo Reina Sofia and say '*That* is Picasso's *Guernica*', or similarly, point to the tank(s) full of formaldehyde and say 'That is Hurst's *Mother and Child (Divided)*' (http://www.damienhirst.com/mother-and-child-divided-1). However, we are typically reluctant to similarly appeal to ostension in picking out the theory of relativity, say. The black marks in my photocopy of Einstein's 1905 paper do not *constitute* the theory, any more than do the marks in Einstein's original draft or the wind-blown squiggles on the beach. And, as we saw in Chapter 2, it is precisely this point that is drawn upon by those adherents of the Semantic Approach who urge that theories should be regarded as *extra*-linguistic entities. But if they are so regarded then the distinction between marks that were intended by Einstein to represent certain phenomena and marks carved into the sand by the wind, appears to evaporate.

Of course, we would only read the squiggles as '$E=mc^2$' in the first place because we already possess the prior intention to use markings of that form to represent the relevant phenomena. Someone not trained or educated in the appropriate way would have no such intention and would indeed see nothing but marks in the sand. However, this has to do with a different, albeit related issue, namely that of supplying an *interpretation* (French 2003). Thus, van Fraassen considers *Le Dejeuner sur l'Herbe* and notes that although Manet's intention was to portray the young men as arrogant, in order for us to *grasp* that intention we must interpret their stances, facial expressions, clothes, etc. (1994). Such an interpretation is dependent on certain cultural factors and if these are absent, quite a different interpretation might be constructed under which the painting ceases to represent what Manet intended it to, just as in the case of *Guernica* above. A similar issue arises in science: by offering a different interpretation of the relevant equations than Einstein's, Minkowski can be understood as having demonstrated that they can represent something else, namely the structure of relativistic spacetime rather than the behaviour of rods and clocks.

Which set of intentions, if any, should then be privileged? Positivist philosophers of science maintained for many years that special relativity is indeed about rods and clocks.[64] Realists argued, on the contrary, that it should be understood as about space-time, particularly given the subsequent development of general relativity (see Nerlich 1994). Present day antirealists may then agree that the theory can be *interpreted* as about this entity, space-time, but that we cannot *know* whether space-time, as conceived by the theory, actually exists. I don't want to get into this debate here but would just like to emphasize two things: first that the role of intentions in establishing what it is that is being represented may not be straightforward and secondly, that these intentions ride on the back, as it were, of the relevant formal relationship.

Even in art it is not so clear-cut that in order to be able to answer the question 'what does *Le Dejeuner sur l'Herbe* represent?' we must have at least a smattering of knowledge in the history of art (or the cultural mores, dress codes, attitudes of nineteenth-century Paris). It can be argued that any piece of art enters into a *multiplicity* of representational relationships, so the question of what it represents shifts with the cultural context. In that case, the intentionality enters in the observer's construction of the object, or set of objects, which the painting is taken to represent: in Manet's case it might be a group of arrogant young men, in the case of a working-class physics student with no background in history of art, it might be a group of charming young people enjoying the sun, in the case of a Bororo tribesperson it might be something quite different altogether. Of course, some might be tempted to ask 'OK, but which is the "correct" representation?' And again some realistically inclined artists and commentators might well respond, 'The correct representation is that which the artist intended', whereas others might insist that all art is subject to a multiplicity of representations, with no one to be favoured over the others. But this is a separate issue. The point is, intention is not invoked as a *constitutive* aspect of the representational relationship per se but in order to pick out one such relationship over the others.

The analogy with theories and interpretations is obvious but here we are typically even less inclined to invoke Einstein's intentions in order to settle the issue of what the theory of special relativity represents. The choice of relationships, or better, sets of objects, phenomena, whatever, is determined by other factors, well known from the realist-antirealist debate. And again, although a hard-line realist might insist that some such additional factor—such as simplicity for example—is *constitutive* of the representational relationship, that is an awfully hard line to hoe.

In sum, then, I agree that in order for us to see a painting as about a certain scene, a certain person, or whatever, we must be able to interpret the splashes of paint a certain way, and that will certainly depend on the broader context. But we should not effectively incorporate a particular intention into the mechanism of representation itself, such that the latter becomes fixed, in the sense that

[64] Howard (2017) argues that this mischaracterizes Einstein's own view.

Le Dejeuner sur l'Herbe must represent a set of bourgeois young people, or Einstein's special relativity *must* represent phenomena associated with rods and clocks. Rather, both artworks and theories enter into a multiplicity of representational relationships, and if the intentions of the artist or scientist are not to be privileged in a problematic way, then we must allow for pragmatic or broadly contextual factors to play a role in selecting which of these relationships to focus on. 'Building' particular intentions into the representational mechanism would then be disastrous.

Criticism: Isomorphism Is Neither Necessary nor Sufficient for Representation

It has also been claimed that neither similarity in general, nor the likes of isomorphism in particular, partial or otherwise, can be *necessary* for representation, since the latter may occur without any relevant similarities being in place. So, recalling the claim that *anything* can be taken to be similar to anything else in certain respects, some 'criterion of relevance' (Suárez 2003, p. 235) is taken to be needed to identify those similarities that are meant to underpin the representational relationship. Identifying such similarities is not straightforward and it is here that we find examples of apparently problematic artworks being introduced to support the argument: in Picasso's *Guernica*, for example, none of the similarities that can be seen—'a bull, a crying mother holding up a baby, an enormous eye' (Suárez 2003, p. 236)—appear to be directly relevant to the targets of the representation, which can be identified with the bombing of Guernica or, more abstractly, the rising threat of fascism in Europe. Likewise, it is claimed (ibid.), and shifting back to science, when an equation, written on a whiteboard, represents a system, the marks on the board of course have no relevant similarity with the system being represented.

Furthermore, there exist examples of artworks that, it is asserted, cannot be held to be isomorphic to *anything* (Suárez 1999). Consider *Guernica*, again and the question of what that painting represents. This is ambiguous: on the one hand, it represents the 'concrete pain' of the inhabitants of Guernica, but on the other it also represents the 'more abstract threat' noted above. Hence, it is argued, *Guernica* cannot stand in any one-to-one correspondence with (and thus be related via isomorphism to) those things that it is taken to represent (ibid., p. 78). Or take certain of the works of Mondrian, such as 'Composition with Colour Panes' (http://www.theartstory.org/artist-mondrian-piet-artworks.htm#pnt_2): to focus on representation in such cases seems to miss the point of the art in that the aesthetic and emotive responses that this type of abstract painting induces do not arise in virtue of its 'representing' anything (ibid., p. 79).

Again, however, care must be taken when it comes to evaluating the evidentiary force of such examples. As already noted, it cannot be the case that similarity *in general* is not necessary for *scientific* representation, since in the absence of any similarities at all, it is not clear how the usual practices of interpretation and inference could get off the ground to begin with (Chakravartty 2010; Shech 2016). And as I've already said, even if we grant the force of these examples from the world of art, it is not at all clear that similarly forceful examples could be found in science. Indeed, how could one claim that a model represents a target system if there is *no* similarity between the model and the system at all?! And of course, once such a similarity is alleged to hold, it can be formally captured via the kind of framework sketched here. It is in this regard that scientific representation differs from the artistic and hence putative counter-examples such as *Guernica* lose their value when it comes to the former, at least.[65]

There is a further worry that the possibility of multiple interpretations suggests that theories themselves should not be regarded as representations *at all*, unlike lower-level models which have their intended uses built in, as it were. Suárez, for example, argues along these lines:

> When a model is presented, its intended use is given by implicit or explicit reference to the object of its application. Thus... a model is a representation, as it is essentially intended for some phenomenon; its intended use is not an external relation that we can choose to add to the model, but an essential part of the model itself. (1999, p. 6)

A theory's intended uses, on the other hand, are not an essential part of it (as in the example of special relativity above) and it is this absence of intended uses that makes a theory non-representational:

> If a theory operates *as if* it had no intended uses—if it operates as a pure form, devoid of empirical content—then the theory is not working representationally. (ibid., p. 8)

An obvious question to ask is whether theories *ever* operate as if they had no intended uses—how would they be distinguished from pieces of mathematics in such cases?—but Suárez is surely right when he states that, at the very least, theories are 'not so tightly linked' to their intended applications as models. As he says, 'we don't typically construe the discovery of a new application of, say, quantum mechanics, as fundamentally changing the theory in any way' (ibid., p. 6). Indeed, as we recall from Chapter 2, it was precisely to avoid such a construal that

[65] It is not even clear that *Guernica*, for example, is a particularly forceful counter-example (French 2003; Bueno and French 2011).

adherents of the Semantic Approach rejected the Syntactic view's incorporation of correspondence rules, which by both tying theories to their applications and, crucially, forming a constitutive part of the theory itself, implied that a new application meant the introduction of a new correspondence rule and, strictly speaking, a new theory. According to the Semantic Approach intended applications are captured by relationships external to the theory itself as represented by partial isomorphisms. These can precisely capture the way in which theories are linked, but 'not so tightly', to their applications.

Moving back now to the artistic realm, if intentions have to be *built in* for something to count as a representation, then *Guernica* cannot count as representational, which reinforces my point that it cannot serve to undermine accounts of representation in science. Nevertheless, it could be pressed that there has to be *some* relation between the marks on the canvas and specific objects in the world in order for our understanding of what *Guernica* might represent to get off the ground in the first place (and that relation can then be represented formally). That understanding may vary with our historical, political, and social sensibilities, but the basic mechanism of representation involves the kinds of relations I have highlighted here and that can be captured within Budd's account. In cases such as this, these sensibilities may drive the symbolic dimension of the artistic representation, requiring the inclusion of a 'key' or 'code' in our account of representation, but as I've already said, in the sense in which this dimension is understood in art, it is not as central to science.

What about examples of artworks that are clearly not representational at all (e.g. Davie's *Figure Mask No. 2*; see http://library.leeds.ac.uk/art-gallery-paintings-highlights#activate-image10)? Again, the obvious response is to deny their evidentiary force here simply because qua artworks they are not analogous to scientific theories or models and hence their 'falsifying' power cannot be carried over. Even in such cases, however, things are not always straightforward. So, consider one of Pollock's work, such as *No. 5 1948* which is often presented as a fine example of abstract expressionism and insist that this is clearly not representational at all but, as the genre label suggests, is intended to *express* something. Even here one might push the line that it does represent something, namely Pollock's movements when drizzling the paint.[66]

What about the works of Mondrian, such as *Composition No. 10* (http://www.wikiart.org/en/piet-mondrian/composition-no-10-1942), as already mentioned. To focus on representation in such cases is to miss the point of the art, or so it is claimed, and although it may elicit various kinds of responses, it does not do so by *representing* anything. Of course, Mondrian himself was notably thoughtful and reflexive about his own work and famously rejected this representational/

[66] I am grateful to Otàvio Bueno for pressing this point.

non-representational dichotomy, arguing that his work *was* 'representational' in a broad sense that encompassed the *representation of pure relationships*. Indeed, Mondrian himself makes it clear, for example, that his use of striking horizontal, vertical, and diagonal lines can be traced back to the geography of the Netherlands, with its flat horizons, lone trees, and windmills and his commitment to representation (in some form) is made explicit in statements such as the following:

> Through the very culture of representation through form, we have come to see that the abstract—*like the mathematical* [his emphasis]—is *actually* expressed in and through all things, although not determinately.... Through painting itself the artist became conscious that the appearance of the *universal-as-the-mathematical* is the essence of all feelings of beauty as pure *aesthetic* expression... He learned to represent *exactly* what is *vaguely perceptible* in nature, he *reduced and destroyed the concreteness* of appearance (by simplification), yet he did no more than carry the *conception of art to its logical conclusion*. And so our age arrived at *abstract-real* painting. The new plastic is *abstract-real* because it stands between the absolute abstract and the natural or concrete-real. It is not as abstract as abstract thought and not as real as tangible reality. (Mondrian 1917)

What he means by the 'new plastic' is just the 'direct expression' of the universal, where the universal is, at least in part, 'pure relationship' (ibid.). The word 'plastic' here is a translation of the Dutch 'beelding' which can also mean 'structure' and signifies form giving and image creation.[67]

Thus, even apparently 'abstract' art may be broadly representational. More generally, care must be taken in presenting putative counter-examples to accounts of representation, particularly when they are being exported from one domain, the philosophy of art, into another, the philosophy of science (see for example Muller 2011, p. 110, n. 33).[68]

Conclusion: Representation Does Not Require Reification

The exchange of examples, devices, moves, and manoeuvres between the philosophy of art and the philosophy of science needs to be handled carefully on a case-by-case basis. On the one hand, the kind of formal structuralist approach that nicely accommodates representation in science can also be applied, with some adjustments, to representation in the arts. On the other, as we've just seen, not

[67] Interestingly—for me at least!—Mondrian may have been influenced by M. H. J. Schoenmaeker's *The New World Image* in which he presented the 'absolute underlying structure of the universe'.

[68] Further criticisms of the semantic approach to representation are disposed of in Bueno and French 2011.

every counter-example faced by the latter can be straightforwardly exported to be used against the former.

Furthermore, we need to keep in mind that what we are doing with such formal frameworks is characterizing at the meta-level of the philosophy of science or art the kinds of representational practices that are engaged in at the level of science and art themselves. I am not, of course, demanding that *artists or scientists themselves* adopt these kinds of formal approaches at the level of their practice and neither am I insisting that the Semantic Approach is the only framework for representations that *philosophers of art or of science* must adopt. Certainly one might advocate a different framework and it may well be the case that different kinds of relationships require different formal frameworks. However, the challenge then would be to show that such frameworks offer the same advantages as the approach indicated here when it comes to the relationships we encounter in science in particular. As we have seen, the Syntactic Approach, for example, may have the resources to meet this challenge, although as things stand, that has yet to be demonstrated.

This then feeds into an old concern, namely that by describing, *as philosophers of science*, theories and models in terms of such a framework and its associated set-theoretical structures, we are committed to regarding theories and models *as* such structures. However, there is no requirement for advocates of this approach to be committed to such an ontological claim (da Costa and French 2003). This presumed reification has generated considerable criticism but it is time it should be laid to rest. Such structures provide a useful meta-representational (or better, perhaps, *descriptive*) device at the level of the philosophy of science. What theories and models *are*, qua objects, is then a further issue; let us now consider it.

4
Theories as Abstract Entities

Introduction

In the previous chapter we examined how the manner in which both artworks and scientific theories and models represent their targets can be formally captured, subject to certain caveats. However it would be bizarre to then insist that such an artwork, a painting say, much less a musical work, should be *identified* in terms of such a formal framework. Likewise, granted that a theory's or model's representational capacity can be accommodated within the Semantic Approach, we should resist the temptation to *identify* that theory or model in terms of this approach.

This then brings our original question back onto centre stage: what *is* the ontological nature of theories and models themselves? When it comes to paintings, at least, one might think there is simply no issue here as to *their* ontological status—we can point to, or even touch (apologies to any curators reading this) Picasso's *Guernica* or Turner's *Rain, Steam and Speed* and ostensively individuate the artwork concerned. Of course, sometimes things are not always quite so straightforward, as the debate over indiscernible examples of a particular artwork testify (for an overview, see Livingston 2013). But the point is, this is not the case with scientific theories or models, or at least not all of them. Although we can, indeed, touch and flip through the pages of Newton's *Philosophiæ Naturalis Principia Mathematica*,[1] in which his theory of mechanics is presented, we would be hesitant, at least, to likewise point to that particular presentation and say *that* is Newton's theory.[2]

Of course, we can, or could, point to and touch Crick and Watson's tinplate-and-wire model of DNA[3] but here it is not the materiality or physical presence of the model itself that carries its scientific significance, or at least not in the way that the physical object that is Picasso's painting is imbued with or carries its artistic significance. In the Crick and Watson case, what was, and still is, significant, of course, is the structure that the model represented DNA as having and that structure would be equally manifested by or presented by another example, built out of Bakelite and string, say, or indeed, an exact replica of the original model, built

[1] As indeed some of us did during a trip to the museum of the Royal Observatory in Edinburgh some years ago!

[2] That the theory is understood to be distinct from a particular written presentation of it is precisely one of the motivations of the Semantic Approach, of course.

[3] Again for a useful discussion of the construction of this model, see Schindler 2008.

There Are No Such Things as Theories. Steven French, Oxford University Press (2020). © Steven French.
DOI: 10.1093/oso/9780198848158.001.0001

from precisely the same materials. In that case, I suspect, scientists would not be particularly bothered about answering the question as to which is 'the' Crick and Watson model; or at least and again, they would not in the context of its scientific significance (they might in the context of its historical significance, of course).[4] But contrast that (suspected) attitude with those evinced in the debate over indiscernible copies and forgeries within the philosophy of art (again, see for example, Livingston 2013). Suppose I were to reproduce down to the last detail, using the same kind of canvas and the same paint colours, even the same paints (rediscovered in some long forgotten warehouse or atelier, or even reconstructed via chemical analysis), *Guernica* or *Rain, Steam and Speed*. Would I have (re-)created the same artwork? The standard view (on which the relevant law is based were I to attempt to sell my effort as Picasso's *Guernica* or Turner's *Rain, Steam and Speed*) is that I would not; on the contrary, what I would have created would be deemed a forgery. The difference, from an aesthetic (and also, it seems, legal) point of view has to do with the context: first of all, '[r]elational knowledge plays a huge role in fixing a work's aesthetically relevant features and how we should appreciate them' (Kieran 2010, p. 371). Thus, knowing that the copy was produced by me, rather than Picasso or Turner, affects its aesthetically relevant features. Furthermore, the role of 'creative originality' is also important (ibid.) and once again, intentions come to the fore—the intentions being different in the two cases, the aesthetic qualities would be different (see also Kieran 2005).

We shall come back to these features shortly, but now compare that to the case of a theory or model. Here the scientifically relevant features are not fixed in the same way via relational knowledge of the provenance of the work, nor does creative originality play the same role. Of course, historians and philosophers of science, as well as scientists themselves, may engage in vigorous disputes over whether Einstein, Lorentz, or Poincaré was the 'true' 'discoverer' of special relativity (for a useful overview of the debate, see http://en.wikipedia.org/wiki/Relativity_priority_dispute)—and I'll touch on this issue again in the next chapter when we look at 'simultaneous discovery'—but most would agree that this has no bearing on

[4] One can of course find replicas of the Crick and Watson model in many museums (for example: http://www.sciencemuseum.org.uk/online_science/explore_our_collections/objects/index/smxg-146426) and can even buy examples. Interestingly, but perhaps not surprisingly, physical three-dimensional models in biochemistry were replaced during the 1980s by computer generated images which are now produced in such a way as to overcome some of the earlier difficulties in viewing these in 3-D (see: http://www.ncbe.reading.ac.uk/dna50/models.html). I said interestingly because here one might see the ontology of scientific models drawing closer to that of artworks, or at least those that are created digitally. Thus, in the case of digital artforms, created on the computer, iPad, or whatever (consider for example, Hockney's 'iPad art': http://www.hockneypictures.com/current.php), artists, critics, and philosophers of art are likewise less concerned about the question as to which of the various copies that can exist (simultaneously) is *the* artwork concerned. Benson (2015) refers to digital photographs as 'simulacra', which echoes Cartwright's view of scientific models (Cartwright 1983).

the theory's scientifically relevant features.[5] In this regard theories and models may seem closer to *works of literature.*

Consider, then, Borges's famous short story 'Pierre Menard, autor del Quijote' in which Menard sets out to rewrite, without simply copying, Cervantes's great work. Here the 'without simply copying' is cashed out in counterfactual terms, with Borges noting how Menard's text could take a different turn or move in a different direction than Cervantes's. Is what Menard ends up writing, the 'same' artwork or is it different from Cervantes's? Again, many have argued that it would be a mere copy or replica and therefore aesthetically deficient (Livingston 2013); others have questioned whether an attempt to rewrite but not copy is even psychologically coherent! But then consider the million-monkeys-on-typewriters scenario in which someone, somewhere, with no knowledge of Cervantes's book, writes a work that is word-for-word identical to it. Again one might want to insist that in terms of its aesthetically relevant qualities, it would be a different work (the intentions of the authors being different, for example). But in terms of its story-telling qualities, in terms of the way it represents the conflict between systems of morality, or world-views more generally, or even the way it leads us to examine the nature of narrative and story-telling in general, it could be argued that the rewrite is the same as Cervantes's. *In these terms*, who cares who actually wrote it? What is important is the message(s) it conveys, the feelings and emotions the story invokes, and so on.

Here we might draw an obvious comparison with scientific theories: suppose I set out to rewrite, without simply copying, Newton's theory of mechanics. In this case, I don't need to construct a fantastical story as Borges does, since every time a physics student answers an exam question 'describe Newton's theory of mechanics' or her lecturer presents it as part of a course on classical physics, they are engaged in just such a rewrite. And of course there is nothing surprising or contentious in the claim that the result of the rewrite *is* Newton's theory (assuming the student has got it right). Indeed, it would seem bizarre to say that the rewrite is a 'mere' copy or is somehow scientifically deficient! The point is that unlike a painting, we are interested only in the 'content' or the 'message', as suggested above, and it is this that motivates the idea that the theory, embodying that content, exists 'over and above' in some sense, any particular written presentation.[6]

One way of capturing that sense of its existing 'over and above' its particular presentation is to regard theories and models as abstract entities in some sense.

[5] Galison regards the debate as tedious and fruitless, since the issue depends on what one takes to be essential to the theory (2003). That there may be no clear answer to that last issue obviously meshes with the overall thesis regarding the nature of theories I am defending here.

[6] Scientific theories and models are not typically compared to works of literature. Hughes does a nice job in comparing Aharanov and Bohm's classic paper on the effect that now bears their names with *commedia dell' arte* (2010). As we shall see, Popper places both theories and works of literature such as *Hamlet* in his 'World 3' and more recently, a number of philosophers of science have compared theories and models to fictions (we'll come back to this as well).

Thus, as we have seen, it has been argued that because models involve abstraction and idealization, they themselves must be regarded as abstract objects, and as such cannot stand in relations of similarity with physical systems (Hughes 1997). However, as we recall, the motivation for this last move is not clear. It might be grounded in a restrictive view of similarity such that the abstracting away of certain properties removes the basis for relating in this way the idealized and non-idealized models that possess them. So, to draw on our much-used example again, it might be insisted that by abstracting away friction and air resistance, the (idealized) model of a simple pendulum cannot then be related by similarity to an actual pendulum. But, again we recall, the model and the actual system share other properties—most notably, length of string, mass of bob, and so on—via which they can still be held to enter into a similarity relationship. The possible response that an idealized length that encounters no friction at point of contact, or a mass that encounters no air resistance, cannot be held to be similar to their actual counterparts might strike many as implausible and would rule out many accounts of representation.[7]

Alternatively, the above move might ultimately be grounded in the view that similarity cannot be a cross-category relation, such that it cannot be taken to hold between abstract and actual objects. However, this again seems implausible. Consider, for example, the relation between Jody and Ian such that Jody is taller than Ian. This is similar to the relevant relation between the numbers that correspond to Jody and Ian's respective heights. There seems to be no objection to maintaining the existence of such a similarity in this case, and hence I likewise see no objection to taking similarity to hold between theories, taken as abstract, and systems, taken as concrete.[8]

Nevertheless, the view of theories and models as abstract objects has gained considerable currency in the philosophy of science literature, particularly following the rise of the Semantic Approach, which, as we have seen, takes as central the idea that they are 'extra-linguistic'. But now a tension arises between such a view and the process by which we construct theories or models. A similar tension has been highlighted in the case of musical works—with which scientific theories have been compared—and the argument can be carried over to scientific theories, as we shall now see (the following is based on French and Vickers 2011).

Abstractness and Creativity (Art)

Using the example of musical works, the tension can be captured in the following form (Cameron 2008a):

[7] Or, at the very least, it requires an appropriate account of property identity.

[8] Again, my thanks to Otávio Bueno for this point.

M1: Musical works are abstract objects.
M2: Musical works are created.
M3: Abstract objects cannot be created.

Each of these claims seems plausible, yet it cannot be the case that all three can be held together. The motivation for M1 comes from the multiple instantiability of musical works (a feature they appear to share with scientific theories, of course). Consider for example, Beethoven's *Fifth Symphony*, sometimes called 'the symphony of symphonies' (https://www.theguardian.com/music/tomserviceblog/2013/sep/16/symphony-guide-beethoven-fifth-tom-service), typically regarded as one of the most played of all symphonies and certainly played thousands of times since its first performance in December 1808. It has been played in different venues by different orchestras and recorded in different media, from shellac and vinyl disks to magnetic tape, laser-read 'compact disks' (ah, how we remember them fondly), to computer hard drives—the First Movement is even included on one of the 'Golden Records' carried by the Voyager I spacecraft launched in 1977 and currently almost 13 billion miles away in interstellar space.

How might this multiple instantiability of 'the' work be accommodated? One (nominalistic) option would be to regard a musical work as simply the sum of, or as composed of, the set of all its performances, scores, and other relevant concrete particulars. That presents an interesting mereological challenge, exemplified by the case of Beethoven's symphony above (presumably the playing of the First Movement by some alien species across the galaxy thousands of years from now would also constitute the symphony as an entity). But more acutely perhaps, how are we to make sense of a claim such as, 'Tchaikovsky's violin concerto in D major is unplayable' (purportedly made by the court violinist Leopold Auer in 1878)?[9] Such claims and others appear to refer to the work itself, rather than any set of actual or possible performances and for this sort of reason, many have dismissed this approach as a non-starter (see Kania 2014).

The obvious alternative is to accept that musical works are abstract objects, in some sense, that can then be instantiated and played in different locations, by different orchestras and recorded and played back using different media. Dodd, for example, has argued that musical works are 'sound structures', in the sense of structured types that have sound types as their constituents (2000; see also his 2002). As types, musical works are eternal existents (see Dodd's summary of his

[9] Three years later, the work was played (although a well-known critic reported that 'beaten black and blue' rather than 'played' was more appropriate—see http://www.bbc.com/culture/story/20150317-the-worlds-most-difficult-music) and it is now apparently regarded as a standard if technically demanding piece in the professional violinist's repertoire. Given its level of difficulty, one might expect that at least some performances of 'it' will contain some bum notes, which on the simple nominalist account would imply that the work itself contains some bum notes—an odd claim to make, at the very least (see Kania op. cit.)!

argument for this in 2002, pp. 381–2) and as such—that is, as things that cannot come into or out of existence—they cannot have been created by anyone. Thus, he embraces M1 above, and resolves the tension by dropping M2.

Now, that's obviously a radical move! After all, Beethoven certainly *composed* his *Fifth Symphony* and what else could such composition be but an act of creativity? Dodd has a response (2000, pp. 427–34):[10] composition should be understood as a kind of 'creative discovery' in which composers 'come across' something that is already 'there', in some sense, but that manner of 'coming across' requires imagination and creativity. Thus, the creative aspect is not tied to any bringing into existence but rather is associated with the act of discovery—to the exploration of a kind of sound space, if you like. However, we might wonder whether the notion of 'discovery' is really appropriate in this context. Consider: unlike, say, Wylie's famous discovery of a proof of Fermat's last theorem, there is no possibility of Beethoven having made a mistake with his 'discovery'—that is, Wylie could have got it wrong, but Beethoven couldn't (Sharpe 2001). In general, musical works seem to be conceptually bound up with the creative act in such a way that the latter could not lead to anything other than that work—something that is not the case when it comes to the discovery of mathematical proofs, for example.

Dodd's response is interesting: it is only *certain kinds of discovery* that allow for getting it wrong and these fall under 'discovery by enquiry', where the nature of the enquiry is constrained by certain criteria of success (2002, pp. 386–7). And to determine whether such criteria are met typically involves some form of comparison between that which results from the enquiry and some description of what one was hoping to end up with: in Wylie's case, that would be Fermat's Theorem, so that if his putative proof had not yielded it, he could not have claimed to have discovered a proof of it. But Beethoven did not set out to discover something answering to a certain description that he had beforehand; rather, he came to form a clear conception of the *Fifth Symphony* and that in itself was the act of composition. Thus, there is no gap between *conception* and *search* as there is in the Wylie example (ibid., p. 387).

Nevertheless, there is another kind of gap here, namely between the composer and the musical work and consideration of it reveals the extent to which we regard such works (and artworks in general) as 'modally flexible', in the sense that had circumstances, including the composer, been different, the same work would have been produced. So, let us ask: could someone other than Beethoven have composed the *Fifth Symphony*? Now, we'll come back to this sort of question later, when we'll explore the claim that some artworks, at least, are not sufficiently flexible modally speaking to allow such a possibility, but Dodd bites the bullet and acknowledges that it is indeed metaphysically possible for 'Beethoven's' *Fifth* to

[10] He captures the tension here in a version of the above argument (Dodd 2002, p. 381).

have been composed by someone else and in a different musical or cultural context. This is not to say that that the actual context does not have some role to play, only that it is limited to the process of formation of a 'clear conception' of the work and thus may be considered as a kind of heuristic device that helped Beethoven in coming to that clear conception. These contextual factors effectively situated Beethoven in the appropriate region of musical space, from where he was able to form such a 'clear conception'. Nevertheless, that is not to say that the context is essential to the work or otherwise underpins its identity—that is fixed, of course, since the work is an eternal type. And someone else—Luigi Cherubini, say—could have formed a conception of it, but that would have required something akin to the relevant contextual factors to place them at the right spot, as it were, in musical space—something that is not metaphysically out of the question, but that would be extremely unusual and anomalous, given that composer's own context (ibid., p. 388).

Now, granted that it is not just anyone who can get to that position in musical space, as it were, and form such a clear conception, the process of this 'creative discovery' remains unclear. How does the creative faculty of the composer and the contextual factors that structure the 'sound space' work together to 'lead' the likes of Beethoven to the work? We know that Beethoven composed the *Fifth* over four years, with the first rough 'sketches' produced in 1804. We know that, for example, the third movement was influenced by the final movement of Mozart's *Symphony No. 40* and it has been suggested that the famous four-note motif with which the symphony opens was derived from the song of a yellowhammer, heard as Beethoven walked through a park in Vienna (Hopkins 1977). On Dodd's account, these heuristic factors effectively propelled Beethoven to a certain portion of the sound space but what about his early 'sketches'? Are they there too? If so, it's not hard to see that this sound space will become pretty crowded! Perhaps this ontological inflation can be avoided by insisting that such sketches are not associated with any 'clear conception' but then what is the measure of clarity here? According to what metric might they be ruled out?

As we'll see, similar questions will recur but perhaps even more sharply when we try to import such a view into the philosophy of science (indeed, Dodd himself draws a comparison with Einstein's 'discovery' of special relativity).

An alternative would be to retain M2 but deny M1, whilst accommodating musical works' multiple instantiability. Thus, Collingwood, for example, argued that '[i]f the making of a tune is an instance of an imaginative creation, a tune is an imaginary thing. And the same applies to a poem or a painting or any other work of art' (1938, p. 139). We shall briefly touch on the role of imagination in art and science in Chapter 5 but as simply stated there's an obvious objection to this claim, which is that it renders artworks inter-subjectively inaccessible—in effect, there will be as many musical works labelled *Beethoven's Fifth Symphony* as there are people having the relevant imaginative experiences, whether as a result of hearing a live performance or a digital download (Kania 2014).

Collingwood's response is quite radical but it does generate a useful comparison with science, as we'll also see later (see also van Fraassen and Stigman 1993). At the core of his account lies the distinction between the artist as a craftperson[11] seeking to stimulate certain emotional effects in an audience and as a 'proper' artist, aiming only to create art. In the former case, what results is something 'real' in the sense that it is made out of some material, such as sound, or stone, or whatever. In the latter case, what is obtained is something that exists only in the artist's head, where it exists as something 'already complete and perfect' (Collingwood 1938/71, p. 139). Of course, when the artist arranges for a piece of music, say, to be played, there comes into existence a 'real tune', which is a sequence of noises, but this is not the work of art. Crucially, '[t]he noises made by the performers, and heard by the audience, are not the music at all; they are only means by which the audience, if they listen intelligently (not otherwise), can reconstruct for themselves the imaginary tune that existed in the composer's head' (ibid.). Thus, Collingwood denies, in effect, multiple instantiability: what might be taken to be instantiations of a work are *different entities* entirely—intelligent reconstructions of something imaginary.[12]

We are urged, then, to move beyond merely hearing the noises in a musical performance, and so also beyond the sensual pleasure those noises may induce, and *listen*, in a way that engages the imagination to produce a 'total imaginative experience' (ibid., pp. 144ff.). And this engagement is both positive, insofar as the imagination fills in sounds that our ears do not catch, just as when we look at a drawing, we take a series of lines as a shadow, and negative, in that extraneous noises at a performance, such as the performer's aural nemesis, the rustling of a sweet wrapper, are 'disimagined'. Collingwood goes even further, noting that when one really listens to a piece of music, one imaginatively experiences all kinds of visions, movements, and so on that go beyond mere noises.[13]

There is much more to say here, on, for example, the extent to which Collingwood's view is a form of idealism, (see Ridley 1997) and how what he himself describes as this 'preliminary account' underpins his complete philosophy of art (Davies 2008). But it offers an interesting response to the objection from lack of inter-subjective accessibility: the nature of the imaginative experience that results from appropriately listening to a musical performance is such that we may reconstruct and thereby reproduce in our own mind the musical work that is in the artist's. Of course, we might be sceptical that a complete reproduction

[11] Or as Collingwood characterizes her/him, as magician or purveyor of amusement (Collingwood 1938/71, p. 139).

[12] A useful collection of papers on Collingwood's philosophy can be found in 'Collingwood and the Philosophy of History', *Journal of the Philosophy of History* 11 (2017).

[13] There is, then, an obvious element of elitism in this view, since to obtain such an imaginative experience, one must be 'properly qualified' (Collingwood 1938/71, p. 148).

could ever be achieved—after all, even if we accept that feelings, emotions, and so forth can be shared, we might doubt that all the nuances and complexity that were in the artist's mind during the creation of the piece could be shared in their entirety and even if they were, how would we know it? But perhaps this is to confuse what goes on in creation with what the artist ultimately intends the piece to convey and if we can 'get' that, we could legitimately be said to 'have' the musical work in our mind.[14]

Alternatively, we might just bite down hard on this particular bullet and accept that the work exists only in the mind of the artist and that all we can aspire to are reconstructions that may be more or less pale imitations of the work. That may seem like a counsel of despair, since it acknowledges that there can be no inter-subjective accessibility, but a Collingwoodian can insist that via appropriate listening, those reconstructions are sufficiently rich and nuanced that we can talk about, exchange opinions on, express views regarding 'the' piece—even with the artist—in such a way that we do not find ourselves talking at cross-purposes or disagreeing too often or to such a degree that we feel we are talking about different works (although that may happen). The point is, lack of such access may be no impediment to critical engagement.[15]

Again, we shall return to this view when we consider how it might be applied to theories but it seems that we are faced with the possibility of a kind of ontological proliferation, once more, only this time not in some abstract space, but in the minds of all the listeners of a performance: in the mind of the composer we have 'the work', as an imaginary entity, and in the minds of all the listeners we have a plethora of more or less intelligent reconstructions, none of which are strictly identical to that work. Perhaps when it comes to musical works this can be made palatable, by, for example, emphasizing the point above that it is sufficient for those reconstructions to be intelligent enough to allow the listener to grasp what it is that the composer intended to convey; but it is difficult to see how that

[14] Cf. Davies 2008, pp. 170–1.

[15] Interestingly, Collingwood (1938/71, pp. 323–4) also argues that '[t]he work of artistic creation is not a work performed in any exclusive or complete fashion in the mind of the person whom we call the artist'. To suppose otherwise is to fall prey to the delusion of individualistic psychology. Instead, aesthetic activity should be thought of as a 'corporate' activity belonging to a community consisting of not only the artist but all those who have influenced her, together with the performers, who, in the case of a musical work, say, effectively collaborate with the artist and, finally, the audience, which also collaborates in the creation of the work. It is not clear to me how one might reconcile this view of art with the claim that artworks are imaginary. Collingwood himself insists that this 'corporate' stance is not inconsistent with the view that aesthetic activity goes on in the artist's mind, because for a speaker to have the experience of being listened to there must be a listener but that doesn't quite address the issue. One could insist that it is this latter stance that represents the more mature account, with the first relegated to preliminary musings (see Davies 2008) but that also does not do complete justice to the claims noted above. There is also the even more radical stance that a 'tune' is indeed an imaginary thing but the imagination involved is that of the community as a collective!

will cut much ice in the case of a scientific theory (again, Collingwood himself compares a musical performance to a scientific presentation: 1938, pp. 140–1).

What about M3? Could we maintain that musical works are both abstract and creatable? Bluntly put, most metaphysicians insist that abstract objects *cannot* be created, on the grounds that creation requires some kind of causation (in the form of a causal chain, at least, from creator to created) and abstract objects are causally inert (see for example Walhout 1986). However, it could be denied that creation requires causation. We could perhaps insist that at the same instant that a composer comes up with a particular musical work, the corresponding abstract object just comes into being. But then that opens the door to speculation as to *when*, precisely, it comes into being—as we've noted, Beethoven took four years to complete the *Fifth*. Did the symphony partially come into being during that time? Or only when Beethoven put down his quill for the final time? Or did an assortment of partially formed versions of the symphony come into being at various stages of the process? (But again, when precisely??)

Yet again ontological inflation looms, but a more pressing problem concerns the relationship between the thought (the conceiving of the musical work) or the practice (the writing of the score) and the corresponding abstract objects. If this is not causation, what is it? It remains a mystery.

In the next chapter, we shall look at one possible way of dissolving the mystery through Thomasson's proposal that artworks should be regarded as 'abstract artifacts' (see Thomasson 1999, pp. 132–4). The core idea here is that musical works are created by and depend for their continued existence on certain human intentional states but are not to be identified with either the imaginary creations of individual minds or physical objects. However, the ontological inflation involved in this proposal is again a concern. In addition, as Thomasson openly acknowledges, it also requires us to accept a new hybrid category of entity that lies between those of concrete individuals and Platonic abstracta. As far as she is concerned, this is a price we have to pay in order to resolve the ontological problems generated by works of art. However, again as we shall see, there may be cheaper ways of dealing with these issues.

A closely related option would be to argue that although creation *does* require causation (in some sense), and we can't causally interact with abstract objects, this in itself doesn't mean that we can't create abstract objects. After all, during the process of creation the abstract object doesn't yet exist, so one is free to tell a story where we manipulate things we *can* causally interact with, and the abstract object comes about as a result of our manipulating these things.[16] But again, quite how the abstract object comes into existence remains a mystery.

[16] This option was suggested by a commentator on an early presentation of the ideas in this chapter.

Abstractness and Creativity (Science)

Turning now from art to science, it seems that a similar tension can be generated when it comes to *scientific theories* and *scientific models*. Consider a similar set of three statements:

S1: Scientific theories (and models) are abstract objects.
S2: Scientific theories (and models) are created.
S3: Abstract objects cannot be created.

Clearly scientific theories differ from musical works in many important respects, some of which we shall explore in some detail shortly. But crucially, they share many similarities. For example, they are 'multiply realizable' in a certain sense: just as there can be many performances of Beethoven's *Fifth*, so there can be many presentations of Einstein's special theory of relativity. Just as the former can be recorded in different media, so can the latter be presented—in conference presentations, journal papers, textbooks, and so on. And just as the musical performances can be subtly, or perhaps not so subtly, different, as the conductor and the performers bring out different aspects of Beethoven's work, so different presentations of Einstein's theory can emphasize different features or aspects of it. And very similar questions of identity can arise too: how different can a performance or presentation be before it is no longer the *Fifth* or the special theory? And in both cases, there doesn't appear to be anything 'concrete' that we can identify the given musical work or scientific theory with, which suggests we should accept S1 (although we'll come back to that choice shortly). But then how do we resolve the tension with S2? Here I am going to quickly sketch some possibilities, before delving into the details of two well-known positions, namely that theories are *abstract entities*, in some sense, inhabiting Popper's 'World 3' perhaps (Chapter 5) and that theories are not abstract, but are *fictional*, also in some sense (Chapter 6).

So, an obvious way of dealing with the above set of claims would be to insist that in the case of science, the grounds for talking about the 'creation' of theories is far weaker than it is for musical works. Thus, some might take S2 to be a prime candidate for rejection on the basis that it is inappropriate to say that scientific theories are *created*; rather we should talk in terms of their *discovery*. Theories are 'already there', in some sense, waiting to be discovered, in some theory space (perhaps fleshed out in terms of Popper's World 3, as we'll see in the next chapter). But, if S2 were to be replaced with,

S2′: Scientific theories (and models) are discovered

there still remains an obvious tension with S1, revealed by the question: by what means would scientific theories, qua abstract objects, be discovered? One option might be to appeal to some kind of Gödelian 'platonic' sense, in terms of which we 'perceive' such entities in theory space, but even if that is plausible for mathematical objects, it seems much less so for scientific theories, whose heuristic development can be much more straightforwardly traced. And this is crucial: in the absence of further elaboration of the nature of the access to such entities, the role of the kinds of heuristic moves that have been well-documented in scientific practice then becomes murky.

One way of responding and dissipating the tension would be as follows: just as those who regard artworks as abstract objects could argue that what we call 'discovery' involves a form of selection, or mediated access to the relevant type that allows it to be tokened, so access to scientific theories as abstract objects could be understood as mediated via the very heuristic moves by which they are 'discovered'. Such 'moves' differ considerably in nature and kind (see Post 1971) and it is difficult to see what these have in common both between themselves and with the sensory modalities in terms of which we gain access to observable objects. Of course one might argue that the relevant comparison should be with our indirect access to unobservable entities, but again it isn't clear that there are any similarities with the inferential moves that are made there.

Alternatively, we might profitably import Dodd's device of 'creative discovery' leading to a 'clear conception'. Thus, the creative aspect would be embodied in the scientist's heuristic moves around that theory space by which she 'comes across' and thereby discovers the theory concerned.[17] Both the relevant contextual factors can then be accommodated and the achievement of the scientist acknowledged, since, just as with the composer of musical works, it is not just anyone who can get to the relevant position in theory space and form such a clear conception. Indeed, as just noted, Dodd himself draws a comparison with Einstein's 'discovery' of special relativity:

> Einstein discovered the facts that constitute the Special Theory of Relativity; he did not invent them. This, however, does nothing to undermine our sense of Einstein's peerless brilliance and creative thought; only someone very much like Einstein could have uncovered such facts. Equally a possible world in which the Special Theory of Relativity emerged in 1850 is a world light years away [ha ha!] from the actual world. Einstein's discovery was contingent upon other discoveries before it, and upon Einstein's internalization of a vast amount of physical theory, so all of this background would have to have been in place too. But, of course, it all might have been. (Dodd 2002, p. 389)

[17] We recall Schiffer's (2003) deflationary account of propositions which takes them to be 'pleonastic' entities in the sense that they are involved in moves that allow us to deduce statements about certain kinds of entity from statements that make no reference to those entities.

Nevertheless, he insists, the comparison should not be taken too far, since Einstein was 'aiming to *find things out*' (ibid.; sic) and might have been mistaken. As a result, and as already noted, Einstein's achievement falls under 'discovery by enquiry', constrained as it is by the relevant heuristic moves.

Actually, however, it seems that one could take the comparison further than Dodd thinks. First of all, it is not that certain facts 'constitute' special relativity; rather, the theory *represents* (according to the realist) certain facts about the world (the behaviour of clocks and rods or the local Minkowski structure of space-time, depending on one's stance). Einstein, of course, did not 'invent' these facts but he might plausibly be said to have 'invented' special relativity. Secondly, however, insofar as Einstein might be said to have discovered or 'uncovered' these facts, it is only in a second-hand sense, once the theory has received sufficient confirmation[18] and has been accepted (as at least a partially true representation of the world). Thus, and thirdly, what we should be interested in, at least from the perspective of the current work, is Einstein's 'discovery' of the theory itself. Whether that discovery was *successful* or not, according to the relevant criteria, likewise seems to be a secondary issue and here we see a difference between this and the example of Wylie's proof of Fermat's Theorem: indeed, Wylie may have been mistaken, in the sense that after chugging through all the lines of his proof, he may have found that he did not end up with Fermat's Theorem at the end, or that there was a impassable lacuna somewhere,[19] but Einstein could not have been 'mistaken' in *this* sense in his formulating of special relativity. It is only later, further down the confirmational line, that one can say whether the theory was successful or not. Indeed, it seems odd to say that Einstein could have been mistaken at all, since it is not as if he had an already prepared description and then went out looking for *the theory*, with the possibility that he might not have found it.[20]

Thus, if one can draw, even roughly, a distinction between the domains of discovery and justification within what Dodd calls the 'domain of enquiry', one can extend, it seems, his account to theories: Einstein 'discovered' special relativity by virtue of being able to form a clear conception of the theory, as an eternal type, existing in some theory space, perhaps. And he was able to do so because he was sitting, as it were, in a certain place in that space, thanks to the relevant contextual

[18] And of course, when that is may be subject to debate—indeed, tests of special relativity continue (see Witze 2014)—but certainly it was not in 1905 when Einstein proposed (or formed a clear conception of) the theory!

[19] Indeed, Wylie's initial proof was found to contain an error and it took a further two years of work before a successful proof was produced, amounting to over 150 pages of very complex mathematics.

[20] Of course, in many cases what scientists have in mind is some description of the relevant phenomena or some low-level phenomenological theory and there is then a question of whether what they discover matches this. In Einstein's case it has long been held that he did not know of the details of the relevant phenomena (namely, the lack of aether drift made apparent by the famous Michelson-Morley experiments) prior to 1905, although the discovery of recent transcripts of some lectures he gave in Chicago in 1921 appear to put the lie to this (see van Dongen 2009).

factors, which include his background knowledge, various heuristic devices, and so forth. These all helped to get him to that point in theory space from which he could form such a clear conception.

But now the questions asked about musical works return, with renewed force; in particular: what is meant by 'clear' in this (scientific) context? So consider again the dispute over whether it was Lorentz, Poincaré, or Einstein who discovered special relativity. Taking the most well-known and best supported line—that it was Einstein—we might then say that Lorentz and Poincaré occupied different places in theory space, such that they were unable to form clear conceptions, in some respect, of special relativity. This then absolves us of the need to envisage that theory space as being cluttered up with proto-theories, heuristic prequels, cranky 'pseudo'-theories, or what have you—an issue that will come to the fore in the next chapter, as we'll see. But then Poincaré seemed to have a pretty clear conception of his aether-based theory (Poincaré 1904/6), so on what basis are we to judge that Einstein had the clearer view? One can do this *post hoc* of course, but then clarity comes down too close to empirical adequacy, and that can't be what it's about at all.

And, given that, how is a scientist's 'place' in theory space delineated? What contextual factors go into that delineation? Poincaré could have 'discovered' special relativity because the theory is about certain objective features of the world (in realist terms) and so, even though he was French, and had a different background than Einstein, it seems plausible to conceive that he could have arrived at the same theory (perhaps after a chance encounter with a Swiss patent clerk who convinced him to drop the aether and reconceive space and time!). This seems more plausible than the counterfactual claim that Luigi Cherubini could have composed the *Fifth Symphony*, precisely because of the existence of relevant world-based constraints. But then the issue of how those constraints structure theory space or guide the scientist through it becomes absolutely crucial. In effect one would have to posit an appropriate relationship (one of similarity if not of mirroring—a partial isomorphism, perhaps!) between all the different kinds of heuristic moves made in practice and that structure. Now, although its not entirely implausible to insist upon such a relationship, it does bring with it further ontological costs: not only is theory space populated with all these abstract entities, but it too is complexly structured, in a way that relates appropriately to scientific practice. As we'll eventually see, there are ontologically cheaper ways of accommodating such moves and such practices in general.

Scientific Theories Are *Not* Abstract Objects

Despite the motivation noted above, a case could be made for rejecting S2, the statement that scientific theories and models are abstract objects. Granted that just as musical works cannot be identified with the score, so a theory cannot be identified with what is written down and just as a musical work cannot be identified with a particular performance (by an orchestra, say) or the totality of such, so a theory cannot be identified with all or any of its presentations, but that doesn't compel us to take them to be abstract entities. What if they were to be identified with the thoughts in some person's head, or indeed the thoughts in a number of persons' heads? Obviously there's the concern about inter-subjective access but we've already seen how Collingwood's approach can deal with that when it comes to music.

So, lets consider the extent to which such moves might be imported into the philosophy of science. We recall that Collingwood himself compares a musical performance to a scientific presentation (1938, pp. 140–1), noting that although they are obviously dissimilar in terms of what we are trying to 'get out of' them, there is the crucial similarity that, just as in the case of the musical performance, what we are trying to get out of the presentation is more than just the noises the speaker is making (or at least one would hope). Furthermore, the focus is not, or should not be, on the sensual pleasure we might get out of the tones of the lecturer's voice (although that might be a bonus). Rather, of course, the lecturer's intent is to develop a scientific thesis and '[t]he noises are meant to assist us in achieving what he assumes to be our purpose in coming to hear him lecture, that is, thinking this same scientific thesis for ourselves' (ibid., p. 140). This is achieved not by the noises effecting the imparting of the thesis to our minds but by their enabling us to reconstruct and thereby reproduce that thesis via 'active thinking' (ibid.).

Here too one might argue that insofar as scientific thinking is imaginative thinking, the products of that thinking—theories and models—are imaginary things.[21] Again, such a view faces the problem of inter-subjective accessibility but Collingwood offers the same way around this: we reconstruct and reproduce the theory concerned via this engaged and informed thinking. Indeed, he insists that if we do not make efforts of the right kind, it will remain 'forever inaccessible' to us (Collingwood 1938, p. 141). It might also be suggested that in this case, at least, there is less nuance to capture or that reproducing the core idea encapsulated in a scientific theory is more straightforward and thus more plausible than reproducing the emotions tied up in a piece of music, so one might be less sceptical about achieving such a reproduction.[22] And, as is clear from the quote above, Collingwood

[21] Despite his emphasis on the imagination, Collingwood's view is very different from the sort of fictionalism that may be familiar to philosophers of science and that we shall consider in Chapter 6. The latter is equivalent in his terms to imagination acting under the censorship of desire, namely the desire that the situation imagined were real, whereas the imagination required for 'proper' art is indifferent to the distinction between real and unreal.

[22] Of course, the sceptic might press the point that a major scientific theory, at least, such as general relativity, or quantum mechanics, is just as nuanced and complex in its implications as Beethoven's *Fifth*, say, and, furthermore, one might well have reasonable doubts as to whether anyone can reproduce what was in Einstein's mind, say, no matter how well trained or appropriately educated!

considers that what is reproduced is indeed 'the same' scientific theory, not some imitation, pale or otherwise.[23]

As we have already seen, issues of what counts as 'the same' now arise. Let's go back again to the example of special relativity: if I read Einstein's 1905 paper can I plausibly be said to have the same theory 'in my head' as he did; or more interestingly, perhaps, could he be said to have the same theory in his head then, as those of us have in our heads now, post-Minkowski?[24] The latter case raises even more acute issues: when Minkowski recast special relativity in terms of spacetime, rather than clocks and rods, we would have to say that he was not, in fact, recasting Einstein's theory but rather his own reconstruction, which would have to be treated as strictly a different theory. We would certainly have to reconsider what was meant by 'theory development' and in effect rewrite the history of science (if, at least, we wanted to be true to our ontological commitments!).

Again, we shall come back to these issues in later chapters but let us now consider the final alternative—dropping S3.

Abstract Objects *Can* Be Created

S3 is identical to M3, of course, and so we can run much the same discussion as in that case. Again, at the same instant that a scientist comes up with a theory, the corresponding abstract object presumably comes into being in theory space. And again, we can ask if it comes into being at the moment the scientist conceives of the theory in her mind, or when she writes down the relevant statements, equations, etc., or if it 'emerges' during such a process. We'll come back to this issue. Furthermore, although the idea of a theory 'coming into being' might broadly mesh with the 'lightbulb' or 'Eureka!' view of scientific discovery, that view has been widely rejected as too naïve in not accommodating the role of heuristic factors mentioned previously. Of course, one could allow for such factors by modifying this approach so that in conceiving of a hypothesis, say, or writing down an equation etc. on the basis of making the relevant heuristic moves, one thereby creates the corresponding abstract element. And just as some combination of such elements in practice would be taken to compose the theory, so, paralleling this, at the abstract level, we would have abstract objects corresponding to such elements composing the abstract object corresponding to the theory. Once again, even those with strong metaphysical stomachs might feel a little squeamish over the ontological inflation involved, but perhaps more crucially, how are we to

[23] Cf. 'the material work is the temporally persisting focus of a sequence of imaginative creations, which can alternately be viewed as a single imaginative work evolving through centuries, liberated from the confines of any one individual mind, including the artist's' (van Fraassen and Stigman 1993, p. 95).

[24] And also post-conventionalism, post-debate over substantivalism, ...

understand the relationship between the thought (the conceiving of the hypothesis) or the practice (the writing of the equation) and the corresponding abstract objects?

As I've said, in the next chapter we will consider Thomasson's proposal to regard artworks as 'abstract artifacts'. Applying such a view to scientific theories, we might then say that insofar as the creation of such a theory involves the intention of the scientist, it is this intention alone that brings the theory into existence. But again, as we shall explore in more detail, this raises questions as to how such a stance meshes with the heuristics of scientific discovery. At the very least anyone adopting this proposal would have to acknowledge that these intentions would have to be constrained in some way. Perhaps you could say that any intention to produce a theory, no matter how bizarre, unjustified, or out of step with current science, creates the corresponding abstract artefact but that only the sub-set of the resultant plethora of such artefacts that meet the relevant heuristic criteria would count as 'theories'. And again you would have to accept certain relations between the various artefacts corresponding to the different stages of the development of the theory, paralleling the relations we discern between their material counterparts in practice. As in the case of musical works—indeed, perhaps even more so—the ontological inflation involved is daunting.

Each of these ways of resolving the tension carries significant costs. In the following chapters we will explore these approaches and the associated costs in more detail. However, I shall go on to suggest that we can avoid these costs by importing into the philosophy of science a further device from metaphysics, by way of the philosophy of art, which provides us with an entirely new exit from our tripartite dilemma.

5
Theories as Abstract Artefacts

Introduction

In the previous chapter we touched on the claim that theories and models are abstract entities in some sense. Here I want to flesh out this idea a little further, and explore some of its consequences through a comparison with not only musical pieces but also works of literature.

Let's begin by recalling the core idea of the Semantic Approach from Chapter 2, namely that theories should be regarded as families of (set-theoretic) models. According to Giere (1988, pp. 80ff.), for example, the latter are generated from the relevant laws, such as Newton's laws, Maxwell's laws of electromagnetism or the laws and principles of special relativity. However, he insists, these should not be regarded as vehicles for making empirical claims, given their considerable generality.[1] If we *were* to insist on taking them to be genuine statements about something, we have to say what it is they are about and Giere argues that the only candidates are 'highly abstract objects', in the sense of 'an object that, by definition, exhibits all and only the characteristics specified in the principles' (see for example Giere 1988, p. 80ff.; ibid., p. 745). The principles are then true, but only trivially so, of such objects and act as templates, with the aid of certain conditions, for the construction of more specific abstract objects, namely the relevant models. To get from these to actual empirical claims we must then 'designate', or denote, a particular concrete system, as when we apply the model for simple harmonic motion to an actual pendulum, for example.

Hence, models must be abstract as that is the only kind of entity that scientific laws, understood as genuine statements, can be 'about'. But, what does it mean to be an abstract entity on Giere's view?[2] Setting aside the obvious but not terribly helpful contrastive claim that theoretical models, for example, are abstract in the sense that they are not concrete (Giere 2008a), it is worth noting that Giere takes such entities to be 'human constructions' (Giere 2004, p. 747, n. 7),[3]

[1] Such claims are embodied in theoretical hypotheses about the models' 'fit' to specific things or target systems.

[2] Its worth noting that according to Psillos one can still be a realist while holding such a view—cf. Psillos, forthcoming. I shall return to the realism issue in Chapter 9.

[3] Thus, he also suggests that as products of the human imagination they are akin to fictions, but differ from the latter with regard to their function (Giere 2008a)—another claim that I shall return to shortly.

There Are No Such Things as Theories. Steven French, Oxford University Press (2020). © Steven French.
DOI: 10.1093/oso/9780198848158.001.0001

whose construction is made possible by the 'symbolic artefacts' of language and mathematics. However they are not to be identified with these artefacts; neither are they purely formal but are created already interpreted, as when we plan a shopping trip (ibid.). As an abstract entity, then, a model should be no more mysterious than a planned shopping trip.

However, even if we take the plan to be a kind of model of the trip, as Giere suggests, it is not clear that appeals to such 'homely' examples dispel the mystery. At the time of drawing up the plan, the trip is not a real, actual, whatever... entity; it itself is only possible. Of course, once the trip is undertaken we can say that the plan represents an actual series of events but this is, at the very least, vastly more ephemeral than the kinds of phenomena that science deals with. In these cases the models concerned represent something actual not possible and to which we may, if so inclined, adopt a realist stance. That something purportedly abstract can represent such an actual, concrete entity, or set of entities, brings into sharp focus again the question: what is the nature of these abstract representational entities?

More usefully, then, Giere goes on to suggest that scientific models are abstract in two ways:

> First, they are abstract objects like numerical relationships or geometrical figures, square roots, perfect squares and circles, or never constructed buildings described in architect's drawings. They are not physically realized. Second, they are abstract in that they are not fully specified. Newton's Laws refer to forces, masses, accelerations, velocities, positions, and times, but not to any specific such objects or quantities. (Giere 2008b, p. 5)

The second sense of abstraction just has to do with their place in the hierarchy between data models and theories and the notion of abstraction, although much debated within the philosophy of science and usually in the same context as that concerning idealizations (ibid.), does not directly pertain to the models' ontological status. However, by again comparing models to other abstract entities, the first sense of abstract does compel us to consider the above question and invites us to consider others, such as: do models, like geometrical figures or numerical relationships exist in some Platonic realm? If they do not, what sense can we make of their purported abstract nature?

Someone who famously, faced up to these questions was, of course, Popper. In effect his notion of 'World 3' offers a kind of space, not just for theories, but also artworks in general and as we shall see, we can further develop the relationship between the theories, as abstract entities in this space, and the scientists concerned, by importing the view of these theories as abstract *artefacts*.

The Surprising Life of a World 3 Entity

For Popper, positing his 'World 3' offered a way of securing the objectivity of knowledge without having to invoke mental states such as beliefs, as in the standard account of knowledge as justified true belief (Popper 1972).[4] In contrast to the world of physical states, comprising 'World 1' and that of mental states, namely 'World 2', 'World 3' is:

> the world of intelligibles, or of *ideas in the objective sense*; it is the world of possible objects of thought, the world of theories in themselves, and their logical relations; of arguments in themselves; and of problem situations in themselves. (ibid., p. 154)

Thus, he conceived his 'World 3' as the world of the products of the human mind in general, occupied by scientific theories as well as works of art (including, notably, musical works) and social institutions (Popper 1978):

> Examples of world 3 objects are: the American Constitution; or Shakespeare's *The Tempest*; or his *Hamlet*; or Beethoven's Fifth Symphony; or Newton's theory of gravitation. (Popper ibid., p. 145)

There are a couple of features worth noting here. First of all, there is the initial characterization of World 3 entities as 'intelligibles'; that is, as entities that have the *potential* to be grasped or otherewise *understood* (ibid., pp. 115–16). Secondly, World 3 encompasses not just these entities but all their logical relations, including, of course, all the logical consequences that could be deduced from them. We'll come back to this point, as it underpins a major criticism of Popper's account. Furthermore, these entities should be regarded as independent of us and, consequently, as *real*—for reasons I will discuss shortly—and, also, as causally effective. Having said that, they are also related to entities in both World 1, the world of physical things—in particular, of course, the 'instantiations' of such entities, as listed above—and World 2, the world of mental states—in particular, our beliefs about such entities.

The nature of that relationship depends on the nature of the object concerned.[5] A sculpture such as Michelangelo's *The Dying Slave* (http://en.wikipedia.org/wiki/Dying_Slave), exists both in World 1 as a block of marble and in World 3, as the creation of Michelangelo's mind (ibid., p. 144). Likewise, a painting such as *Rain,*

[4] Some of the following is taken from French and Vickers 2011 and I'm grateful to Pete Vickers for letting me appropriate this work.

[5] There are also interesting things to say about the relationships between Worlds 3 and 2, where there is a kind of 'feedback' effect: 'Our minds are the creators of world 3; but world 3 in its turn not only informs our minds, but largely creates them' (Popper 1978, p. 167). So, for example, the very idea of 'the self' depends on certain views of time that underpin self-identity and exist in World 3.

Steam and Speed exists as a painted piece of canvas in World 1 and the result of Turner's imagination in World 3. One might see this as offering an interesting twist on the infamous problem(s) of material constitution (see Wasserman 2015). A well-known approach to nailing down the relationship between the statue and the marble from which it is formed is to insist that the former possesses properties—in particular aesthetic or more generally, non-categorical—that the latter does not. These properties can then be grounded in certain relational facts regarding the relationship between Worlds 3 and 1: *The Dying Slave* is a statue rather than just a mere lump of marble because it exists in both worlds, whereas prior to being shaped by Michelangelo, the marble existed only in World 1 (cf. Baker 2000, pp. 35–46). The 'obvious' concern that this gets the grounding the wrong way round—*The Dying Slave* acquires aesthetic properties and is admired as such *because* it is a statue and not vice versa (Wasserman, 2015)—begs the question here: it is only because it exists in World 3 that it can be said to have such properties to begin with.

There is of course the further issue of what happens when a statue is destroyed, such that it can no longer be said to exist in World 1. Consider Michelangelo's early piece, *Sleeping Cupid* (https://en.wikipedia.org/wiki/Cupid_(Michelangelo)#Sleeping_Cupid), created as part of a scam and subsequently presumed destroyed in the great fire of the Palace of Whitehall in 1698: does the artwork nevertheless still exist in World 3? It would defeat the point of Popper's distinction if *Sleeping Cupid*, as a World 3 entity and thus product of Michelangelo's mind, ceased to exist when Michelangelo did. But does it still exist there when its World 1 counterpart has gone? Of course, we might try and recreate it, if, for example we were to discover a sketch of the piece (see http://www.michelangelo-gallery.com/michelangelo-drawings-list.aspx), but then *that* statue would not be the creation of Michelangelo's mind and a different World 3 object would be created, however similar to the 'original'. To maintain that artworks persist in World 3, no matter the vicissitudes of World 1, is to grant them more of a Platonic nature than seems compatible with Popper's view of theories, for example, but to deny this raises certain obvious issues that I shall come back to shortly.

When it comes to 'repeatable works of art, such as literature, the relationship is more akin to that between types and tokens: there are many different copies of *The Tempest*, scattered about 'World 1' but insofar as they contain the same text, they are all the embodiment or physical realization of 'one and the same book' that 'lives' in 'World 3'. One can then extend this more or less straightforwardly to theories—think again of Newton's *Principia*, for example. Significantly, given what we just considered with regard to Giere's stance, Popper draws the distinction as follows:

> One can, if one wishes, say that the world 3 objects themselves are *abstract* objects, and that their physical embodiments or realizations are *concrete* objects.
> (Popper 1978)

However, the relationship between these two kinds of objects is not to be conceived of in straightforwardly Platonistic terms. First of all, the objects of 'World 3' are not timeless; indeed, Popper takes it as crucial for his overall epistemology that theories, qua 'World 3' objects, be seen as tentative[6] and hence as subject to change. He writes,

> In opposition to Plato and Hegel I consider *tentative theories* about the world—that is, hypotheses together with their logical consequences—as the most important citizens of the world of ideas; and I do not think (as Plato did) that their strangely non-temporal character makes them eternal and thereby *more real* than things that are generated and are subject to change, and to decay. On the contrary, a thing that can change and perish should for this very reason be accepted as prima facie real; and even an illusion is, *qua* illusion, a real illusion. (Popper 1972, p. 300)

Secondly, these objects interact *causally* with those of the other worlds:

> [T]hey [the objects of 'World 3'] may be real in that they may *have a causal effect* upon us, upon our world 2 experiences, and further upon our world 1 brains, and thus upon material bodies. (1978, p. 150)

Clearly, then, World 3 objects are different from mathematical entities, as traditionally conceived Platonistically, since they can be created, and are not causally inert.

As much as one might balk at this extension to what we typically take to be 'real', it is important to recognize Popper's motivation here: the only sensible alternative to 'World 3' would be to take theories to be mental representations, whether conceived of physically in terms of memory engrams, or 'mentalistically' in terms of mental experiences, but that is obviously problematic. Interestingly, his rejection of the mental representation account is based on the attempt to apply it to musical works such as Beethoven's *Fifth Symphony*: in such cases, he argues, unless these works were taken as real, we would have no way of objectively judging what is a good or bad performance of them.

Now, it could be said that we simply have to accept that, in fact, we *don't* have a way of objectively judging what is a good or bad performance of Beethoven's *Fifth*, but have to rely on our own past experiences, our own aesthetic judgement, and what other people tell us. More significantly, perhaps, we might worry along lines already touched on previously, that, even if such an argument were acceptable when it comes to Beethoven's *Fifth*, it represents an inappropriate importation of considerations from the philosophy of art into the philosophy of science.

[6] They are, of course, conjectures, subject to refutation and hence falsifiable.

Here we might recall our discussion of Dodd's distinction between musical works and theories and the relegation of the latter to 'discovery by enquiry', precisely because we don't have good or bad 'versions' of the special theory of relativity, for example. At any rate we *can* judge whether we have the theory 'right' (in whatever sense), by working out whether what we have provides the predictions and explanations the theory is supposed to achieve.[7] In a sense this is just a reiteration of the standard point: in the case of theories, we have, at the very least, empirical adequacy by which to judge them.

Perhaps it is better, then, to look to the positive arguments for the existence of theories as independent, real, 'World 3' objects. The first appeals to the coordination of understanding about a given theory that we find in scientific practice: consider a group of scientists discussing such a theory. They may, of course, differ in their assessment of it, or of what it implies, etc., but, it is claimed, they take themselves to be discussing the same *thing*:

> How is this (rough and imperfect but undeniable) coordination of their understandings of the theory to be explained if it is not there to be understood, there being nothing but their respective subjective conceptions? (Watkins 1974)

With the identification of the theory with a particular set of marks on a whiteboard, say, ruled out, the only alternative explanations are some deity or 'sublimal advertising' (ibid.)![8] But of course, one might object that these are not the only possible explanations—an obvious alternative would be to appeal to social factors to explicate the 'coordination of their understandings' (see for example Bloor 1974).[9] One way of undercutting such an appeal would then be to appeal to the *causal* role of theories in explaining the coordination.[10]

Indeed, the second argument invokes a causal criterion of reality, as already indicated:

> [W]hat is real or what exists is whatever may, directly or indirectly, have *a causal effect* upon physical things, and especially upon those primitive physical things that can be easily handled. (Popper 1978 p. 153)

Thus, what Popper calls his 'fundamental argument' for theories as World 3 objects invokes the way in which science has changed the (physical) world, or

[7] There might be questions of better or worse *interpretations* of the theory, of course.

[8] This is strongly reminiscent of a similar move made in the scientific realism debate: we note certain phenomena and posit some unobservable entity or process as the *best explanation* of that phenomena.

[9] In effect this is to argue that the appeal to theories in World 3 is not the *best* explanation of the relevant phenomena.

[10] And again the parallel with moves made in the realism debate is obvious.

more particularly, the way in which 'conjectures and theories' are used as 'instruments of change'. Hence,

> [S]cientific conjectures or theories can exert a causal or an instrumental effect upon physical things; far more so than, say, screwdrivers or scissors.
>
> (ibid., p. 154)[11]

Now, in defending this claim against the opposing view that it is the mental representations—whether conceived of physically or mentalistically—that exert causal effects, Popper argues as follows. He distinguishes between the subjective thought processes that led to, for example, the special theory of relativity, and the objective *content* of that theory by noting that,

> There are many important logical consequences of the Special Theory of Relativity which Einstein did not think of in 1905; and there may be important logical consequences of this theory which nobody has thought of so far, and which perhaps nobody will ever think out. (ibid., p. 158)

In particular, mental experiences, however conceived, can stand in causal relationships with one another, but not in logical relationships. Thus, he insists,

> These purely objective logical relationships are characteristic of the entities which I have called theories, or knowledge, in the objective sense. (Popper 1972)

That these logical relationships are *objective* can be inferred from the fact that, with regard to special relativity for example, there are many such relationships that are not conceived of when the theory is first proposed, or that are never conceived of. However, one can agree to this whilst disagreeing with the claim that theories cannot be mental entities for *this* reason. If one accepts that two beliefs can be inconsistent one must accept that at least some mental entities can stand in logical relations.

However, Popper continues his argument as follows,

> [A] full understanding of a theory would mean understanding all its logical consequences. But these are infinite in a non-trivial sense: there are infinitely many situations of infinite variety to which the theory might be applicable; that is to say, upon which some of its logical consequences may bear; and many of these

[11] Furthermore, Popper takes it to be a characteristic that they can be improved by co-operative criticism, in the sense that people who had nothing to do with the original idea can work on it and develop it. This provides a further reason to view World 3 objects as objective and as being causally efficacious (they cause people to think!).

> situations have never been thought of; their possibility may not yet have been discovered. But this means that nobody, neither its creator nor anybody who has tried to grasp it, can have a full understanding of all the possibilities inherent in a theory; which shows again that the theory, in its logical sense, is something objective and something objectively existing—an object that we can study, something that we try to grasp. (Popper 1972, p. 299)

The key element of this passage is the claim that theories *cannot* be mental objects, because theories include their consequences, but some of these consequences do not exist in any mind.

Even if the Syntactic Approach were rejected, the argument still goes through as long as theories are thought to have as constitutive elements explicit parts and unknown consequences of these explicit parts. But if these unknown consequences do not exist in concrete form or in any mind (at time t at least), then where do they exist? In World 3, says Popper.

Of course, we could question the key element, arguing that insofar as deducible consequences are obtained from their premises, if the latter are 'in the mind' then so must be the former. Or, relatedly, we might question the lack of logical omniscience on which this argument appears to depend; or, more radically perhaps, simply insist that the logical consequences in question are not proper parts of the theory *per se* (that is, we would deny that a theory is constituted by its deductive closure). Or, as I hope to convince you, a less radical alternative is to reject the underlying assumption that theories *exist*, in whatever sense, and therefore need a home, whether in the mind of World 3 to begin with. Whatever causal efficacy accrues to them should, then be placed elsewhere—indeed, where it has always been placed, with material objects.

However, there is still the idea that theories are 'something we can study, something we can try to grasp'. In this context Popper argues that theories, as described above, have a property that *only* existing things could have: the element of *surprise.*[12] He writes,

> Such a theory, or such a system, is infinite, and may be full of surprises. Thus it must have been a surprise for Einstein when he found, shortly after writing his first paper on Special Relativity, that the now-famous formula $E=mc^2$ could be deduced from it as a theorem. (1978, p. 162)

Of course, you might be tempted to think that the role of surprise in science is well-known, in the guise of novel predictions as discussed in the realism debate. But here the sense of novelty is less psychological, as might be associated with

[12] Cue Monty Python *Spanish Inquisition* sketch….

surprise, and more to do with whether information about the phenomenon predicted was used in the theory making the prediction which might thereby block that prediction from having any confirmational force (see the discussion in Psillos 1999, p. 71). Einstein's 'deduction' can of course be equally regarded as a prediction and it was certainly novel in *this* sense[13] (see Einstein 1905b).[14]

For the scientific realist, novel predictions are powerful motivators for her stance (we'll return to this in Chapter 9): the core idea being that it would be highly improbable for a theory to make such predictions, and get them right, unless it were somehow 'latching onto' reality. However, here the element of surprise is serving a different function: Popper intends it as a mark of the *reality* of something. So, just as physical objects surprise us as we discover more about them, so too do scientific theories. It is surprising (ha ha) that the nature and role of surprise in science is so little discussed in the philosophy of science.[15] Here again we can draw on work in the philosophy of art.

Consider the issue whether our own imaginings can surprise us. According to Stock, Sartre, for example, thought not, arguing 'that the image teaches nothing, never produces an impression of novelty, and never reveals any aspect of the object' (Stock 2006, pp. 173–4; she refers to Sartre 1972).[16] As she goes on to note, according to Wittgenstein this is because imagination is subject to the will (the contrast being that 'real' objects, of course, are not). Or as White says 'one can't be surprised by the features of what one imagines, since one put them there' (White 1990, p. 92).[17] Now, of course, this begs the question against the Popperian view but there's a similar strain of thought, credited to Wittgenstein, that runs as follows: 'the reasons why people are surprised lie in their limitations: a proof is too long to keep all its steps in mind, so something is lost from purview' (Simons, unpublished).

If theories are characterized in terms of the Syntactic Approach, then Wittgenstein's argument would seem to apply: the only reason that the deduction of $E=mc^2$ from the principles of special relativity surprised Einstein, and everyone else, is that not even the great Albert was logically omniscient! Given the deductive opacity of the theory, we should not be surprised that anyone was

[13] The other sense is known as 'temporal novelty' and, as the name suggests, is concerned with whether the phenomenon in question was known *before* the theory was constructed. Famously, Einstein's formula was known before special relativity, albeit deduced on a very different (that is, classical) basis (see https://en.wikipedia.org/wiki/Mass–energy_equivalence#cite_note-inertia-2). As with the recent 'observation' of a black hole, we can relate this to the issue of 'multiple discovery' in science, something we'll discuss in Chapter 9.

[14] Einstein writes, 'The results of the previous investigation [i.e. that of Einstein 1905a] lead to a very interesting conclusion' (1905b).

[15] Heinz Post made this point (numerous times!) during my PhD studies.

[16] I'm grateful to Alice Murphy for bringing this paper to my attention.

[17] Stock (2017) herself goes on to argue that we can in fact learn from the visual imagination, not least because, for example, such imaginings can mislead us and we can discover that the associated beliefs are false.

surprised. Now, even as an adherent of the Syntactic Approach, a Popperian could resist the Wittgensteinian line by insisting that when we think of theories what we are afforded is a glimpse into World 3 and the surprise we feel is occasioned by the intrusion into our mental picture of further features of the theory as an entity in that world. What the deduction is doing, then, is in effect mapping out the contours of those features. Our lack of logical omniscience is merely a reflection of our lack of complete knowledge of the relevant abstract entity. Again, to insist otherwise would be to beg the question.

Such a stance would be boosted were the Popperian to adopt the Semantic alternative. Of course adherents of the latter do not eschew logical deduction when it comes to exploring the consequences of theories, but with theories regarded as something other than logico-linguistic entities, it seems even more plausible to suggest that such deductive consequences are merely revelatory rather than constitutive.[18]

The upshot, then, is that Popper's 'surprise' argument for the existence of World 3 entities may be more difficult to undermine than one might initially think. Nevertheless, as in the case of purported causal efficacy of such entities, this element of surprise can be accommodated in a way that is less ontologically committing, as we shall see in Chapter 7. However, before we come to that, let us consider how we might render Popper's core idea more plausible by drawing further on the philosophy of art, specifically the set of moves associated with the claim that artworks are 'abstract artefacts'.[19]

Models and Theories as Abstract Artefacts

The issue of how to regard the ontological status of 'repeatable' works of art, such as pieces of music or literature, is of course central to aesthetics, where the delineation of the persistence and identity conditions of artworks has been discussed more extensively than is the case for theories in the philosophy of science. One option is to regard such works as abstract types with multiple tokens, or 'norm-kinds' having correct and incorrect instances, for example. This seems a plausible approach when it comes to musical works, where not only can we distinguish 'correct' and 'incorrect' performances but also place them on a spectrum of better or worse. However, as we have already noted, this just doesn't seem to be appropriate when it comes to theories. After all, if I were asked to sketch London's

[18] There is a further useful comparison to be drawn in this context with recent discussions over whether thought experiments and simulations may surprise us (see Brown 2004; Norton 1991 and 2004); again I am grateful to Alice Murphy for drawing my attention to these discussions.

[19] Weisberg (2013, p. 52) mentions this as a possible approach to theories but dismisses it on the grounds that it is not discussed in the philosophy of science literature. That's not entirely true (see French 2010)!

theory of superconductivity on the whiteboard I might well get it wrong (indeed that's quite likely) but rightness or wrongness is surely not the issue. Rather it has to do with the relationship between those tokens, say, of the theory in textbooks and papers—indeed in London's own book—that get it right, and the theory itself, if we're minded to distinguish it from 'its' tokens. Is that relationship akin to that between a musical work and one of its instances? More generally, what is the nature of this instantiation?

Here the role of the artist or scientist intrudes, respectively. 'Abstractist' views of art that take artworks to be abstract objects face the obvious challenge of accounting for the role of the creator of the artwork (see Thomasson 2006, p. 247). Thomasson has responded to this challenge by characterizing certain artistic objects, such as works of music and literature, as 'abstract artefacts' in the sense that although they lack a spatio-temporal location, and hence can be regarded as 'abstract' in this sense, they are still created, come into existence, change, and may cease to exist (Thomasson 1999 and 2004; other works such as painting and sculptures, obviously, count as concrete artefacts). It is through their creative activities that artists bring such artworks into existence, and their continued existence depends on the artists' intentionality.

However, this generates an obvious dilemma. Consider the characters in a literary fiction, say, taken to be abstract entities that somehow encode the properties that the characters exemplify in the relevant story (Thomasson 2003). How do we reconcile the apparently abstract status of such fictional characters with their created and contingent nature. Thomasson's solution is to argue that 'a fictional character is an abstract cultural artefact created at a certain time by the acts of an author writing a work of fiction' (ibid., p. 139). Like other created artefacts, it is contingent—so, consider the character of Frodo from *The Lord of the Rings*: if Tolkien's onerous examination duties had been even worse, he might never have finished or perhaps even begun the book and Frodo as a character would have been very different or not even have existed at all. But of course, Frodo and other literary characters lack any spatio-temporal location and thus are not concrete.

This approach best preserves, she argues, the 'common sense' (or 'folk') view of such characters, although as she acknowledges, certain elements of that view may still have to be revised or abandoned altogether. Again, take the suggestion that Frodo might not have existed had Tolkien become more involved with university administration. Such claims appear to be in flat contradiction with common sense—'Of course Frodo does not exist', you might say; an assertion that may appear to carry even more force than in the case of Sherlock Holmes, say, where we might point to certain real historical people on which the character was based. However, Thomasson avers, there is no contradiction, as long as we keep the distinction between persons and fictional characters to the forefront. Nevertheless, there is some revision involved, since we now have to accept that Frodo qua fictional character does indeed exist, *but as an abstract artefact.*

The question then arises as to what establishes or pins down the nature of such characters? Thomasson is clear: 'the very nature of fictional characters is determined by the beliefs and practices of those who competently deal with works of literature' (ibid., p. 142). So, to ask whether Frodo 'really' exists after it has been established that there is a relevant literary practice concerning him, as manifested in *LOTR*, is simply a mistake. *All that there is to be said about Frodo is embodied in that practice.* As she notes, this is not quite the same as taking such practices as being constitutive of *particular properties* of a given character; rather, she is interested in the meta-level claim that 'literary practices are constitutive of the general *ontological* nature of fictional characters as a kind' (ibid., n. 12). Thus, such practices determine the characters' metaphysical status and in particular, are constitutive of the existence of Frodo, say, so that questions as to what it might take for him to exist over and above such practices simply reveal a fundamental confusion:

> [O]nce we see that the existence conditions for fictional characters are determined by practices, and recognize how minimal those conditions are, it becomes difficult—and unnecessary—to deny that there are fictional characters, so understood. (Thomasson 2003, p. 151)[20]

According to Thomasson, given this dependence of the status of fictional characters on practice, the role of the philosophy of fiction must be 'to extract and make explicit the principles that are embodied in our practices, assess what these commit us to (or what principles they tacitly presuppose), and how we can best make a consistent, coherent theory that accommodates them' (Thomasson 2003, pp. 146–7). Any philosophy that is so revisionary that it conflicts with those practices must then be regarded as at best a view of what the relevant characters might or should be (which may still be illuminating) rather than how they are.

Can we appropriate aspects of this stance for the philosophy of science? Yes we can! Indeed, Thomasson herself has done so (forthcoming). Thus, she invites us to consider a typical element of scientific discourse regarding models, such as 'consider a simple harmonic oscillator' and suggests that 'utterances or inscriptions of sentences like these, made in the appropriate theoretic context, in a way that enables us to follow implicit rules for determining further features of the model system with the aim of aiding in acquiring knowledge of a target system, may also *entitle* us to introduce reference to the model-system itself. For that,

[20] One might worry that this could be used to support the existence of all sorts of entities, including, for example, unicorns, elves, gods, and the like (and hence could form the basis of a *reductio* response) but Thomasson emphasizes that the existence of these sorts of 'things' requires further substantive conditions that we have every reason to think are not met, nor can they be met merely by telling stories or constructing scientific theories (Thomasson 2003, p. 151). Of course, we can say that the character Daenerys Targaryen exists, qua character.

according to ordinary and scientific standards, is all it *takes* to develop a model' (ibid., p. 15). And she continues, 'historical, theoretical and critical discourse about model systems can then be read straightforwardly...as claims about the relevant model system, considered as an abstract artifact' (ibid., p. 16).

I'll return to this suggestion in Chapter 7, prior to presenting my own view that lies close to Thomasson's when it comes to the significance of the relevant practices but insists we can do without the commitment to abstract artefacts. However, let us first note some obvious concerns.

Concerns

Thomasson herself raises and address a number of immediate worries about this extension of her approach from fictions to scientific models. One has to do with the attribution of properties and identity conditions to models, understood as abstract artefacts. Here she refers to her previous work on fictional characters, cited above (1999; 2003) and expresses the hope that a similar move will work in the scientific case. And there seems good reason for such hope: we can take what she says above about fictional characters and similarly suggest that if we take the identity conditions for scientific theories and models to be determined by the relevant practices, and recognize how minimal those conditions truly are, then we have no grounds for denying that there *are* such theories and models.

There might be a lingering doubt, however, that has to do with the differences in the relevant practices associated with Frodo, say, and the model of the simple harmonic oscillator. In the case of the former, we might expand the remit and include *The Silmarillion* and the various drafts and miscellaneous documents that have emerged over the years, but these might nevertheless seem less complex and multifarious than the plethora of practices associated with the simple pendulum, stretching from lectures and textbooks on classical mechanics to those on quantum field theory. Even if we take a different example, such as that of James Bond, who has been written about in authorized novels by eight other authors since the death of Ian Fleming, we might still feel that, by contrast, the practices in science are too diverse to unequivocally determine the relevant identity conditions for the model of the simple pendulum. But of course, perhaps constraints can be delineated in the latter case that parallel those that are presumably insisted upon by Ian Fleming's estate: just as James Bond's preference for his martinis to be 'shaken and not stirred' might be laid down as canon, so a simple pendulum must exhibit sinusoidal motion (given the relevant idealizations). Whether this determination of identity conditions can plausibly be extended to theories is something we shall come back to in Chapter 7.

However, a more serious concern has to do with introducing this 'strange kind' of objects that are abstract artefacts. Thomasson herself is dismissive: such objects are in fact commonplace and include theories, stories, symphonies, and so forth, so we should not balk at taking fictional characters and models to be the same kind of thing (forthcoming, p. 19). However, that might seem a little quick. First of all, one of the prime motivating features behind this view, taken as a fundamental requirement on any such ontology, is that works of literature and music are brought into existence at or by a particular moment in time via creative activity. Now, it is not being suggested that such coming into being is instantaneous (!) or even takes place over a very short space of time. Although Mozart was famous for being able to improvise 'on the spot', there is apparently little or no evidence that these improvisations led to more well-known and established compositions (https://en.wikipedia.org/wiki/Mozart%27s_compositional_method).[21] When it comes to such works, he apparently took two days to write his Twenty-Fifth symphony, for example, while Beethoven spent six years on his Ninth. *The Lord of the Rings*, famously, took many years to complete, starting in 1937 and eventually being published in 1954/5.[22] The point, of course, is that there comes a time when the composition is deemed to be finished and can then be said to have been brought into existence.[23] Can something similar be said about theories and models?

Now, we might well balk at the suggestion that general relativity just came into existence when Einstein thought it up. But some obvious care needs to be taken. First of all no-one is suggesting—at least not here—that *curved space-time* popped into existence when Einstein came up with general relativity! What, then, about the 'laws' of the theory, embodied in the famous tensor equations? Here things get a little (but only a little) more tricky: if you are a realist about laws and take them to be modally informed features of the world that 'govern' the phenomena, in some sense, then again, it is not being suggested that these likewise 'pop' into reality through Einstein's creative endeavour. If you are a Humean, on the other hand, then you might be inclined to think they do pop into reality, inasmuch as the theory could be said to do so, since on this view laws or, better, law statements, are just shorthand descriptions (within some 'best' system) of repeatable phenomena (which, again, no one is suggesting is doing any kind of popping). With that out of the way, the balking might subside—we are talking about

[21] This bears on the issue of creativity and heuristics in music—the view of Mozart in particular as composing through some impulsive, muse-driven craeative act over which he had little or no control is now seen as a product of nineteenth-century mythologization of the composer.

[22] Although it was published as three volumes, Tolkien in fact intended it to be a single book.

[23] Of course, when this point is reached may still be debatable. In a famous incident, included in the film by Mike Leigh, when Turner viewed his seascape 'Helvoetsluys' hung next to Constable's more colourful 'The Opening of Waterloo Bridge', he daubed a red blob in the middle of his sea (Constable's painting featured figures in resplendent red jackets as well as red standards flying) and fashioned it into a buoy, thus completing the painting while it was being exhibited (http://www.gettyimages.co.uk/detail/news-photo/an-employee-looks-at-turners-helvoetsluys-hanging-alongside-news-photo/90996762).

whether the theory, *as an entity in itself*, could be said to come into being through Einstein's creative acts. And on the face of it, there seems to be a reasonable degree of similarity between artworks and theories in this regard. So, just as musical compositions can be said to be created over a period of time, we can think of Einstein's theory as being developed from 1907 but finally coming into existence in 1915, with his presentation of the field equations to the Prussian Academy of Sciences. And, heading back in the other direction, just as we can trace the development of the theory through Einstein's notes, papers, letters, etc., thereby extracting the heuristics moves he took, so we can do something similar with Mozart or Beethoven (just as we can with Picasso, as we noted in Chapter 3).

However, this is again a little too quick as it stands. Returning to Popper and his World 3, this account has been criticized on precisely this point of *when* the relevant theory or hypothesis can be said to come into existence (Church 1984). Consider the following example: a computer is programmed to churn out tables of logarithms. According to Popper, the mathematical content of such tables exist in World 3 and would continue to do so even if there were no human beings around to read them (Popper 1972, pp. 115–16). But then the obvious question can be asked: '*when* in the encoding process does the new World 3 denizen appear?' (Church 1984, p. 381). Does it appear when the last element of a table is printed out? Or when enough has been printed as to make things understandable? And if the latter, what is required for understanding? Such questions seem to have no definite answer but this points to a serious problem, illuminated by the following variant on the classic 'monkeys and typewriters' thought experiment (ibid., p. 382):

> Imagine that someone has programmed a computer to print out one line after another in English, automatically selecting for each line a random combination of letters of the alphabet, punctuation marks, and numbers. The programme includes an injunction forbidding any line of symbols to occur more than once. Assume also that the programme calls for each line to be no longer than a certain specified length (say, 65 spaces). (ibid.)

Granted that most such lines will be gibberish, some will not only make sense but, given enough time, will encode certain propositions, such as:

> All Greeks are daffodils and Socrates is a Greek (ibid.).

Call this proposition *P*. This encodes or instantiates a definite World 3 entity. If we now suppose that this is a 'stand alone' entity, in the sense that *P* has never been written down or thought of before, it has never been deduced from some other proposition and so on, then we can say that it is only by virtue of the computer spitting the statement out at a certain time (and place) that the entity is manifested in World 3. Now, this manifestation encompasses not only the entity itself but all the consequences that can be deduced from it, including the 'unintended interrelationships and interactions' (Popper 1974, p. 1051) between this entity and others. Thus, if another proposition, *Q* say, was encoded, in some book for example, prior to *P* being encoded, then when *P* is manifested in World 3, so is '*P&Q*'. By the law of excluded middle, the total set of propositions in World 3 must either entail '*P&Q*' or not entail it. But then, at any given time, World 3 must either contain *P* or not contain it. But *that* means that *P* cannot gradually be manifested in World 3 and hence it must appear there instantaneously. That then raises variants on the above obvious questions: why does it appear at that particular instant and not at one an infinitesimal fraction just before or just after?

> On this point the Three World theory is, to put it mildly, highly arbitrary and implausible. For the theory implies, if taken literally, that a written inscription's capacity for being deciphered must leap into existence where less than one ten-billionth of a second before no such capacity existed. Yet what possible physical change—i.e., what possible change of coding—could our hypothetical computer inflict on a written text instantaneously that would serve to alter its meaning, or bestow a meaning where none existed before? If we agree, as I think in fact we must, that it is physically impossible for a computer, as described, to make a significant English-language change on a written page instantaneously, then it follows that Popper's account of World 3 is empirically false. (Church 1984, p. 384)

The refutation hinges on the disparity between the coming into existence of the relevant World 3 entity and the physical process of its being encoded, by which we come to understand or interpret it (ibid.). Now, the Popperian could try to wriggle free in various ways. She could, perhaps most plausibly, deny that the instants of time are arranged in a dense continuum, so that it's simply not the case that there are an infinite number of possible other instants either side of the one at which *P* is manifested (ibid.). And of course, according to certain, albeit still speculative, theories in physics, time should be conceived of as discrete, rather than continuous. But the level of granularity here is still such that the relevant difference (between *P* being manifested at this granule of time or the next one in the series) will still lie far below any possible human level of awareness and so the problem remains.

And of course this issue can be writ large in the scientific context: how do we achieve a reconciliation between the gradual process in practice by which theories are conceived, developed, and interpreted, with their instantaneous emergence as entities in World 3? We could solve the problem by excising the practice based processes and adopt a Platonist view according to which the World 3 entities are not created at all, but exist eternally. However, that raises its own set of issues as to how we come to know them and in what sense theories can then be said to be discovered (see also Popper 1976, pp. 185–6; Church 1984, p. 387). Taking theories to be artefactual, in the sense indicated above, suggests a way of allowing for creation and discovery but in turn raises a general issue: how does engaging in a certain practice, artistic or scientific, or following a certain heuristic move, *create* an artefact that is abstract? In the case of a concrete artefact, such as a painting or a physical model, we can answer this question, not least because the practices and moves considered will embody certain causal relationships that we can understand as effecting the final product, namely the painting or model. We can see how Picasso's studies of the gored horses in the bullring relate to the horse in agony on *Guernica*, where that relationship is effectively mediated by the physical movement of his hand with the brush against the canvas and so on. Likewise, we can track the various moves that led Crick and Watson to construct their tinplate and wire model of DNA (see for example Schindler 2008), where again these moves are translated into their positioning the tinplate and wire in a certain way, at a certain angle, of a certain length, and so on. But how is that effectivity established in the case of abstract artefacts? Here we face the Benacceraf problem of the causal efficacy of abstract objects all over again and for all that it might be insisted that we must not beg the question here and that what is being articulated is a new category of entity, some account is owed of how we can bridge that final gap, between the practice and the entity itself. And the same holds, of course, when it comes to scientific theories and their status as World 3 objects.

One way to do that would be to say that insofar as the creation of such a theory involves the *intention* of the scientist, it is this intention alone that brings the theory into existence. Again consider the case of fictional works: it is one thing to say that to *produce* such a work is to intend the reader to imagine something but quite another to claim that the intention of the author is both necessary and sufficient for fictional *content*.[24] A defence of the latter claim has been mounted by Stock (2017) who argues that, with regard to word-based, single-authored works of fiction, 'there is a certain sort of imagining which readers appropriately extend to the contents of fiction and the contents of fiction are determined by the author's intention that they have particular imaginings of this sort' (ibid., p. 20). However, in spite of the impressive array of arguments deployed, we can immediately see how this runs into problems if we were to extend it to the case of scientific theories[25] (even putting to one side the thorny issue of the prevalence of multiple co-authorship in this context): Einstein produced the special theory of relativity intending the reader of his 1905 paper to imagine rigid rods and clocks behaving in a certain manner[26] but to insist that the *content* of the theory is determined by that intention would be to concede a certain interpretation[27] and discard the

[24] For more on intentions in art, see Wimsatt and Beardsley 1946.

[25] Not for nothing is this view known as 'extreme intentionalism'! As a reader has suggested, whether intentions are necessary and/or sufficient for content may depend on the context.

[26] For discussion of this invitation to imagine, see French forthcomingb.

[27] This is typically taken to be a positivist interpretation, but again see Howard 2017.

post-Minkowski understanding of the theory in terms of space-time. At best we would have to acknowledge that Minkowski, with his different set of intentions, effectively produced different theoretical content, causing, on the Popperian view, the manifestation of a different entity in World 3. And we can obviously multiply such examples, citing, say, Brown's recent interpretation that eschews the role of Minkowski space-time in explaining relativistic phenomena in favour of an explanation in terms of the relevant dynamics (Brown 2005). It seems that on this view we are forced to accept that when it comes to the determination of theoretical content, World 3 is densely populated.

Relatedly, the relevant intentions would have to be constrained in some way, on pain of allowing further over-population of this abstract realm—after all, do we really want to say that every half- or quarter-baked proposal thought up by anyone, not just those trained in science but cranks and crackpots, day-dreamers and lay-people, counts as an abstract artefact? If we're not careful our World 3 will become cluttered with not just discarded theories but also those that didn't quite make it, like Einstein's of 1913, as well as those that developed into those that 'made it', the blind alleys and false starts, the detours and digressions....[28]

You could try to mitigate this by imposing a set of filters, either on the intentions or on the entities they bring into existence, as already suggested. So, taking the first option, you might insist that not just any intention can bring a scientific theory or model into existence, but only the *right* intentions. Now, granted that we might want to be more relaxed about what the right intentions are when it comes to artistic artefacts as compared to scientific, even then it seems reasonable to insist that only those intentions that are appropriately embedded in artistic practices lead to the coming to be of artworks. And even more so when it comes to theories and models: only those intentions that arise as a result of the kinds of heuristic moves previously touched upon generate these sorts of artefacts. We could then say that those intentions not so entwined with the right sorts of practices, such as those associated with 'lightbulb' moments so beloved by cranks and amateurs, are either impotent in that they do not generate any artefacts at all, or generate the 'wrong'—i.e. non-scientific—kinds of artefacts. However, all the work here is being done by the heuristic practices—it is only these that pick out the right sorts of intentions and thereby, albeit indirectly, yield the theories and models.[29]

[28] As one of the readers has noted, the same concern arises with regard to Thomasson's account of fictional characters: as is well known, Frodo did not appear until the third draft of the writing of the first chapter of *Lord of the Rings* and had a different name ('Bingo') and different parentage. Presumably with each draft a new abstract artefact is created and similarly when it comes to Frodo fan-fiction (yes, that is indeed a thing).

[29] There is also the concern that the relevant practices are just too vague and diverse to support the coming into being of an entity in any clear or well-defined manner. The questions begin to proliferate: which practices? Over what time-frame? How are their entwined interrelationships reflected ontologically in the artefacts created? And so on.

But even if we allow for such constraints by acknowledging that in conceiving of a hypothesis, say, or writing down an equation etc., *on the basis of making the relevant heuristic moves*, one thereby creates the corresponding abstract element, variants of the questions raised above must be faced: does the abstract artefact only come into being at the end of the heuristic process? Can we pin that down? In Einstein's case was it really when he presented the general theory of relativity to the Prussian Academy of Sciences? That seems somewhat arbitrary, as if the theory could not be considered to be in existence when he carried his notes or paper into the lecture theatre and then, again, popped into being when he presented it (again, when he began or when he finished?). Or did it come into being when he wrote the final sentence or the final full stop? Or when he completed it 'in his head'?![30]

And what about all the steps leading up to that final moment? Suppose some led to something akin to the final theory but were not quite there yet. We know that in Einstein's case, the infamous hole argument[31] led him to abandon his focus on general covariance, even publishing a relativistic theory of gravitation that was not covariant in 1913, only for him to return to this requirement when he took the argument (erroneously) to be a trivial error (see Norton 1984). More generally, we can find other examples in the history of science of 'proto'-theories, of not just speculations but theories that are almost in their final form but aren't quite 'there' yet. Are we to say that all of these 'almost rans' are also created in World 3 via the scientists' intentions as a result of the relevant practices or heuristic moves being enacted? And even if we could find a principled way of excluding them, we would surely have to acknowledge that some combination of such elements in practice can be taken to compose the theory concerned, and so, paralleling this, at the abstract level, we would have abstract objects corresponding to such elements composing the abstract object corresponding to the theory.

Alternatively, perhaps it could be argued that any intention to produce a theory, no matter how bizarre, unjustified, or out of step with current science, creates the corresponding abstract artefact but that only the sub-set of the resultant plethora of such artefacts that meet the relevant heuristic criteria would count as 'theories'. This allows for any kind of intention, no matter how odd, to yield an abstract artefact but imposes some sort of filtering mechanism such that only certain of these count as scientific theories and models. Thus, just as those who regard artworks as abstract artefacts could argue that what we call 'creation' of a

[30] Likewise, did *Moby-Dick* come into existence only when it was published with that title in New York in November 1851 and not when it was published as *The Whale* in London the month before? Given that the London publisher cut passages perhaps we *should* take this as a different entity but Melville himself made last-minute changes—should we take this as being made *to* 'the' work or as each creating a different literary entity?

[31] Much discussed in the philosophy of physics in the context of space-time substantivalism; for a useful summary, see Norton 2014.

work of art involves a form of selection, or mediated access to the relevant type that allows it to be tokened, so access to scientific theories as abstract artefacts could be understood as mediated via the very heuristic moves by which they are 'discovered'. But again, note that these 'moves' differ considerably in nature and kind. It is difficult to see what they have in common between themselves such that they can be considered to grant the relevant kind of 'access'. If it is just the fact that they are the kinds of heuristic moves practised by scientists, then again, all the relevant work here is being done by the practices, with both the coming into being and the selection as merely metaphysical 'add-ons'. Furthermore, you don't have to be an adherent of Laudan's (currently neglected) 'reticulated' model of scientific progress (Laudan 1984; 1990)[32] to appreciate that these moves have changed and evolved over the years: the aforementioned role of symmetry principles in modern physics has become much more prominent over the last century or so with the intrusion of group-theoretic techniques into both space-time physics and quantum theory (for a brief overview, see chapter 4 of Bueno and French 2018). This adds further complexity to the relationship between such selection devices, the relevant intentions and the theories/models qua artefacts themselves.[33] And again one would have to accept certain relations between the various artefacts corresponding to the different stages of the development of the theory, paralleling the relations we discern between their material counterparts in practice.

Furthermore, the sense in which we gain 'access' to such artefacts that are brought into being via whatever intention one might have remains unclear. There are obvious differences between the sorts of heuristic moves we find in science (or modern science at least) and the sensory modalities in terms of which we gain access to observable objects. Of course it might be suggested that the relevant comparison should be with our indirect access to *unobservable* entities, but again it isn't clear that there are any similarities with the inferential moves that are made there. Some, for example, have argued that the appropriate sense of access is a kind of extension of sense-based observation (Shapere 1982) but then the previous point obtains. Others, especially those of an antirealist persuasion with regard

[32] According to this model, aims, methodologies, and theories are inter-dependent such that changes in theories (e.g. from classical to quantum mechanics) may drive changes in aims (e.g. from having deterministic theories to indeterministic) which then drives changes in methodology (e.g. a greater emphasis, at least, on probabilistic outcomes and confirmation), which then yields different theories... and so it goes (think of the Manx triskelion or 'Ny Tree Cassyn'). In this context, we can insert heuristics into the picture, such that the shift from classical to quantum mechanics led to a greater emphasis on symmetry as both a heuristic and justificational device which then fed into the search for new theories in high energy physics.

[33] You might think that such concerns bite less severely when it comes to artworks but in addition to the examples already mentioned, note that there are at least three different early versions of *Hamlet*, for example, that a raised fist, as a sign of resistance, was included in an early version of *Guernica* but was subsequently removed (http://www.isreview.org/issues/52/guernica.shtml), and of course, what was to become the statue of David remained 'badly blocked out' for many years before Michelangelo was given the commission to complete it (https://en.wikipedia.org/wiki/David_%28Michelangelo%29).

to such entities, maintain that this extension is blocked at some point (see Bueno 2008) and that any such claims to access must be based on other kinds of moves that can be put into epistemic doubt. Are there analogues to such moves when it comes to theories qua abstract artefacts? At the very least, further work is required to articulate how the post-intentions access to theories and models is akin to whatever access we have to unobservable entities. At best, until this mode of access to theories is spelled out in terms of the specific heuristics, the proposal remains just a suggestion.

Alternatively, we could drop Thomasson's device and appropriate Dodd's account: theories exist fully formed in Popper's World 3 and scientists come to form a 'clear conception' of them via the kinds of heuristic moves we find in their practices.

Of course, that doesn't sit too well with Popper's own views of the matter. In his classic statement that relegates heuristics to the realm of psychology, Popper himself talks of 'the act of conceiving or inventing a theory' and of ideas 'occurring' to a scientist (Popper 2002/1959, pp. 7–8). Here Popper clearly doesn't mean 'discover' in the sense of 'already there waiting to be found'. Indeed, his language is suggestive of 'creation' more than 'discovery', and the fact that he directly compares how we come up with new musical themes and new scientific theories (and of course places both in World 3) further supports the thesis that Popper, at least, was not using the word 'discover' in a literal sense.

Putting that historical point to one side, insofar as having a clear conception involves having some kind of access to these abstract entities, the same issues as outlined above recur. Furthermore, it may appear particularly difficult to hold on to the idea that scientific theories subsequently revealed to be *false* are discovered, either in general or in Dodd's sense. The suggestion that *true* theories can be discovered has some initial plausibility at least. After all, didn't we discover that the earth orbits the sun, and not the other way around? In fact even this doesn't support the claim that we discover theories: we discover *that* the earth orbits the sun, we don't discover the *theory* that the earth orbits the sun. But when we move to false theories we don't even have this initial plausibility. And it is hard to understand in what sense one might have a 'clear' conception of such theories. Furthermore, to insist that even false theories are 'out there' in whatever sense again has obvious inflationary consequences.

On the other hand, a realist who maintains that appropriate commonalities can be found across even quite radical instances of theory change may suggest that it is these common features that support claims of approximate or partial truth that are discovered, but this would be to again confuse the discovery of that to which these features refer (on a realist view) with the discovery of these features as referential or, more broadly, representational devices. Where there are few if any such commonalities, such suggestions appear even less plausible. Consider the rather wild theory once put forward that electrons in an atom orbit the nucleus in hexagon-shaped trajectories. How could this feasibly be 'discovered' rather than created, in some sense?

It might be argued that if theories were to be identified with sets of propositions (closed under some appropriate logical operation), then the claim that *false* theories can be discovered may appear more plausible. On this view, false theories are just sets of propositions that contain a false proposition, which should be no more ontologically troubling than sets of true propositions. And if propositions are taken to be effectively 'out there', in some sense, then *if* (a big 'if') true theories can be discovered, then so can false ones. As we've seen, the *identification* of theories with either sets of propositions or set-theoretic structures faces well-known problems. But even setting these aside, pushing the above argument takes us away from the view that *theories* have ontological status as things. Of course, one could then take propositions themselves to be eternally existing abstracta. Hence via a two-fold identification (first of theories with propositions and then of the latter with abstracta) one could maintain that false theories are just as discoverable as true ones, since both would now be identified with sets of propositions, in turn understood as abstracta. But then the above problems recur, albeit on a smaller metaphysical scale perhaps: either our theory space is littered with a myriad false propositions or we have to accept that somehow we can have a clear conception of that which is not true.

Perhaps these kinds of worries can be swallowed when it comes to artworks but they do give us pause in thinking of scientific theories in this way. And that pause may be deepened, by a further feature that distinguishes theories from artworks.

Multiple Discovery and Modal Flexibility

The phenomenon of 'multiple discovery' has been repeatedly discovered itself by historians and sociologists of science, as Merton famously noted (Merton 1963).[34] Here my focus is on the multiple or even simultaneous discovery, or creation, or whatever, of *theories and models*, rather than phenomena, substances, or features of the world in whatever sense. Among the latter we might include the discovery of oxygen by Lavoisier, Priestley, and Scheele, or of sunspots by Harriot, the Fabricius father and son team, Galileo, and also early Chinese astronomers, or even of pulmonary circulation by al-Nafis, Servetus, and Harvey…the list is

[34] Here Merton is operating at the meta-level in noting that this rediscovery of multiple discovery in science is caught in a static condition (again, at this level) 'as though it were permanently condemned to repetition without extension' (1963, p. 238). He then considers the reasons for this resistance to examining the phenomenon by scientists themselves, concluding that these have to do with complex patterns of behaviour associated with, on the one hand, ethical and other issues concerning priority and, on the other, the influence of the 'Eureka!' view of scientific discovery. But of course, this resistance can be, or should be, easily overcome by philosophers, or even sociologists, of science!

extensive. In all these cases the relevant identity conditions are reasonably clear, so that the 'fact' of multiple discovery can be, equally reasonably, established (although there may be some issues of dispute in certain cases). That's not so when it comes to theories, where it is not clear, at the very least, that we have the relevant identity conditions and where considerations of what the theory is 'about' may bear on multiple discovery claims.

So, in addition to the (possible) case of Hilbert vs Einstein and general relativity, consider the following examples, taken from different sciences: the special theory of relativity, again, the theory of evolution, and the theory of the asymmetric carbon atom. The first was co-discovered or so it has been claimed, by both Einstein and Poincaré (Zahar 2001), the second by Darwin and Wallace (van Whye 2013), and the final example, by van 't Hoff and LeBel (Gay 1978). One can give various explanations for this 'phenomenon': a sociologist, such as Merton, might point to the relevant historico-cultural context as a kind of common cause of such multiple discoveries (or at least those that are, or nearly, simultaneous), whereas a philosopher of science might seize on certain 'internal' factors, having to do with the initial choice of problem, the limited range of heuristics available, and, more generally, the deployment within a given field of a more or less agreed (if only tacitly) methodology (involving both heuristics and justification) over a period of time.[35] And, going further, a realist might tie in this form of multiple discovery to the kinds I just dismissed, arguing that given the way the world is, we should not be surprised that more than one scientist comes up with the same, or similar ways, of describing it.[36]

Certainly, it raises obvious problems for the view that the creation of theories qua abstract artefacts is mediated by the intentions of the scientists concerned. It hardly seems plausible that the intentions of Poincaré and Einstein, or Darwin and Wallace, van 't Hoff and LeBel, could be the same as one another and thus, even if we could dispel the above mystery, we face the further problem of accounting for how different intentions could produce the same theory (do the intentions become entwined in some way?!). Even if we downplay the role of intentions and fall back on practices—perhaps following Merton's suggestion that this can explain the phenomenon—we're not out of the woods. Although it might seem plausible that the historical context and associated practices that Poincaré and Einstein were engaged in were sufficiently similar as to lead, somehow, to the creation of the same theory, it stretches the point to claim that the same holds

[35] Thus, recalling our earlier discussion of the London–London theory of superconductivity, it turns out that the idea of an analogy with diamagnetism was 'in the air' at the time (see Potters 2019).

[36] An antirealist may indulge in the time-honoured manoeuvre of pointing to cases of underdetermination in response, or, more interestingly, might note that even if we grant that there is a way that the world is, this does not compel its description in the same terms. This again touches on the issue of how we identify the same or different theories and the question whether, in such cases, theories that describe the same unobservable features in different terms are merely nomological variants of one another. Again, we'll come back to this.

for Darwin and Wallace, given their different social status, cultural context, geographical travels, and so on. As for van 't Hoff and LeBel, both the rationales and inspirations were different: the former was primarily concerned with isomerism, and was following a line established by Kekule, whereas the latter set himself the problem of explaining optical activity and followed Pasteur (Gay 1978, p. 222). Of course, one might try to give a more sophisticated account of the relevant practices that would reveal what they had in common. Thus, as already indicated, there may be methodological commonalities (ibid.) that we could highlight but insofar as these are intended to cover other theories as well, they cannot be appealed to in order to account for the creation of these ones in particular.

Alternatively, we could accept that the intentions, practices, heuristic devices, etc. *were* different in each case and that as a consequence, different theories were in fact created. In that case, *if* we were to continue to maintain that, for example, the theory that Poincaré, for example, created was identical *on the face of it*, to the theory that Einstein created (and that's a big 'if' as we'll see), then we would obviously have to maintain that the relevant identity conditions are not what they appear to be 'on the face of it'. Thus, we could insist that the different intentions, practices, heuristics devices themselves form part of those identity conditions and thus the theories are, in fact, different by virtue of all those elements and factors that led to their creation. This obviously suggests a different conception of what a theory is than the standard picture, insofar as we can identify such a picture and equally obviously meshes with a view of theories as historically situated or practice-based entities.

Interestingly there does not seem to be a phenomenon similar to 'multiple discovery' when it comes to works of art; that is, we do not find, it seems, that as well as Picasso, another artist created *Guernica*, or a painting indistinguishable from the latter, or that someone other than Melville wrote a work identical in content to *Moby-Dick*, or that as well as Beethoven, someone else also composed the *Ninth Symphony*.[37] Of course, there are precursors or works that deal with similar themes, or yield similar insights, or whatever.[38] And there are of course, regrettably, numerous examples of plagiarism in literature (see, for a disgraceful example from poetry, http://www.nytimes.com/2013/04/28/books/review/nice-poem-ill-take-it.html?pagewanted=all&_r=0)[39] and many examples of forgery from the

[37] Thus, we might follow Wolterstoff (1980) and understand an artwork as a 'norm kind' in the sense that the artist selects a certain set of properties in creating a work such that those properties become normatively associated with that work. So, in the case of literature, a certain fixed sequence of words will determine what counts as a correct instance of the relevant work.

[38] So, for example, Jeremiah Reynolds's account of an aggressive (justifiably) white whale in his 'Mocha Dick: Or The White Whale of the Pacific: A Leaf from a Manuscript Journal' (https://en.wikipedia.org/wiki/Mocha_Dick) is known to have influenced Melville.

[39] In many cases, the plagiarised copies are said to be 'almost' or 'virtually' identical to the originals, the plagiarist perhaps trying to assuage their guilty consciences at least a little by changing a word here and there; see for example, http://www.theguardian.com/books/2013/may/22/plagiarism-scandal-poetry.

world of painting, as we've noted already (for an interesting use of a 'replica' painting to provoke public discussion, see http://www.dulwichpicturegallery.org.uk/whats-on/exhibitions/2015/february/made-in-china-a-doug-fishbone-project/; for a useful discussion of forgery more generally, see Lenain 2011). However, we recall the point that an artwork's aesthetic features—and hence its nature, as that artwork—can be regarded as determined *relationally* (Kieran 2005). As a result, a forgery of one of van Gogh's sunflower paintings is not the same artwork as that painting. There are also 'appropriation artists' such as Elaine Sturtevant, who 'challenges the concept of originality' by replicating the works of other artists, such as Lichtenstein, Warhol, and Duchamp (http://www.nytimes.com/2014/11/14/arts/sturtevant-double-trouble-a-career-retrospective-at-moma.html?_r=0). Nevertheless, these works are not exactly the same as the 'originals', since Sturtevant, for example, would usually introduce a 'signature' difference.[40]

Furthermore, not only does multiple discovery seem implausible when it comes to artworks, no one has claimed (at least not as far as I know) that Einstein was a scientific forger or that he 'appropriated' Lorentz's or Poincaré's work in the above sense. Whittaker, as mentioned earlier, gave priority to Lorentz when it came to relativity theory, with Einstein mentioned only in passing as having 'completed' the former's equations (Whittaker, 1910; in the 1951 version Einstein is mentioned much more, but primarily for his work on general relativity and quantum theory, with the chapter on special relativity entitled 'The Relativity Theory of Poincaré and Lorentz').[41] And the Nazi-affiliated scientists Lenard and Stark famously excoriated Einstein's work as 'Jewish science' (Gratzer 2001, ch. 10) and urged that it be removed from physics courses. But those extreme examples aside, even in cases of such priority disputes, or of apparent multiple discoveries, there are no analogues to the above examples of forgery and appropriation.[42] It is tempting to conclude from this that the authorial relationship plays a diminished role, at best, when it comes to scientific theories as compared to artworks. And in that case, the extended abstract artefacts account further loses purchase.

We can further explore the issues here through the question: *could someone else have done that?* Given the relevant history, it seems plausible to answer 'yes' in

[40] Having noted that, if they are accepted as art—and practice, as reflected in both art museums and the auctions houses, suggest that they are—then they clearly undermine the ruling that every aspect of a work of art must be the result of some decision by the artist (Gombrich 1995, p. 32). But then, if so little of the resulting artwork is the result of an appropriation artist's decision, obvious questions arise as to the nature of the authorial relationship (see Irvin 2005).

[41] A useful summary of the priority issue can be found here: https://en.wikipedia.org/wiki/Relativity_priority_dispute. In an exchange with Born, who lamented Whittaker's refusal to give credit where it was due, Einstein replied: 'I would not consider it sensible to defend the results of my work as being my own 'property', as some old miser might defend the few coppers he had laboriously scraped together' (Born, Born, and Einstein 1971, p. 200); thanks to https://skullsinthestars.com/2008/07/15/einstein-vs-whittaker-with-born-in-the-middle/ for pointing me towards this!

[42] Of course, there are very well-known examples of scientific forgeries, such as Piltdown Man (for an entertaining list, see http://www.strangescience.net/stfor2.htm) but that's a whole different kettle of (dubious) fish!

the case of special relativity, or indeed, of the theory of evolution: Poincaré or Lorentz, in the former case, and Wallace, in that of the latter, not only could have discovered/created these theories, respectively, but according to some, they did. And if they did, others could have. This plausibility adds further impetus to the view of theories as entities that are, in some sense, 'out there', waiting to be tripped over.

But in the case of artworks, such as Beethoven's *Fifth*, the relationship between the plausibility of a positive answer to our question and the view of such works as abstract entities seems to be inverted. As we saw, Dodd answers 'yes' to the question when it comes to musical works *because* of his view of such works as existing in some abstract soundspace. It is by virtue of its situation in such a space that the *Fifth Symphony* might have been accessed by someone other than Beethoven. But now we have to judge the plausibility of that latter claim and without the relevant history to call upon, as in the case of science, what was a modus ponens becomes a modus tollens.

Thus, consider the further specifications of the above question: could Manet have created *Guernica*? Could Tolkien have written *For Whom the Bell Tolls*? Here, positive answers seem implausible. Why? The lack of plausibility in these cases seems to hinge on the nature of the possible world, or to use a less metaphysically loaded term, 'scenario' under consideration. To consider whether Manet might have painted *Guernica* requires a stretch of the imagination that takes that possible world or scenario *some considerable distance* from the actual one. But now suppose we consider alternative questions: could Braque have painted *Guernica*, or Tolkien *The Chronicles of Narnia*? These seem to be more easily imaginable and positive answers correspondingly more plausible, at least on the face of it: we can imagine Braque having been commissioned to produce a mural for the *Exposition Internationale des Arts et Techniques dans la Vie Moderne* (although perhaps not by the Spanish Republican Government) and reading the horrific eye-witness reports of the bombing in the Times or New York Times and we can likewise see, in our mind's eye, Tolkien sat in the Eagle and Child pub, reading passages from one of the Narnia books to Lewis and his friends.[43] Here the plausibility of our imagining is supported by, not just our understanding of the character of the artists involved, but also the 'closeness', in some sense, of their actual work with the possible.

But of course, imaginability, easy or otherwise, or more generally, conceivability, does not imply 'genuine' possibility. There is an issue as to how naturalistic we should let our modal metaphysics be,[44] but for our purposes, as soon as we start

[43] Actually that one is still stretching it a bit, given Tolkien's opinion of his erstwhile friend's work!

[44] So, consider an example from science: in thinking about intrinsicality, metaphysicians often deploy the possible scenario of a lone object, such as a particle. But can we have a possible world in which there is just one proton, say? Insofar as that which makes the proton the particle that it is bound up with the Standard Model, it is actually not clear that the question even makes any sense (French and McKenzie 2012; 2016).

to flesh out some of the details of these 'imaginings', their plausibility fades (Rohrbaugh 2005, p. 215): would Braque have had the detailed knowledge of bull fighting to be able to produce those anguished animal figures? Or if we chose some lesser-known Spanish painter, would he or she have had the skill and relevant artistic background?[45] Appealing to what we can imagine in order to pursue this issue runs into obvious problems.

One explanation for these difficulties and of the difference between artworks and scientific theories when it comes to multiple discovery in general has to do with the apparent necessity of authorship in the case of the former, as already noted. Under what conditions does this hold? To answer this, consider the extent to which artworks might be said to be 'modally flexible' in the sense that they could be different in some way yet still be considered 'the same' artwork (Rohrbaugh 2005); or in other words, the extent to which we should tighten up the relevant identity conditions. If Picasso had painted *Guernica* so that the lightbulb/eye had fifteen lashes/spikes rather than fourteen, would it still be *Guernica*? Intuition—always a sure guide, of course!—suggests yes. And if the lightbulb/eye had been replaced by an actual human eye, or a candle, our intuitions might push us towards a 'no' answer.

Now, consider Rauschenberg's *Erased de Kooning's Drawing* (see https://www.sfmoma.org/artwork/98.298/).[46] Here it is not possible that two artists, Rauschenberg and Warhol, say, could produce the same artwork, in the same world, since the two productions would compete for the same raw materials (the de Kooning drawing). Again the modal flexibility of the artwork is crucial:

> Is it essential to *Erased de Kooning Drawing* that it be made by erasing this particular de Kooning drawing, or could another have done just as well? If Rauschenberg could have used a different de Kooning drawing in his production process, much as Picasso could have used different tubes of paint or different paintbrushes, then that production is compossible with Warhol's. If not, then the two productions are not compossible and it is open whether Warhol could have made *Erased de Kooning Drawing.* (Rohrbaugh 2005, p. 226)

It is in cases where artworks have essential features that are not sharable that these features will give rise to production processes for which there is no necessity of authorship (ibid.).[47] Where there is sufficient modal flexibility or the artworks

[45] Likewise, those who have read his letters and essays would balk at the idea that Tolkien, a Catholic, with his dislike of allegory and his insistence that a mythical world must be constructed via the bedrock of language, could have written a work featuring a Christ-like lion and a mix of mythological creatures from different cultures!

[46] Rohrbaugh attributes the example to my colleague Aaron Meskin (Rohrbaugh 2005, p. 226).

[47] Recalling the earlier point about priority in the case of theories, such a feature might be the time of production of an artwork, such that it is essential to its identity that it be the first of that kind (Rohrbaugh 2005, p. 227). In that case, the necessity of authorship does not hold. The very nature of priority disputes suggests that theories are not identified in this manner!

concerned do not possess unshareable features, then the necessity of authorship holds.[48] Thus, when it comes to the questions whether Manet could have painted *Guernica* or Tolkien written *For Whom the Bell Tolls* given that *Guernica* plausibly has no unsharable essential features of the above sort we can conclude that no possible production of Manet's could have resulted in *Guernica*, and hence Manet could not have painted it (ibid., p. 228). And the same goes for Tolkien!

Can we apply this analysis to scientific theories? We've noted that the authorial relationship seems to play a stronger role when it comes to artworks than for theories. And now, by probing deeper into the issues associated with the question, 'could someone else have done that?' we can discern the conditions under which a yes–no answer can be given: *where there is sufficient modal flexibility or no possession of unshareable features, then the relationship between the artist and the artwork may be necessary*. So, do theories or models possess unshareable features, such that the production/discovery of 'the same' theory or model would be blocked? It is hard to see how one could answer 'yes', at least on first iteration. Of course in the case of concrete models, such as the Crick and Watson DNA helix, one could presumably come up with some outré example that is similar to Rauschenberg's artwork, whereby the nature of the model is such that its construction uses up all known quantities of a particular material so that no other examples of that model could ever be built, but I know of no such cases in the history of science (thank goodness!).[49] And all other kinds of theories and models are such that materiel considerations do not apply.[50]

The more interesting issue is whether theories and models might be said to be sufficiently 'modally flexible', in some relevant sense. *If* the existence of cases of multiple discovery were accepted, then we could sidestep all the discussion of flexibility by noting that the question 'could someone else have done that?' (or, better perhaps, could someone else have discovered that?) *seems* to have a clear positive answer: after all, in this, the actual world, it is claimed, someone else *did* come up with special relativity, the theory of evolution, and so on. But that very fact of course undercuts the necessity of authorship in the case of theories, since, unlike the case of Manet and *Guernica*, a given theory could be created/discovered by distinct scientists within a possible world. Again, then, theories and models *seem* to be different from artworks and in such a way that the kind of authorial relationship that might sustain the extension of the abstract artefacts account simply does not hold.

However, this is too quick. We recall that the plausibility of a positive answer to our question, 'could someone else have done that?' was underpinned by certain

[48] Rohrbaugh himself suggests that works of art should be regarded as 'historical individuals' that persist through time in an ontologically dependent manner (see Rohrbaugh 2003).

[49] Suppose it were necessary that your model were made of painite, of which until recently only twenty-five crystals were known, so that all the painite in the world was used in the construction of your model....

[50] Again, one could insist that priority be associated with such an unshareable feature but again, again, that would make it hard to understand the vehemence behind priority disputes.

historical claims. However, many if not most historians and philosophers of physics would now agree (contra the likes of Whittaker) that whatever it was that Poincaré came up with—theory, proto-theory, hypothesis...—it was *not* the special theory of relativity. Thus, Poincaré and Einstein had different understandings of absolute motion—the former taking it to be undetectable, the latter as non-existent—and of the Lorentz transformations—the former taking them to be merely calculational devices, defined only with respect to the aether rest frame, the latter as representing physical relationships between coordinate systems (for an overview see Adlam 2011). More importantly, perhaps, Poincaré did not see the relativity principle as explanatory in and of itself but rather as an empirical summary that could be used to confirm or refute various hypothesese (ibid.; Katzir 2005). Einstein on the other hand, took the principle to be a fundamental constraint on the form of the relevant laws and this represents a fundamental shift in what was taken to require explanation in this context, akin to that which marks the difference between the Aristotelian and Newtonian approaches to what we now call inertial motion. And there is the further, related, difference regarding the associated ontology, with Poincaré retaining the electromagnetic aether, and Einstein dispensing with it, presenting the theory initially in terms of rods and clocks, before reluctantly accepting Minkowski's space-time formulation. Relatedly, Brown has argued that a Lorentzian approach offers a better explanation of relativistic effects (such as time dilation and length contraction) by virtue of appealing to the dynamics rather than space-time structure (Brown 2005), but that simply reinforces the point that these are different theories.[51]

Likewise, although Darwin and Wallace themselves thought their theories were identical, various historians of science have noted crucial differences. Thus, Nicholson has argued that whereas Darwin's theory was concerned with *competitive* selection between members of the same species, such that the less 'fit' lose out to the more 'fit', Wallace's theory was about *environmental* selection, whereby an individual has to survive the rigours of a particular environment (Nicholson 1960). Bowler, on the other hand, insists that Darwin's theory is different from Wallace's because the former is about competition between individuals and the latter about competition between varieties (Bowler 1976). More recently, Bulmer has argued that the theories differ in that according to Darwin's, an advantageous variation would increase and its parental form decrease in frequency until it became extinct, whereas according to Wallace's, both would exist, albeit with the former as more numerous than the latter, until some environmental change forced the

[51] Obviously, a core issue here has to do with what counts as an acceptable explanation (my own view is that structural accounts fit the bill) but that only bears on the concerns discussed here if it can be related to the nature of the theories concerned. One might argue that insofar as Poincaré-ian SR offers a better home for dynamics-based explanations of relativistic effects as compared with Einsteinian or Minkowskian, then this marks a significant difference between the two such that we cannot say that Poincaré and Einstein discovered the same theory.

extinction of the parental form (Bulmer 2005). Again, we don't need to get into the details, or the dispute between historians of science; what is clear is that an argument can be made that the theories are different, despite appearances and indeed, the views of the discoverers themselves.[52]

I could continue, but I won't. Already one begins to suspect that the claims about multiple discovery are over-played. Once we strip away the irrelevant cases, such as the multiple invention of blast furnaces at one end of the spectrum, and the co-discovery of the calculus at the other,[53] there are precious few, if any, cases left standing. But then it seems that our conclusion that theories are different from artworks in this respect is not so firm and we may have opened the door to the necessity of authorship argument when it comes to the former, just as for the latter. That would then give a boost to the extension of the abstract artefacts account.

Nevertheless, someone who thinks that theories, such as special relativity, are different from artworks in that they are not created via the scientist's intentions but are 'out there' in some abstract space, just waiting to be found, might still balk at this. We recall Dodd's idea that the composer or scientist has to be in the right 'position' in this abstract space in order to find the artwork or theory, respectively. And so our friend might insist that all that the above discussion of multiple discovery shows is that the likes of Wallace and Poincaré were not in the right 'position' to find the theory of evolution or special relativity, respectively—they were close, however, and so found theories that were very similar. Perhaps all they needed was to have been situated in the right *context* and they would have discovered the theories attributed to Darwin and Einstein. Let's explore that suggestion a little further.

Contextuality

So, let us grant that Poincaré, or Lorentz, did not, in fact, produce the special theory of relativity. Now, consider the question again, '*Could they have?*'. Suppose that Poincaré, say, had lived and worked in the same context, scientific and otherwise, as Einstein, and, crucially, had changed his attitudes in certain ways—by, for example, dropping his commitment to the mechanical aether, and changing his

[52] Thus, Bulmer notes that in his autobiography, and also in other accounts of his discovery, Wallace explicitly identified his theory with Darwin's (2005, p. 134). This yields a nice example of a scientist's restropective re-evaluation and, in this case, re-identification of their work that further muddies the waters when it comes to identifying theories.

[53] Leaving aside issues to do with both men's debts to others such as Barrow (Hall 2002), a mathematical Platonist will obviously have a ready answer to how such multiple discoveries could be possible—indeed one might wonder why, on this view, there aren't more—while the nominalist, who sees mathematics as no more than a representational device, will presumably point to the similar context of applications for her explanation.

attitude towards the Lorentz transformations—then it might seem plausible to imagine that the theory he would have presented in, say, the *Revue générale des sciences pures et appliqués*, would have been identical to Einstein's. But that seems just the same as insisting that if Braque had attended a lot of bull-fights and had developed the same techniques as Picasso and had acquired other artistic attributes, then he could have created *Guernica*. Again, we have to ask, is that plausible? And again, we come back to the issue of the 'modal flexibility' of theories and artworks and underlying that, the question of their identity conditions.

In the case of Poincaré and Einstein, that plays out as 'what would it be to produce a theory 'identical to Einstein's?'. As already noted when it comes to forgeries, in the case of *Guernica* it would be to produce something identical in form, composition, colour, hue, saturation, brushstroke, and so on, to Picasso's painting. But if Einstein had written his 1905 paper in French instead of German would it have been the same theory? Clearly yes, a response that the Semantic Approach draws upon as we have seen. And thus if the author of that paper, written in French, had been Poincaré we would still be inclined to say the theory, in this possible scenario, is identical to Einstein's in the actual world. What if it had not had the Lorentz transformations? Equally clearly, no—these are a necessary component or feature. But what if this possible theory had kept the aether? Again, it depends on how we characterize or delineate the theory. Whittaker, of course, would say yes; indeed, he did say yes, that theory that Poincaré and Lorentz produced was the theory of relativity. Most historians of physics today say 'no', but adjudicating in such cases can be a tricky business, dependent as it is on pinning down what 'the' theory in question *is*. And in the absence of such adjudication, on what basis can we claim that theories are 'out there', waiting to be discovered by whomever?!

Where are we then? The plausibility of Dodd's idea—that to 'discover' a theory, the scientist needs to be appropriately situated in the relevant theory 'space', or more broadly, perhaps, Popper's World 3—seems to depend, again, on issues of theory identification. On the other hand, an obstacle to the extension of the abstract artefacts account from artworks to theories—namely, the possibility of multiple discovery—appears to have collapsed. Nevertheless, we recall that this account faced problems when it came to the 'creation' of theories via scientists' intentions. As we shall now see, it faces further problems when it comes to their persistence, since it is not just the case that they are brought into existence through intentions, they are also *maintained* by them.

Persistence

Once created, it is claimed, such artefacts are not independent of us nor do they exist eternally, but rather, they depend for their ongoing existence on the relevant intentions being maintained. Now, it is not entirely clear what that last phrase amounts to. Consider a piece of music, such as one of the 'lost' compositions of the young Mozart (a number have been rediscovered, in old notebooks and the like; see for example http://www.bbc.co.uk/news/entertainment-arts-17493329). Suppose, first of all, that all it takes for one of these compositions to be brought into existence as an abstract artefact is for Mozart, having composed it 'in his head', say, to have the relevant intention to play it, again 'in his head' (indeed, he may have mentally played it through, hearing all the sounds on his piano). Is it the case that the piece of music is then maintained in existence by Mozart thinking about it? That would surely render it too fleeting and ephemeral! But suppose it remains in existence as long as Mozart can recall it, or bring it up from memory (and we might wonder, knowing what we do now about memory, whether what is 'brought up' is identical to what was originally composed)—presumably that means it ceases to exist when Mozart permanently forgets it, perhaps through illness, and certainly, it would seem, when he dies.

Suppose that what it takes for the music to be brought into existence as an abstract artefact is not just that Mozart composes it in his head but that he also writes it down as a score. Again one might wonder about the metaphysics of the relationship between intention, score, and artwork but let's leave that to one side. One might then insist that the artwork persists in existing as long as the score exists, since the latter provides the means of producing a performance of the music. One might further extend this from the score to a recording or just a memory (again, doubts might arise) or some combination of these. But questions as to the relationship between the score, the peformance, the relevant intentions, and the artefact intrude once again. Suppose the notebook containing Mozart's childhood composition had not been found and indeed, was never found. Would the artefact persist in existing even though the composition was never played? In which case, what's the role of the score in this relationship? Suppose the score gently moulders away into dust—does the artefact likewise fade into the dusk in World 3?

Or suppose that what it takes is not just that the music has to be composed and written down or noted in some way but it also has to be performed in order to sustain the existence of the artefact. Now, the question is, does it just take one performance to sustain it? And for how long? Does it cease to exist after a certain period? (Surely not.) Or when the means to perform it no longer exist and/or can never be re-created (for example, following the worldwide banning of all pianos or the means for crating them!). Or perhaps the piece, as an abstract artefact, is brought back into existence each time it is played? Again, ephemerality, not to mention absurdity, beckons.

Many of the same questions arise when this view is taken to embrace theories and models. So, consider Einstein's 'lost' precursor to Hoyle's steady-state theory of the universe, written on notepaper during a trip to California in 1931

(Castelvecchi 2014).[54] Again, we can ask what it was that sustained this theory as an abstract artefact—was it the mere thought? The writing down of the relevant equations? (Or the equations plus an interpretation??) Did it remain in existence all the time the manuscript was lost? Or did it pop back into existence when the manuscript was found? Or, more fine-grainedly, when not only was the manuscript found but when it was understood (after all, it could have been discovered by a musician!)? Again, it is impossible to begin to answer these questions until we have a better idea of the nature of the purported relationship between the intentions, the theory, and its concrete manifestation.

Here, we might appeal to the different epistemic status of theories and artworks—the former are deemed, at some point, to be right or wrong, correct or not, true, or approximately so, or false, usually not approximately so (although recall what was said about models and falsehoods), empirically adequate or inadequate, confirmed or falsified and so on. In some cases, even, indeed *often*, before experiment or observations are brought to bear, they are ruled out because of inconsistency, incoherence, or obvious incompatibility with experience. Thus, in Einstein's case, the manuscript concerned was probably put aside and 'lost' because Einstein realized that he had made a mistake in his calculations and the theory simply wouldn't work (as revealed by his crossing out part of the relevant calculation; see Castelvecchi ibid.). Now, one could insist that some musical works, for example, are likewise put aside because of internal incoherence, or the melody just isn't strong enough or the overall theme just doesn't work but the force or pressure involved is different in the case of theories—one can of course try to fix the problem via some ad hoc manoeuvre, but there are well-known concerns about such moves (see Forster and Sober 1994) and in general, constraints that are grounded in empirical features apply that are different from those we find in art.

This offers both an opportunity and further challenges. So we might seize on this point and suggest that theories as artefacts go out of existence when they are abandoned by their authors, for any of the reasons indicated above. Unlike artworks, this might provide an appropriate 'end point' for the theory concerned. But this raises further concerns: does a theory have to be abandoned only by its author or by everybody to cease to exist? Or conversely, what about cases where only the author clings to the theory, in the face of opposition, only to be subsequently vindicated? Is that desperate belief enough to sustain its existence as an abstract artefact? Here we might think of Wegener's continental drift hypothesis—although it is perhaps too crude to say that Wegener was alone in maintaining his belief in the theory, the majority of geophysicists rejected it through the 1940s and '50s, particularly in the United States (see Frankel 1987).

[54] According to this theory, there was no 'Big Bang'; rather the universe undergoes continuous expansion with constant density maintained via spontaneous particle creation.

What about cases where a theory is generally abandoned, only to be rediscovered and revived? Does it come back into existence? Or does it cease to be when the flaw is discovered, whether by the author or someone else? In that case, could we have a situation where a theory as abstract artefact continues to exist for many years, perhaps forever, even though internally incoherent say, as long as no one discovers that fact? And the blurred boundary between this kind of process and the domain of heuristics raises further issues since not only may someone have the idea for a theory only to almost immediately discover that it is flawed, leading to a flickering existence at best on this view, but tackling a perceived flaw in a new theory is often part of the heuristic process,[55] as 'the theory' is honed and shaped, typically prior to publication or presentation. Again, do we have just one artefact that is shaped and refined during such a process or, in line with what is suggested above, does the theory vanish from World 3 when the flaw is discovered and a new flawless artefact subsequently created?

Furthermore, the idea that the general theory of relativity, say, could 'cease to exist' just seems *odd*. Now, this oddness might simply be a function of one's realist inclinations. Certainly, if you were a social contextualist of some stripe, you might be inclined to accept that just as the apparent confirmation of theories amounts to nothing more than the stamp of approval from some socio-politically supported scientific grouping, so theories come into and go out of existence as abstract artefacts according to the whims of such groupings. But if you were a realist, as you should be, of course, you might well wonder about all this talk of theories going out of existence (never mind their being sustained by intentions!) given that theories—or at least those that are internally coherent, empirically adequate, and determined to be approximately true in some sense—*represent* phenomena, systems, reality, or whatever, as we discussed in some detail in Chapter 3. And talk of such representations going out of existence then seems to be *beside the point*; after all, even if Einstein had abandoned his general theory, or his original paper had been lost, or physics had taken a very different turn and devoted all its resources to other research, the relevant empirical phenomena, or features of the world (to do with space-time and gravitation) would still be there to be described and represented, if not by Einstein, then someone else (Hilbert perhaps).

Here, again, we have to be a little careful and remind ourselves that a realist commitment to the entities mentioned by a theory does not entail realist commitment to the theory itself *as an entity*, whether abstract artefact or whatever. This also applies to the laws and symmetries and related features of the theory: whether we regard these as merely summaries of empirical regularities or take them to embody—whether dispositionally or primitively—necessity 'in the world' (see French 2014, ch. 10), the relevant features of the world will still be there, whether

[55] Or tackling the flaw in an old theory, another of Post's list of heuristic moves.

or not anyone is around to compose the laws and symmetries representing them. Again, to say that a law exists 'in the world' is not necessarily to say that the corresponding theory (which 'contains' in some sense a statement of that law) exists, whether in this world or the Third one (and once again, we'll come back to this later).

Of course, things look a little differently to the non-realist (but not a social constructivist). She might allow that more than one theory explaining a given phenomena can be empirically adequate (this being a case of underdetermination of course, which we'll also come back to). From the perspective of such a stance, theories can be regarded as descriptions of 'ways the world could be' and it is only if you have a realist view of such possibilities that you might have the same inclination to think that talk of theories going out of existence is beside the point because the ways the world could be will always be there to be described by someone or something. If you do not adopt such a realist view of modality, but hold the latter to be 'in the models', in the sense of it being a feature of our representational devices then the ways the world could be spring from the theories and models of course, so 'no theories' implies 'no such ways'. But of course, our antirealist friend could still accept that there is a way the world *is*, just that we cannot know it and of course, the going out of existence of some abstract artefact is not going to change that!

There is also the further point that unlike most musical works, or artworks in general, there may be questions as to where one theory ends, as it were, and another begins. Thus, consider again the Bohr model of the atom and Sommerfeld's extension of it via the quantization of angular momentum, which rendered the electron orbits elliptical rather than circular, thereby allowing for quantum degeneracy (see Eckert 2014). This is typically portrayed as the pinnacle of the 'old' quantum theory, before the new matrix and wave mechanics of Heisenberg and Schrödinger changed the landscape completely. Now, was this merely an adjustment to or at best an extension of Bohr's model, or did it constitute a new model in its own right? It is referred to in both ways and there's no obvious criterion by which to determine whether it is one or the other.

Or consider Bohmian mechanics, sometimes called de Broglie-Bohm theory, the pilot wave theory, or the Bohm interpretation (https://en.wikipedia.org/wiki/De_Broglie–Bohm_theory). Standardly and originally, the core formalism is obtained by transforming the Schrödinger equation, yielding two coupled equations that invite the well-known interpretation of particles moving along classical trajectories under the influence of the 'quantum potential' (other derivations have been proposed, all of which proceed from some aspect of the already accepted formalism of quantum mechanics). Is this a different theory from standard quantum mechanics or merely an interpretation of the latter, one among a number (such as the so-called 'many worlds' interpretation, the Ghirardi-Rimin-Weber spontaneous collapse interpretation, and so on)? On the one hand, the same formalism is involved, so one can argue that all we have here is a different interpretation but on the other, that interpretation supplies a completely different ontology (insofar as one can say that the 'standard' theory has an ontology) and hence that makes it a different theory (again, we'll come back to this point later).[56] Likewise, how does post-Darwinian evolutionary theory relate to Darwin's work? Is it merely a development of the same theory, or something else?[57] Or perhaps we shouldn't be talking about theories and models in this context at all, but about research programmes or even, heaven forfend, paradigms, and such. Whose intentions would sustain such a programme?

Do we see anything similar in art? Well, if we think of Bohr and Sommerfeld as effectively collaborating on the 'old' quantum theory, then there are certainly collaborations in art: think of 'The Juniper Tree', an opera based on a Grimms story, with alternate scenes composed by Philip Glass and Robert Moran; or, of course, the book *Good Omens*, co-authored by Neil Gaiman and the late and much lamented Terry Pratchett, just to name a couple that come to mind. More interestingly, perhaps, there are examples of the French group of composers, 'Les Six', who collaborated on the work *L'Album des Six* and who can be compared to the group of mathematicians known by the collective pseudonym, Nicolas Bourbaki. I say more interestingly because if one were a Platonist in philosophy of mathematics, one would presumably have to say something about how any results were discovered, whether by some collective 'eye' scanning the Platonic landscape, or more reasonably, in terms of some more fine-grained analysis of who among the group were responsible for which discovery, or which aspect of a discovery, or which line in a proof, and so on. If one were to adopt the mathematical version of the abstract artefact view—let's call it 'Platonic constructivism'—then one would presumably face even more acutely the same old problems regarding discovery/creation and the end of the existence of such entities.

A more useful point of comparison might be sequential creations, where one author takes over from another, or develops a work in a way similar to the way that Sommerfeld 'developed' Bohr's theory. An obvious example (the clue lies with the word 'sequential'!) would be comics or graphic novels.[58] As with novels, they are standardly (but not, perhaps, necessarily) 'multiple' works of art in the sense that 'they are repeatable, admit of instances, or occurrences rather than mere copies and in virtue of this they allow for simultaneous but spatially-distinct and unconnected reception points' (Meskin 2014, p. 32). However, unlike works

[56] In which case, Bohm 'theory' yields another case of the underdetermination of theories by evidence (see, for example, Cushing 1993).

[57] Thus, some commentators talk about evolutionary theory having itself 'evolved' (yes, very amusing; see http://evolution.about.com/od/scientists/ss/5-Post-Darwin-Evolution-Scientists.htm). And of course, the very word 'theory' here is loaded in the context of the 'debate' with the anti-evolution zealots!

[58] For a useful introduction to the philosophy of comics which touches on some of the issues considered in this chapter but not, however, this particular one, see Meskin 2011 or for a general reader, see Meskin and Cook 2014.

of literature, comics are 'encoded' rather than 'exemplar based', in the sense that neither the original art nor the engravings or web code or whatever subsequently produced counts as an instance of the comic itself (ibid., p. 38). In this sense, a comic is more like a print and like the latter, is produced in two stages rather than one, as with a novel.

We might draw on this aesthetic device and think of a theory in this way, as 'encoded' in the abstract artefact, or whatever is in World 3, say, from which instances can be obtained via a similar sort of process, resulting in the theory's reproduction in a book, or a scientific paper, or in a Powerpoint presentation. And like comics, theories are, or may be, at least, hybrid, in the sense that they may include linguistic, mathematical, and pictorial elements. But the important similarity here, of course, is that comics offer obvious and numerous examples of works that are initiated by one author but continued by others: so, consider a recent incarnation of the Marvel comic *Moon Knight*, initially written by (the inimitable) Warren Ellis and drawn by (the masterful) Declan Shalvey, replaced by (the wonderful) Brian Wood, still with Shalvey. The character is the same, the artist is the same (except for issue 10), the series is the same, so the general framework, overall plot, issues tackled, hero methodology, etc. are all the same, but of course, certain plot lines are further developed, new ones are introduced, new challenges are offered and so on. So, is the Wood/Shalvey *Moon Knight* the same or a different artefact from the Ellis/Shalvey one? On the one hand, with a different writer, and given the inter-dependent relationship between writer and artist when it comes to the creation of comics, one might be inclined to argue that the change marks the creation of an entirely new artefact. On the other, one could equally well argue that the Wood/Shelvey collaboration represents a development of the comic, as a series, similarly to the way the Sommerfeld model represents an extension of the Bohr system. And as in the case of theories and models, when faced with the choice of line to take one might feel both that the abstract artefact view is, at the very least, unhelpful or even irrelevant with regard to what we might, as philosophers, deem to be more important issues, and also that talking of theories and comics, respectively, as the appropriate units of assessment, is not entirely appropriate—rather we should talk of 'programmes' and 'series', also respectively.

There is comparatively little that has been done in comparing the philosophy of art and the philosophy of science in this regard. Rickles has made a start in this direction by considering whether musical styles might be likened to scientific paradigms (Rickles 2013; also Rickles 2017). However, as he notes, the driving forces for change are, at the very least, different and we don't seem to have anything equivalent to a crisis or anomaly in a musical style that provokes something equivalent to a revolution.[59] We might also draw a comparison between

[59] Although Bueno has suggested that the very lack of novel phenomena might itself function as the generator of crisis in an artistic movement.

Lakatosian programmes and such stylistic movements or movements in painting, such as Impressionism, Cubism, Expressionism, and so forth (for a list, see: https://en.wikipedia.org/wiki/Art_movement). So, for example, we could take as the 'hard core' of the 'neo-plasticist' movement to which Mondrian belonged, his insistence that 'this new plastic idea will ignore the particulars of appearance, that is to say, natural form and colour. On the contrary it should find its expression in the abstraction of form and colour, that is to say, in the straight line and the clearly defined primary colour' (as cited here: http://www.tate.org.uk/learn/online-resources/glossary/n/neo-plasticism). We might even pursue analogies to 'auxiliary hypotheses' and 'protective belts', and chart how such art movements change and evolve but the same disanalogy remains stubbornly in place: there is no equivalent to the motor that drives a scientific programme to progress or the lack of which leads it to become degenerate (although of course the latter word has been used as a pejorative in the case of artworks!).

All of this reveals a further critical problem with the abstract artefacts view as applied to scientific theories and models—it doesn't help us in understanding, and doesn't appear to be able to accommodate, the inter-relationships between scientific works. And the indeterminacy we face in deciding whether Sommerfeld's model is the same as or different from Bohr's further contributes to the feeling that perhaps *this is just the wrong way of thinking about theories and models.*

Conclusion: Heuristics and Sustainability

We began this chapter by recalling Popper's notion that models and theories can be situated in what he called 'World 3', understood as a kind of abstract space, perhaps. We then drew on Thomasson's account of artworks as abstract artefacts in order to further develop this view. However, problems emerge when we begin to think about how this sort of account might mesh with scientific 'discovery' in general and well-known heuristic moves in particular. As we have seen, questions begin to multiply about how such scientific abstract artefacts come to be created and manage to be sustained. It may be, of course, that these questions can be answered. Nevertheless, such answers will add further complexity to this account and furthermore, the ontological cost will almost certainly remain.

Now, we could just chop through this knotty (Gordian) bundle of issues by simply denying the initial assumption, namely that theories are things or entities, abstract or otherwise, to begin with. But before we get to that clear and ontologically brutal point, we need to consider an alternative position, namely that which takes theories and models to be *fictions*.

6
Theories as Fictions

Introduction

We recall that Popper includes in his World 3 works of literature such as Shakespeare's *Tempest* and *Hamlet*, as well as musical works such as Beethoven's *Fifth*. This is because he feels that such works are more than merely ephemeral mental entities and have an objective existence or 'life' of their own in some sense. As we've already noted, we can likewise think of Newton's mechanics as being more than the words and geometrical diagrams in the actual physical copies of the *Principia*, and a number of philosophers of science have taken this idea seriously to suggest that theories and models are akin to works of literature[1] and thus should be regarded as *fictions* (see Frigg and Hartmann, 2018; we'll leave concrete models to one side for now).[2] This view is motivated not just by the claim that models are creations of human imagination, just like novels and plays, but also by the thought, reflected in Popper's placement of them, that like the latter models have a 'life' or, rather, internal structure of their own which can be explored and developed and which, again reflecting Popper, can yield surprising results.[3] I shall consider these twin motivations in turn.

Models and the Imagination

With regard to the first, a number of people have focused on this idea that when modelling a system, or phenomenon, or whatever, scientists are engaged in an act of imaginative construction whereby the model is nothing more than an

[1] We also recall that Hughes has suggested that one may approach scientific texts in a similar way to which critics approach literary texts, thereby bringing into the light the nature of the practices embodied in the text (Hughes 2010).

[2] Some have related this stance to Vaihinger's 'as if' approach to philosophy (Frigg and Hartmann 2006; and Vaihinger's view can in turn can be related back to Bentham's *Theory of Fictions*, and so it goes...), drawing on the role of idealizations in models but as Weisberg notes, the claim that models describe fictional scenarios because they are idealized (Suárez 2009) can also be accommodated by other views.

[3] Thus, it is not just that models are 'are in the scientist's mind rather than in the laboratory and... do not have to be physically realized and experimented upon to perform their representational function' (Frigg and Hartmann 2018) so that it is 'natural' to view them as fictions, but rather that modeling involves an engagement with fictions that is at least akin to that one has with novels (Weisberg 2013, p. 46).

There Are No Such Things as Theories. Steven French, Oxford University Press (2020). © Steven French.
DOI: 10.1093/oso/9780198848158.001.0001

imagined system, phenomenon, or whatever which, *if made real*, would be a concrete or 'flesh-and-blood' system/phenomenon/whatever (Godfrey-Smith 2006; for a summary, see Weisberg 2013, pp. 48–9). This imaginary system/phenomenon/whatever will possess the properties attributed by the act of construction (indeed, such attribution is largely what the act amounts to), which will typically be fewer than or different from the properties possessed by the real system/phenomenon/whatever given the role of idealizations in models. Many possible properties of course will not be attributed at all—as in the case of colour (in the old classical sense) when it comes to atoms, for example—whereas some might not be explicitly included but might be inferable on the basis of those that are.

This then provides the basis of the comparison with literary fictions where '[t]he text only contains some of the details and the rest must be filled in by us in order to make a coherent story' (Weisberg 2013, p. 50). Thus, consider the example of *The Lord of the Rings* again and the property of Frodo's handedness: presumably he must be either right- or left-handed or ambidextrous (ibid.) but Tolkien does not explicitly state which, nor, so far as I know, can this property be inferred (at least, not obviously or directly), although in making sense of the story we must attribute one or the other or the other. Furthermore, once these properties are attributed, there is established a kind of 'internal logic' to the story, similar, it is claimed, to that we find with models. So, to deploy another much used example, consider Sherlock Holmes: once we have established his acute observational and analytical skills, his haughty demeanour, and his attitude towards personal relationships, certain things then follow and others are ruled out. In effect, these properties act as constraints such that if violated, or violated egregiously, we would find the story unsatisfactory or even incoherent.

As Weisberg (2013, pp. 50–1) notes, this sort of account does have certain advantages. So, recalling our discussion in the previous chapter, it helps us get a grip on the relevant identity conditions. The idea is that each imaginary system is a model that can be presented in different ways, using language, mathematics, diagrams, whatever. Obviously there will be an element of underdetermination as a given imaginary system can be presented in more than one way, but this is not a problem and indeed, may even be a further advantage as it allows for the generation of families of such models by imprecise descriptions (ibid.). But of course, the 'abstract entity' account of models offers the same advantage, as does the view that they are, in some sense, set-theoretical. And what this fictions account does not do is solve the problem of distinguishing one model from another, since any indeterminacy in how we delineate a given model will be carried over to the imaginary system and vice versa.

More importantly, the fictions account takes seriously the reports of *scientists* about how they think about models (ibid., p. 51); that is, by first constructing a mental image and then refining it, before writing it down or literally building it, as in the case of the Crick and Watson model of DNA. One can find many examples of this in scientists' own recollections of their discoveries such as presented in Maynard-Smith's evolutionary genetics textbook involving a model of RNA replication (see Weisberg 2013, pp. 48–9). Another is Feynman's report about how, burned out after the effort involved in the atomic bomb programme, he idly watched the wobble of a spinning plate in the university cafeteria, which led him to think about and construct a model of electron orbits in the relativistic context (Feynman 1985, pp. 157–8).

But one can just as easily find examples where mental imagery has led scientists astray to greater or less degree. So, in giving his account of the construction of the Watson-Crick model, Crick notes how when they actually built the model the base molecules (adenine, cytosine, guamine, and thymine) turned out to be bigger than the mental picture he had formed of them (Crick 1988, pp. 70–1). And more generally, the use of mental imagery can be understood as just one of a whole variety of heuristic techniques that scientists deploy, and thus it is not decisive when it comes to the ontological or metaphysical issue of *what sorts of things models are.*

Furthermore, we have to take some care in pressing the similarities between models and fictions. Suppose, as suggested above, that Frodo's handedness is neither explicitly indicated in *LOTR*, nor can it be inferred from anything Frodo himself or any of the other characters says, or from any of the fight scenes or, indeed, any feature of the narrative. Nevertheless, in the act of imagination involved in reading the story, we may find it necessary to attribute handedness to him, even if that is arbitrary both with respect to other people's reading of the text and our own at different stages (so we may imagine that when at Weathertop and attacked by the Nazgul, Frodo draws his sword in his right hand but when the Fellowship is in Balin's tomb and Sting's blue glow indicates the presence of orcs, Frodo is drawing it with his left (contrary to what is shown in the movie)). It is not clear, at the very least, that we or scientists do the same when it comes to models. In that case, while it may be that some inferences are easier to draw than others, given that we are not logically omniscient, and hence some derivative properties easier to attribute, those that are neither explicit nor inferable simply *do not feature in discussions pertaining to the model* (and if they are brought in, then either colleagues would take you to be constructing a new model or that you simply don't understand the current one!).[4]

There is a further difference that is significant. It is telling, perhaps, that the novels of *Sherlock Holmes* and *The Lord of the Rings* are often given as examples of the 'imagined objects of literary fiction' (Godfrey-Smith 2006, p. 735), since here

[4] Here we might recall our discussion of a 'key' or 'code' in the context of a comparison between artworks and scientific representations.

the relevant imaginary worlds are presented in some detail: Victorian London in the former case and Middle-Earth in the latter.[5] However, when we are asked to consider a model, we are not asked to imagine an entire world, or to situate it in such a world. Take again, the model of the pendulum—there is no equivalent to the richly detailed Middle-Earth here! Perhaps it could be said that the 'world' in this case is rather 'thin' and that something like the whole of classical mechanics, as a theory, would correspond to *LOTR*. In that case, the 'world' underpinning the model of simple harmonic oscillation is implicit and the imaginary aspect is closer kin to a character in a novel, such as Frodo, as already mentioned, or Sherlock Holmes himself. But then of course the similarities quickly break down: we can draw certain inferences about such characters, as already noted, but the relationship between models and the underpinning theory seems very different from that between a character and the fictional world they inhabit (models don't change their relationships to parts of the theory for one thing!). And of course if the novel is any good, the character will change and develop in ways that models (typically) don't.

Finally and relatedly, what would it be for a model to be 'made real'. In the case of literary fictions, making Frodo or Sherlock Holmes real in *this* world (consider the example of Homer^3 in the Simpsons episode Treehouse of Horror VI, in which Homer enters the 'hypothetical' third dimension and becomes trapped in the real world!) would destroy the integrity of the character through the removal of at least some of the constraints on its behaviour. Making real the relevant world, as the given imaginary entity in Godfrey-Smith's sense (ibid.), would be an entirely different matter, and different also between *LOTR* and the Sherlock Holmes novels, of course.[6] Turning to models, if the 'making real' operates on the model as it is, then what we would get would be a concrete system that is different, perhaps significantly, from the system modelled, due to the presence of idealizations (which may indeed prevent the system or phenomenon from functioning and thus undermine that 'making real' of the model). Alternatively, the making real of the model could be thought of as incorporating the necessary de-idealizations, yielding not another system or phenomenon, as it were, but the original system/phenomenon itself. Again, there are differences between models and literary fictions in this regard.

However, perhaps the biggest concern regarding this 'simple fictions' account is that it does not appropriately spell out the *metaphysics* of models-as-fictions (Weisberg 2013, p. 51). I shall canvas two fairly obvious ways of undertaking such a spelling out, before returning to the problems with fictionalism in general.

[5] Tolkien often stated that he constructed Middle-Earth as a world for his imaginary but carefully worked out Elven languages.

[6] Not least because in the latter case it would involve making real a past reality.

Models and Possibilia

The first is to take models-as-fictions as *possibilia*, in the Lewisian sense that they would be concrete but non-actual possibilities.[7] This has the obvious advantage that one can then draw on appropriate metaphysical resources from the 'toolbox' presented by current discussions in modal metaphysics (French and McKenzie 2012 and 2015). Thus, for example, we take models-as-fictions to be possible worlds or parts thereof. On this view, the statement that 'The period of a simple pendulum would be proportional to the square of its length, if there were no friction or air resistance and were the angle of displacement kept small' is true about a concrete simple pendulum in a world in which there is no friction etc. This simple pendulum is the counterpart in that world of the simple pendulum in this, the actual one. Since both are concrete objects, albeit in different possible worlds, there is no issue about how to relate the abstract and the concrete. And since the counterpart relation is a similarity relation (so, it is neither transitive nor symmetric), we can straightforwardly apply the Semantic Approach, for example.

However, such resources also carry a certain amount of baggage with them that may ultimately undermine this sort of approach. Although we can sidestep some of the well-known concerns that have to do with mere conceivability as a generator of possibility (and here we recall our earlier discussion in Chapter 1; see, again, Yablo 1993; Chalmers 2002), since the kinds of considerations that factor into model construction function as acceptable constraints on that conceivability,[8] there is the worry that a world in which air resistance is ignored or, more plausibly, where the pendulum bob is treated as a point mass is not a nomically possible world (Contessa 2010, p. 221). An obvious response would be to insist that all we need is *logical* possibility in such cases, but then just as we might wonder what relevance consideration of single proton worlds can have for discussions of the metaphysics or nature and role of laws, for example, in this, the actual, world (French and McKenzie 2012 and 2015) so we might then have a concern about how merely logically possible constructs can inform us about actual phenomena, in the way that models can.

A better response might be to insist that insofar as such idealizations are appropriately justified and well grounded they can be taken to be nomically 'kosher'. So, a world without air resistance is clearly so—it is simply a world in which the pendulum is swinging in a vacuum. There also doesn't seem to be anything nomically

[7] It may seem odd to those who are not metaphysically inclined to describe a situation as non-actual but concrete! However, the idea is to take such situations to be exactly the same kind of thing as the actual world, except they're not actual (see Menzel 2017).

[8] So, the worry here is that many of the 'merely conceivable' counter-examples to various arguments that tend to be thrown up in modern metaphysics are actually insufficiently specified or fine-grained enough to function as such. This is essentially Hacking's complaint about purported 'two globe' counterexamples to Leibniz's principle of identity of indiscernibles (Hacking 1975).

problematic about point masses in this case—classical mechanics can accommodate them just fine (see for example Strauch 2009).[9] Having zero friction between the string and the support is more problematic but if one restricts what one means by nomically possible to exclude the laws of thermodynamics and electromagnetism, so that again we are only considering classical mechanics then it appears that circumstances can be arranged such that there is zero friction (using materials whose coefficient of friction fall to zero as the temperature falls and taking the possible world to be a chilly one, i.e. at absolute zero).[10]

However, there is another more serious problem that this sort of account must face that, ironically, perhaps, derives from precisely thinking about fictional entities as possibilia. So, towards the end of his classic book, *Naming and Necessity*, Kripke suggests that since there is no Sherlock Holmes, 'one cannot say of any possible person that he *would have been* Sherlock Holmes, had he existed' (Kripke 1980, p. 158). And the reason he gives is that it is indeterminate which of any number of possible or actual people who could have performed Holmes's exploits we could say would have been Holmes, had he performed those exploits. Putting this in the form of a well-articulated argument is not entirely straightforward (see Liebesman 2014), but a key premise is that if Holmes is a possible person, then there is nothing further to being Holmes than satisfying the properties laid down in Conan Doyle's stories (we recall Thomasson's account of fictional characters considered in Chapter 5). In that case we can obtain a situation in which objects in different possible worlds, or indeed in the actual world (such as Joseph Bell, a surgeon at the Edinburgh Royal Infirmary, or Sir Henry Littlejohn, Chair of Medical Jurisprudence of University of Edinburgh Medical School, both of whom have been cited as inspirations for the character), may satisfy these properties and thus may legitimately claim to *be* Holmes. But we have no way of legislating between these competing claims and hence must drop our initial supposition, that Holmes is a possible person.

Contessa notes that a similar problem arises for models-as-possibilia (2010, p. 222): consider the Rutherford model of the atom, with electrons orbiting in (classical) orbits. The models-as-possibilia account suggests that corresponding to the relevant figment of Rutherford's imagination (or that of anyone thinking about the model) or the description of it, is an entity in some possible world. But now the Holmes problem bites, as there could be more than one such possible entity or indeed an actual one. So, Contessa uses the latter to argue that agreeing with the mental image, say, or the description, is not sufficient for something to be the Rutherford model: there could be an actual atomic system in the world, unknown to us, that exactly agrees with that description. But we would not call

[9] Of course, that's not so in the quantum case.

[10] Curiously enough in this case, it is quantum mechanics that permits friction-free processes, even allowing for the laws of thermodynamics (see del Campo, Goold, and Paternostro, 2014).

that *Rutherford's* model, since it had nothing to do with Rutherford! (Recall our discussion of the necessity of authorship argument in the previous chapter.) Furthermore, such agreement is not even necessary: take the claim that 'electrons in the Rutherford model orbit the atom along well-defined trajectories'. Although we would agree that there is some truth to that (assuming we're realists of course), no one would take it be literally true (it's a model after all). But if the model were a possible concrete system (akin to a possible person), then the description or, as a façon de parler, the figment, would be *literally true* of it.[11]

One solution is to reintroduce the idea of models as actual *abstract* objects, and take this as complementary to the above models-as-possibilia view. Thus, Contesssa adopts a 'dualist' account according to which a fictional model is an abstract object that 'stands for' one or other of a set of possible concrete systems. So-called 'external' sentences about such models (such as sentences describing their creation, for example), talk of them as actual abstract entities, whereas 'internal' sentences (such as those that refer to the internal composition of the model, for example), talk of them *as if* they were concrete physical systems.[12] The former can be literally true, while the latter are literally false, but insofar as the model can be considered to 'stand in' for a particular concrete system, they can be regarded as true 'by proxy'. Thus, a sentence such as 'The Rutherford model of the atom was created by Ernest Rutherford at the turn of the twentieth century' is an 'external' sentence and whether it is true or false depends on how the actual world is (a historian of physics might declare it false, as Nagoaka seems to have had the idea first). The sentence, 'In the Rutherford model of the atom, electrons orbit around the nucleus in well-defined orbits' on the other hand, is 'internal' and is typically taken not as literally true but only as true 'in some sense' (as quasi-true perhaps; see Bueno and French 2011).

On this account, the creation of a model involves the public description of a possible system in a given context and its proposal as a model of a certain kind of actual system. This 'generative' description of the model can then be further specified or altered in various ways as it is both itself investigated and applied to a range of situations. Unlike other advocates of the view of models as abstract entities who regard such descriptions as complete, Contessa takes them to have an *open-ended character*, in the sense of features that are neither explicitly nor implicitly attributed to it but that are unforeseen at the time of creation but can be subsequently discovered, as in the case of the stability of the Rutherford/Bohr

[11] There is a related worry that just as Frodo's handedness is left unspecified in *LOTR*, so is the length of the string when we imagine the model of the simple pendulum. One could take the latter to refer to different specific simple pendula in different contexts (Contessa 2010, p. 222), but that hardly seems to fit our linguistic of scientific practices (imagine if it were applied to Frodo!). Perhaps the simplest way round this problem is to take the model to be a possible concrete object that is indeterminate in certain respects, such as that regarding the length of the string.

[12] Here we might recall the Suppesian distinction between 'external' or extrinsic and 'internal' or intrinsic characterizations of theories and models (da Costa and French 2003).

model of the atom. And whereas the implicit characteristics can be excavated through logical investigation, these other features that give models their truly open-ended character typically arise when further laws are applied to the objects of the model, as in the case of the Rutherford/Bohr model and the laws of electrodynamics.

So, here we have a dual ontology involving both abstract *and* possible objects that, as Contessa himself acknowledges, is metaphysically quite inflationary. But, he insists, the package should be accepted in the absence of any viable alternative. Below we shall consider two such alternatives that draw on the kinds of considerations that feed into the models-as-possibilia view—namely Frigg and Toon's accounts that draw on Walton's view of fictions in the philosophy of art that we discussed in Chapter 1. However, in Chapter 7 I shall argue that we can avoid all the problems that both the models-as-fictions and models-as-abstracta accounts have to face by adopting an eliminativist approach.

Models and Make-Believe

The core idea here has already been touched upon in our consideration of propositions and suggests that scientific models should be seen as akin to literary works in the sense that both models and fictions involve the construction of *games of make-believe*, so that models can be seen as merely props in such games. Thus, Frigg and Hartman (2018) write that '[i]t seems natural to view [theories and models] as fictional entities', since this 'squares well' with both scientific talk and practice, which take models (and specifically here, non-concrete models) to be objects and also philosophical accounts that focus on their mediating role and manipulability. As we have already noted, models may also *surprise* us, in the sense that they may possess more features than they are explicitly taken to have when initially constructed (ibid.), just as the characters in books may surprise us. But rather than take models as objects in the sense of possibilia, with all the commitments that involves, Frigg takes them to be objects in the sense that Sherlock Holmes or Frodo are objects—namely *objects of our imagination* (Frigg 2010a).

This idea is then fleshed out by drawing on Walton's view of literary fictions as *representational* in the sense of functioning as a props in certain games of make-believe (Walton 1990).[13] According to this account, we recall, a fictional world is associated with a certain cluster of propositions and '[t]he propositions fictional in the world of a game are those whose fictionality is generated by virtue of the principles and props of the game—the propositions which, because of the principles in force and the nature of the props, are to be imagined by participants in

[13] Thus, they are not representational in the sense discussed above, namely that of bearing a relation to some target (see Frigg 2010a, p. 259).

the game' (ibid., p. 59). And a prop within this framework is simply that which 'mandates imaginings' (ibid., p. 69). Thus, *The Lord of the Rings* is a prop that mandates the imaginings of the game of make-believe associated with Middle Earth and its characters, with the book effectively licensing certain rules that allow us to determine where Mordor is in relation to The Shire, for example, or how long it takes Gandalf to ride from Orthanc to Minas Tirith (Weisberg 2013, p. 53).

Likewise, model descriptions should also be seen as *props*, on this approach, with the relevant rules given by that description, together with the relevant background theory, the assumed mathematics, and so on (Frigg 2010a). This description, and the associated rules, may become firmed up or fleshed out or otherwise developed as time goes on, so this account can accommodate the shift from a partially baked idea to a more solid suggestion to something about which the question can be asked, 'does this correspond to the target system?' (ibid., p. 260).

On this account, then, to say that a model possesses certain properties involves nothing over and above saying that 'within a certain game of make-believe we are entitled to imagine the entity as having these properties' (ibid., p. 261). Likewise, truth in such a model can be understood in the same way as we understand truth in a fiction. So, adapting Currie's terminology (Currie 1990, ch. 4), 'intrafictional' statements such as 'on a night of rain Frodo smelled a sweet fragrance on the air and heard the sound of singing that came over the water' (Tolkien 1968, p. 1068), can be distinguished from 'metafictional' statements such as that made by someone who, having read *LOTR*, asserts that 'Frodo passed into the West' and 'transfictional' statements, as when someone compares Frodo to a friend of theirs, or to the character Peter Pevensie in *The Lion, The Witch and Wardrobe*.

The first are statements made within the fiction, of course, and are true insofar as the relevant props and rules of generation prescribe them to be imagined, independently of whether they actually are. Likewise, there are statements that are true in the model if the model description and the associated laws and principles imply them. The truth of metafictional statements then piggybacks on that of intrafictional ones, on the understanding that the assertion that 'Frodo passed into the West' is elliptical for 'In *LOTR*, Frodo passed into the West'. And again, the export of this into the realm of models is straightforward: the statement 'The period of the pendulum is proportional to the square root of the length of the string' gets parsed as 'In the model of the simple pendulum, the period is…'. As Frigg notes (2010a, pp. 263–4), transfictional statements are trickier, because here we are being asked to compare an existent object, say, to a non-existent one. But here again, and specifically with regard to exporting this account into the domain of models, such statements can be regarded as prefixed with a clause setting the context of the comparison, so 'My mate Simon is just like Frodo' becomes 'When it comes to seeing something through, my mate Simon is just like Frodo'. And likewise, in the case of models, this sort of statement is rendered down to a comparison between properties associated with the model and those of the target system, such that we can say the former is similar to the latter in certain respects. Similarly, we can analyse statements relating two models, for example.

Thus, bringing this account of fictions from the philosophy of literature into the philosophy of science yields a broad and robust framework for understanding the epistemology of models. Of course, you might be leery of the role of imagination here, but it is not the case that this account involves some fatal element of subjectivity, since the props, the rules of generation, the game of make-believe in general are all shared by the relevant community and thus are objective to that extent. There might also be a worry about how representation is handled within this framework since both *LOTR* and the model of the simple pendulum now count as fictions but whereas the latter represents a real target system, the former 'represents' an imaginary world and its characters. Here a distinction can be drawn between 'prop-representation' and 'target-representation' respectively and of course, one can begin with a p-representation, which introduces an imagined object that is then claimed to t-represent some system. And although Frigg has his own account of the second, as we have already seen (Frigg 2010b; Frigg and Nguyen 2017), the partial structures version of the Semantic Approach is also perfectly serviceable in this regard.

As for metaphysical baggage, this is pretty light: there are no fictional or abstract entities assumed in Walton's view and likewise, in Frigg's application of it to models, there are no such ontological commitments. Or so it is claimed.

Toon, however, argues that Walton's antirealism about fictional characters sits uneasily with the above account (Toon 2012, pp. 57–9). Consider: according to the latter, model *M* represents some real system *S* if and only if *M* denotes *S* and there is some 'key' that tells us how facts about *M* can be translated into facts about *S* (we recall the significance of such 'keys' in the context of the 'DEKI' approach). But if one imports Walton's antirealist stance and applies it to models, then, strictly speaking, there can be no facts about *M*, and no relevant relation can be established between models and target systems.

One option would be to simply abandon the antirealism (Frigg 2010b, p. 113), but then we need some account of what models, qua fictional entities, actually *are*, and we're back in the ontological mire. Another way forward would be to adopt an across-the-board form of antirealism with regard to representation as well, which would somehow explain away the apparent commitment to fictional entities. However, talk of models denoting real systems or standing in relations with them has now to be construed as merely a 'way of talking', rather than of how modelling actually works (Toon 2012, p. 59).

As a result, Toon takes a different fictionalist path. He sees Frigg (and others, like Giere[14] and Godfrey-Smith) as offering an 'indirect' account of modelling,

[14] Interestingly, given recent interest in this area, Giere suggests that we, as philosophers of science, should refrain from pressing for a 'deeper understanding of imaginative processes and of the objects produced by these processes' (Giere 2009, p. 250).

whereby scientists modelling a system are taken to be describing an imaginary model system, which can then be compared to a fiction. Toon, by contrast, argues for a 'direct' account, according to which scientists simply *imagine* things about real systems. So, consider again what is involved in modelling a bob on the end of a spring undergoing simple harmonic motion (ibid., pp. 38ff.): idealization yields a 'prepared description' of the spring which *represents* it—following Walton's view—by prescribing imaginings about it (ibid.). Just as the passage from H. G. Wells's novel *The War of the Worlds* that imagines St Paul's cathedral having been blasted by the Martians (sorry—should have included a spoiler alert!) represents St Paul's by requiring us to so imagine certain things about it, so the prepared description of the spring, plus the associated equation of simple harmonic motion, represent it by virtue of requiring us to imagine that the bob is a point mass, air resistance is zero, the spring exerts a linear restoring force, and so on. Hence, statements such as 'the bob oscillates sinusoidally' should be understood not as assertions about some model-system ('the bob and spring system') but as about what the model prescribes us to imagine and when we make such utterances we are engaging in certain acts of pretence.

How then should we understand the kind of talk that scientists engage in when they are modelling systems? Take for instance the statement 'the bob oscillates sinusoidally'—on Toon's account there is no object (abstract, fictional, or whatever) about which this statement is true. Thus, we have to construe the statement as asserting that the relevant prepared description and equation of motion make it fictional that the bob oscillates sinusoidally (ibid., p. 48). In other words, the truth of the statement is underpinned by the description and equation and *not* by some object. Thus, talk about what is 'true in the model' is not to be understood as talk about certain objects but rather as prescriptions for what we are to imagine.

I am sympathetic to this move of grounding the truth of statements about models in something other than the models themselves, conceived of as objects. However, there is a further worry with this account that revolves around the question why, then, do scientists talk as if such statements *are* about model systems? In answer we can draw again on Walton's philosophy of fiction: when scientists utter such statements, they are, in fact, engaging in certain games of pretence, where these games are authorized for the model in question (ibid., pp. 49–50). When a scientist asserts, for example, 'the bob oscillates sinusoidally', what she is actually doing is *pretending* to assert that the (actual) bob oscillates sinusoidally.

Still, for all that one can certainly find examples where scientists do seem to be engaging in some kind of pretence and for all that such verbal pretence is but one aspect of scientists' imaginative engagement (Toon 2012, ch. 5), this re-analysis or non-direct understanding of scientists' talk appears to be a cost associated with this sort of account (see also Thomasson forthcoming, for similar concerns), much as the non-literal parsing of such talk was for broadly positivist approaches to science. First we have to understand models, not simply as descriptions but as *prescriptions*,[15] dictating what we are to imagine, and secondly, we have to take talk that is putatively about model-systems as a kind of pretence. It seems to me that if we can avoid reifying models without having to resort to such re-analysis of scientists' talk and practices, then this would represent an advantage—and indeed, that is precisely what I shall try to do in Chapter 7.

These sorts of worry become even sharper when we consider how statements that compare models to the world (exemplified in Giere's 'theoretical hypotheses') fare on this approach. As Toon notes, the previous analysis can't simply be carried over, although he still maintains that such theoretical hypotheses do not require commitment to some model-system. So, to assert that 'the period of oscillation of the bob in the model is within 10% of the period of the bob in the system' is just to compare what the model prescribes us to imagine with what is true of the system (ibid., p. 51). Again, there is no invocation of some model-system with the attendant worry about reification. But the problem now is with explaining scientists' talk, as the above Waltonian line can't be run without some adjustment.

Here we have to invoke the notion of an 'unofficial' game, which is introduced by Walton as a way of accommodating comparative statements such as 'Robinson Crusoe was more resourceful than Gulliver'—in this case we would participate in an unofficial game in which both the books *Robinson Crusoe* and *Gulliver's Travels* function as props, each functioning as they do in their separate 'official' games (Walton 1990, p. 407). More pertinently, the notion also covers situations such as when one says with regard to a performance of the comic opera, *HMS Pinafore*, 'The orchestra is in the water', which should be understood in terms of an unofficial game in which the orchestra pit is fictionally in the ocean off the bow of the ship (ibid.). In such cases we utter statements that combine and compare the real and the fictional, just as theoretical hypotheses do.

Thus, we should understand such theoretical hypotheses as 'invoking an unofficial game in which it is fictional that there exists both the bob and an entity called "the model bob" which, fictionally, has all the properties attributed to the bob by the model' (Toon 2012, p. 52). Claims regarding the accuracy of the model, such as made above, can then be understood as prescriptions of how to speak the truth in this unofficial game and by making such a claim, we effectively assert that the conditions responsible for our fictionally speaking the truth are in place (ibid., pp. 52–3).

However, this paraphrasing of everyday and scientific talk also comes with costs (ibid., p. 53). First of all, we've had to invoke a new fictional entity—the model bob—thus undermining one of the principal virtues of this 'direct' account.

[15] Insofar as such a prescriptivist attitude is neutral with regard to the status of theories as models, this may not seem so distant from the stance that I take (I'm again grateful to one of the readers for pointing this out).

Secondly, consider the statement 'the period of oscillation of the bob in the model is longer than that of the bob in the system'. Under the Waltonian interpretation both 'the bob in the model' and 'the bob in the system' refer to the same thing, so it seems that we have, in fact, pretended to assert a contradiction (ibid.). Now, Toon himself feels that these problems do not undermine his account and that we can stick with the original analysis and simply take theoretical hypotheses to be asserting comparisons of what the model asks us to imagine with what is true. However, as sanguine as I am about inconsistency in science (see for example da Costa and French 2003, ch. 5), I don't think these problems can be evaded quite so easily. At the very least, invoking this notion of an unofficial game, with some attendant manoeuvre to resolve the problem of contradictory assertions, further raises the cost of this sort of account.

These issues then feed into two broader concerns, which have to do with the nature of the relata in the representational relationship and consequently, with the terms in which that relationship can be established. The first is called the problem of 'transfictional propositions' by Frigg and has to do with how we can compare *imaginary* systems with their counterparts 'in' the world. His response is to shift up a level to the features or properties of the systems and to insist that it is these that are compared, rather than the systems themselves. Thus, 'transfictional statements about models should be read as prefixed with a clause stating what the relevant aspects of the comparison are, and this allows us to rephrase comparative sentences as comparisons between properties rather than objects' (Frigg 2010, p. 263; see also Weisberg op. cit., p. 54 for commentary).

Of course, this seems quite reasonable; after all, it's not even clear what it would mean to 'compare' imaginary and real objects 'directly'. Advocates of *non*-fictional accounts would surely agree that insofar as we can be said to be establishing a comparison when talking about representation (and not everyone agrees with this of course), that comparison holds between features or properties, rather than the 'essences' or 'substances' or 'primitive thisnesses', or whatever of the objects concerned. Let us recall the Semantic Approach, cast in the form of partial structures: all the 'action' as it were, takes place with regard to the various R_i, in our formal characterization $<A, R_i>$, that stand for these features or properties. Unfortunately, however, this obvious move doesn't get the Friggian fictionalist off the hook: insofar as the original concern had to do with how we might compare real objects with imaginary ones, the shift to properties just shifts the question to how we can compare instantiated properties with uninstantiated ones (Godfrey-Smith 2009).

Toon offers a far more radical solution: he denies that representation is a relation to begin with. On the above account, we recall, *M* is a model-representation if and only if *M* functions as a prop in a game of make-believe (Toon 2012, p. 62) and thus prescribes the relevant imaginings (see also pp. 81–2). Thus, to state the obvious again perhaps, the appropriate view of representation in science should draw on representation in *works of fiction*, rather than on representation in *paintings*, and consequently moves away from a relational take.[16] In particular, as we have seen, this account denies that there is any object or model-system that 'fits' the relevant prepared description and equation of motion (Thomson-Jones 2010): 'to say that it is fictional that the bob is subject to a linear restoring force is not to say that there is any object of which this is true. It is merely to say that we are to imagine of the actual bob that it is subject to a linear restoring force' (Toon 2012, p. 43). Nevertheless, comparative statements can still be made and we are still invited to understand representation in terms of a comparison of what the model asks us to imagine with what is true of the system, with the relevant pretence now understood to take place within an unauthorized game of make-believe. Hence, the claim that the system is similar to the model in certain respects and degrees invokes an unofficial game in which it is fictional that there is a pendulum bob and a model bob that fictionally has all the properties attributed to the pendulum bob by the model.

Obviously, this accommodates the 'aboutness' of scientific models touched on earlier—that is, the fact that such models are 'about' systems, processes, objects, etc., whether real or not—since on this view it is *all* about the real world system. It also has the advantage of being less ontologically inflationary than alternative fictional accounts, at least at first glance. But on second glance, as Toon himself admits, it seems that fictional entities creep back in as a result of invoking Walton's notion of unofficial games in order to accommodate theoretical hypotheses, as we have noted. Furthermore, there is a certain advantage to articulating representation in relational terms: recall our discussion from Chapter 3 about how the inferential aspect of scientific modelling rides on the back of the representational relationship. And if we capture that relationship in terms of, oh I don't know, *partial isomorphisms*, say, then we obtain all the advantages of (meta-)representing the relationship between typically mathematically presented features, whether these be properties or whatever, also in formal terms.

Now, fictionalist accounts are not devoid of rules or structure. There are 'principles of direct generation' and 'principles of implication', both of which yield fictional truths (Toon 2012, p. 46). The former are conditional on certain features of the fiction—what is explicitly stated in *Lord of the Rings* about the geographical relationship between Mordor and The Shire, for example. The latter, as the name suggests, give us the further truths that are implied by these primary truths—for example, that Rohan lies to the west of Mordor, even though this is not explicitly stated (Weisberg 2013, p. 57). Carrying this distinction over to scientific modelling,

[16] But note that this view does not commit us to the view that models *are* works of fiction; nor does it deny important differences between models and novels. It only asserts that scientific models and novels share the core function of prescribing imaginings. As we shall see, this relates to a certain more minimalist view of the ontological status of such models.

that the bob is a point mass, and the spring exerts a linear restoring force in our fiction would be primary truths, generated directly by the prepared description and equations of motion (Toon 2012, p. 46). Implied truths can then be obtained, such as the bob oscilllates sinusoidally for example, or that the period of oscillation obeys a certain relationship to the mass of the bob, but the principles governing such implications will be case specific (we might discard certain solutions of the relevant equations as non-physical for example; ibid., p. 47).

That is all well and good. Obviously we would expect the relevant principles to be both more explicit and more tightly constrained when it comes to models than for novels. But there is still a worry: in thinking about the model, it would seem that we naturally shift our thought from it to the system being modelled. We might draw out some of the implications of the principal truths, as indicated above, and then consider whether these implied truths also hold of the actual system (a consideration that is explicitly accommodated by the 'immersion, inference, and interpretation' account of Bueno and Colyvan (2011)). This can be naturally captured in terms of a mapping between the model and the system, across which we move, in thought, from one to the other and, indeed, capturing this to and fro explicitly in this manner allows us to schematize not only the way in which models apply to the world, or to each other but also how mathematics is applied (see also Bueno and French 2018). Toon's account obscures, at best, that to and fro, folding it all up into fictional imaginings about the system. Of course, that's not a problem as far as the scientist doing the modelling is concerned; she will continue her practices as always. But it is a problem for us, as philosophers of science, interested in representing, again at the meta-level, those practices in the most illuminating and clarificatory way as possible. And offering a way of making that inferential to and fro explicit confers certain advantages on such a system of meta-representation, or so I maintain.[17]

The second concern has to do with cases where we're not modelling an actual entity. What about models of the aether, for example (see Harman 1982)? The relevant prepared descriptions and laws can still be treated as props in games of make-believe à la Walton: imaginings are still prescribed, only there are no (actual) things these imaginings are about. Here the comparison is with passages from novels like *Dracula*, or any of the Sherlock Holmes stories, where the Count or the detective, respectively, do not correspond to any actual entities. Again, these passages can be regarded as props in the relevant game. However, how are

[17] We might also note that Toon's account is derivative in the sense that models' representational power ultimately derives from that of certain mental states, namely those of the imagination. Yet even if we accept that such states are different from those having to do with belief, say (for a discussion, see Gendler 2013; but see also Stock 2017 and French forthcomingb), they have a certain intentionality associated with them and can be said to be 'about' something whether real or not. Any attempt to appropriately capture this aboutness needs to create a conceptual distance between the relevant state and that which it is about, such that a clarificatory meta-representation of such states can be constructed within the philosophy of mind or psychology.

we to understand statements like 'the aether is always at rest'? As with 'Sherlock Holmes had clear, hard eyes', such assertions appear to licence certain claims about something or someone which/who does not actually exist, bringing the ontological concerns back to the fore.

Furthermore, it is not just models of the aether and other disregarded and discarded entities that fall under the designation 'models without objects'—consider, for example, the postulated 'island of stability' of isotopes of the transuranic elements, such as Unbihexium-310 (see: https://en.wikipedia.org/wiki/Island_of_stability). Although these elements have yet to be detected or synthesized, they can already be modelled—indeed, their existence is predicted on the basis of the so-called nuclear shell model according to which nuclei 'sit' in and fill certain energy shells, or groups of energy levels separated by energy gaps, akin to the way electrons may be modelled within an atom.[18] How can such models be regarded as representations when it is not clear, at the very least, that the systems they are taken to represent actually exist?

One option would be to acknowledge that these models don't actually represent at all, although the scientists involved believe that they do (Callendar and Cohen 2008, p. 81, n. 11). However, this move overlooks the differences between such models and clearly non-representational devices that may use the same or similar elements (Toon 2012, p. 78). Consider again the comparison between the Phillips-Newlyn machine which models the economy by means of water flowing through certain tubes (as discussed in the context of the DEKI account in Chapter 3) and water flowing through the pipes to one's bath or shower (ibid.)—in the former, the water flow signifies something, in the latter it does not. And this problem appears to bite particularly hard for similarity or isomorphism based accounts since it would seem that for such relations to hold we need both relata to exist: models and system being modelled.

Of course, we could take the system being modelled itself to be fictional and yet still exist but this yet again dramatically inflates our overall ontology. More significantly, perhaps, this seems to get the relationship the wrong way round: '[f]ictional entities are dependent on representations for their existence in a way that normal objects are not' (Toon 2012, p. 80). The point is, we have no independent and in particular, *empirical* access to such entities, so all the properties ascribed to them are so ascribed via the relevant model. Indeed, in the case of the superheavy elements it is the model that predicts their existence and the properties they have.

[18] However, in a nice exemplification of *modus tollens*, or Popperian falsifiability, it has been argued that the lack of any evidence for such elements indicates that the model is flawed and should be replaced (see Cook 1990 and 2010). Of course, the current lack of evidence for the existence of such elements should not be taken as evidence for their non-existence and recent results have suggested the island may soon appear on the horizon (see http://www.nature.com/news/element-117-hints-at-island-of-stability-on-periodic-table-1.15189).

Alternatively, we could argue that the models *aim* to represent, in some sense, but that they simply fail to do so, because the appropriate system does not exist. Thus, consider again the 'inferential' conception of representation, according to which a model *M* represents some system *S* only if the 'representational force' of *M* points towards *S* and *M* allows us to draw specific and appropriate inferences about *S* (Suárez 2003). On such a view, we could say that even though the aether does not exist, Maxwell's famous model has representational force that points towards it. However, not all models without objects are like this: '[w]hen we use the Phillips machine to represent an imaginary economy we do not attempt, but fail, to represent any real economy' (Toon 2012, p. 80). Consequently, it is hard to make sense of such models possessing 'representational force'. In the case of non-existent entities, the concept of 'representational force' incorporates the dependence that Toon has usefully flagged up and we are left wondering what is this entity, this 'it', to recall our statement about Maxwell's model, that the representational force is pointing towards.

Returning to the fictionalist camp, Frigg would say that representation in such cases involves a comparison between uninstantiated properties in the fiction and properties that are either uninstantiated or instantiated in something other than the entity being modelled (the electromagnetic field supported by space-time, perhaps, in the particular case of the ether). At least now we don't have to face the ontological disparity noted before! As for Toon, since he denies that representation is a relation, not only need there not exist any object for these prescribed imaginings to be about but the model need not even attempt to represent any object. Thus, Maxwell's model of the ether is representational because it prescribes relevant imaginings within a certain game of make-believe.[19] Likewise the Phillips-Newlyn machine is representational, even if it is not used to represent any actual economy, because its pipes and tanks prescribe imaginings according to certain rules, unlike the overflow pipe from my bath, for example.

Of course, there is still the question: how are we now to understand the content of such imaginings when they are 'about' *non-existent* entities? Toon passes this on as a general problem for the philosophy of mind (which can be generalized to encompass all intentional states) and insists that it is not one he, or we, should have to solve here (ibid.). That's fair enough when it comes to the general issue of how mental states can be about that which does not exist but it still leaves open the issue of how representations can be said to be *about* anything—the generality of this account is bought at the cost of shifting this basic issue into the philosophy of mind.

What about the Semantic Approach? Here, as we saw, representation is definitely relational and involves at its core the comparison of properties or, more generally,

[19] Walton also applies his account to existence claims in theology: the atheist who asserts 'God does not exist' is engaged in an unofficial game of make-believe in terms of which God is fictional (see http://onlyagame.typepad.com/only_a_game/2010/06/walton-on-makebelieve.html).

features, themselves represented at the meta-level by the relevant R_i. Now, even before we get to the concerns listed above, there is the issue of the mathematical/set-thereoretical and therefore abstract, representing the non-mathematical and concrete. This has come up before in the guise of the concern about how the relation of (partial) isomorphism can be said to hold between a set-theoretical entity and a physical one (Downes 1992). The response runs along Friggian lines insofar as we essentially abstract away from the physicality of the latter by representing it in set-theoretical terms (French and Ladyman 1999). This still leaves the more fundamental issue of how the set-theoretical can be related to the physical, or, more generally, how any signifier relates to that which is signified, but we can take this, in the former case anyway, to be an instance of the general issue of how the mathematical relates to the physical.

Thus, we could perhaps see this as analogous to the first concern above, namely of how the imaginary can be related to the real, with broadly the same response being deployed. But what about the second concern? Does a similarly analogous problem arise? Namely, how can a model, under the Semantic Approach, represent entities which are either known not to exist, such as the ether, or which we do not know whether they exist, like the superheavy elements?

Now, certainly, when it comes to entities about which we do not know whether they exist or not, one might not expect these to pose a problem for this sort of approach, given that antirealists such as van Fraassen and Bueno adopt the former stance (across the board), while deploying the latter in their accounts of representation. And the reason there is no incompatibility is obvious: when it comes to what we might call, broadly, theoretical representations, the relatum at the other end of the representational relationship is always 'putative', in the sense that it might exist but, as far as the antirealist is concerned, we can never know (given their understanding of what it would be to know). And the realist can adopt the same general strategy, with the rider that under her conception of what counts as knowledge, the designation 'putative' may at some point be removed.[20]

Still, both the antirealist and realist may agree that the ether does not exist, so it is not in any sense a 'putative' entity, in which case, how can the relevant models be said to represent 'it'? Here one can simply follow Frigg again and insist that what we are relating are properties meta-represented via set theory on the one hand and properties that are uninstantiated on the other (or instantiated in some entity, such as the electromagnetic field plus space-time). But, again, there may remain the niggling worry as to how something that is not instantiated or, more generally manifested in the world, can be compared to or thought to stand in some relation with anything at all, whether described set-theoretically or otherwise.

[20] Of course, realists of different stripes will disagree as to the nature of the entity at the other end of the representational relationship, whether it be an object, structure, or whatever, but most will typically accept that the same cluster of properties is involved, so the Friggian move can still be made.

Again, I'll leave the cutting of this further knot for later but before we leave fictionalism, there are some broader concerns that need to be addressed and which will help illuminate certain issues we shall come back to.[21]

Subjectivity and the Limits of Imagination

First of all, there is a concern that as fictions are basically products of human imagination, if there were to be no humans, or no sentient creatures in general, then there would be no imagination, no stories, no games of make-believe, and thus on this view of modelling, no models and hence no quantum mechanics, no evolutionary biology, etc., etc. (cf. Nolan 2016). Here again we face the following dilemma: either theories exist independently of their creators/discoverers/whatever or they do not. If they do, then of course they would continue to do so even if no sentient beings ever evolved anywhere in the universe to think about them. However, as we have seen, with theories sitting there in World 3, say, some account needs to be given of how access to such entities is achieved and how this relates to the nature and role of the sorts of practices that scientists appear to engage in (particularly the heuristic moves we've already discussed).[22] On the other hand, if they do not exist independently of us, or other forms of sentient life, but are sustained by our intentions, say, then they are as contingent as we are: no life, no quantum theory. And we must face all the issues already covered in the previous chapter.

We might blunt the point of this horn by noting, again, that no quantum *theory* does not mean no quantum *physics*. To say that if there were no sentient life there would be no theories is not to say there would be no *world*. As with laws, the fictionalist can insist that our theories and models are merely descriptions of the way the world is (in both its observable and unobservable respects) and that just as the description (in a limited sense) of the impact of Napoleon's invasion of Russia contained in *War and Peace* would not exist if Tolstoy had not, so special relativity would not if Einstein had not (but space-time would still be locally Minkowskian).

Secondly, one might worry that qua products of human imagination, models, on this view, are somehow rendered arbitrary and unconstrained in nature and in a way that is not appropriate when it comes to the products of scientific practice.

[21] See also Giere (2008b) who lists and rejects three motivations for fictionalism: (1) that scientists themselves sometimes invoke fictional entities when discussing theories and models; (2) that many such models are not actually realizable; and (3) that this view meshes with a fictionalist (and anti-realist) account of science in general.

[22] We face something akin to Benacerraf's problem regarding mathematical objects: how can such mind-independent entities be applicable to the actual world? Once again, I'm grateful to one of the readers for pointing this out.

There is an easy response to this worry: even in the case of literary fictions I am sceptical that these are as arbitrary as this might suggest. Here too I doubt that 'Eureka!' or muse-inspired accounts of creativity are appropriate. But whatever one feels about that, it should be acknowledged that scientific models are definitely constrained, not only in their construction or discovery, in the ways we have already considered, but in what can be inferred from them, as also noted. Putting it rather bluntly: not just any old story can serve as the model fiction in such cases.

However, this feature leads into a further concern, namely that as fictions are products of human imagination, models-as-fictions will differ, perhaps significantly, between different scientists (Weisberg 2013, pp. 56–62). The question now is, how can fictionalist accounts ensure that scientists are not represented as engaging in different games of make-believe about the same system?

Here we can draw on the distinction between 'focal' and 'non-focal' properties of stories (ibid. p. 57): the former include both principal and inferred truths, such as the geographical relationships between Mordor, Rohan, and The Shire, which are certainly important for the story in *LOTR*; the latter might include the number of toes on Orcs' feet (ibid.), which are not so important. Transposing this distinction to the case of scientific modelling, the fictionalist needs to ensure that on her account, there will be a high degree of consensus within the relevant scientific community with regard to the relevant focal properties.

But now a dilemma arises: either the fictionalist insists that the model description and associated equation yields all and only the focal properties; or she accepts that these properties go beyond this description. With the first option, the problem of possible inter-scientist variation is easily solved: everyone who accepts and works with the model description will agree on the relevant focal properties. However, the worry now is that this yields a rather sparse landscape, particularly if the model description is presented mathematically: '[f]ictions then cease to look at all like real-world scenarios, militating against the claim that models can be compared to real systems in a straightforward way' (Weisberg 2013, p. 58).

Unfortunately, the second option is also problematic. If the fictionalist insists that the focal properties go beyond what is given by the model description, then she needs to explain how they are generated. Walton offers the 'mutual-belief' principle (1990, p. 151; Weisberg 2013, p. 59), to the effect that focal properties can be generated by what would be believed in the 'artist's society', given the relevant primary truths. Transposing that principle, the idea would be that those focal properties that go beyond the model description are whatever the relevant scientific community mutually believe to be the case, given that description (ibid.). This certainly offers the possibility of some constraint on variation.

However, this only works if there is sufficient homogeneity within the community. Consider the example of the Lotka-Volterra predator-prey model: here the interactions between two species—one the predator, the other the prey—are modelled via two non-linear first-order differential equations that reveal a characteristic oscillation between the population numbers of the two species. Putting it very simplistically, as the numbers of prey animals increase, so do the numbers of predators, but with a certain time lag; at a certain point, due to the increased predation, the number of prey animals falls and, consequently, so does the number of prey, again with a certain time lag. These oscillations between predator and prey numbers are crucially dependent on a number of idealizations and other assumptions, including the spatial structure of the two species. Now, consider that spatial structure as the relevant focal property. Researchers may imagine the spatial distribution of predator and prey creatures in different ways but only certain of these will yield the crucial feature of oscillation between the respective numbers. The question then is, how could the imagined scenarios that *don't* lead to these oscillations be taken to correspond to instantiations of the model in the context of the Waltonian approach? In response, Weisberg suggests that the mutual belief principle, applied to this example, just isn't restrictive enough.

It's not clear to me that this dilemma is so acute. If the fictionalist is minimalistically inclined, she could accept the first horn, while noting that even if the model description is presented mathematically, that maths will have to be *interpreted* for the model to count as physical. Hence there are no grounds for the claim that comparison with real-world scenarios will be problematic. Alternatively she can buy into the second horn and insist that either there will be further constraints other than the mutual-belief principle or that principle will embody further constraints that can then be made explicit. An obvious such constraint will be empirical adequacy, of course and an equally obvious response to the Lotka-Volterra oscillation case would simply be that those imaginations that don't lead to the observed fluctuations in predator-prey numbers will simply be rejected on these grounds.[23]

There is a further problem that has to do with the *limits* of imagination: although it is easy to imagine the content of finite and deterministic models, say, it may not be so straightforward when it comes to more complex cases, such as indeterministic models or those involving probabilistic interactions in general (Weisberg 2013, p. 63).[24] Given the prevalence of such probabilistic features in many models in current science, from physics to the social sciences, this is a real concern.

There is again more to say about the role of imagination in science, of course (see for example French (forthcomingb), Murphy (forthcoming), Salis (forthcoming),

[23] You could insist that empirical adequacy does not count as a constraint in the sense meant here and that the concern has to do with how scientists' imaginations might be restricted before we reach this stage. However, empirical adequacy does not just feature in the justification stage but in the heuristics phase also, when models are ruled out early on when it becomes apparent that they don't 'fit', even on qualitative grounds.

[24] The issue is that, since any given fictional scenario can only be a single instantiation of such interactions, how can we capture that probabilistic aspect (Weisberg 2013)?

and Stuart (forthcoming)) but it is just worth noting here that the notion of imagining that underpins Walton's ideas of props in games in make-believe, carried over from the philosophy of literature, appears heavily biased towards *visual* imagination, not surprisingly perhaps. Is such a notion appropriate for the scientific context, especially when we move away from the more straightforward examples such as simple pendula or weights on a spring? Here one might appeal to some kind of distinction between 'sensory' and 'conceptual' imagination, with the former associated with forming a mental image and the latter with entertaining certain possibilities (Gendler 2013) but it is not clear how helpful this is—in the context of this review of fictionalism, how is *conceptual imagination* to be demarcated from the kinds of conceptual explorations that advocates of the Semantic Approach or Weisberg's 'mathematical models' approach will also acknowledge?[25] And what would be the role of 'props' in such imagination? Could it really be characterized in terms of a game of make-believe? Here Toon's account seems to be under particular pressure, as such imaginings are directly about the physical system concerned. But when it comes to complex conceptual entertainment or, again, probabilistic considerations, how *can* they be? At the very least, we are owed an extension of the account to accommodate such considerations.[26]

The final concern is related to the above: in their practices, scientists often consider models with properties that appear to be far removed from those possessed by concrete systems. Weisberg gives an example from population biology that involves very general properties of infinite populations (Weisberg 2013, p. 65). We might add the invocation of systems with infinite degrees of freedom in statistical mechanics, where this is claimed to play a crucial explanatory role (see Batterman 2010; for criticism, see Bueno and French 2018), or that old standby, infinite dimensional Hilbert space in quantum mechanics, or anti-de Sitter models in general relativity, or indeed, any model in modern physics where the number of spatial dimensions is greater than three! In what sense can these be regarded as games of make-believe with associated props? Of course, one might

[25] One way fictionalists might respond is to draw on discussions of 'imaginative resistance' (see, yet again, Gendler 2013 for an introduction). This is where someone finds themselves either unable to engage in a prompted imaginative activity, or 'drops out' of one due to some jarring feature. Recent considerations suggest that the extent to which the latter happens may be influenced by genre (as in literary genres) conventions and expectations (Liao, Strohminger, and Sripada 2014). So the suggestion would be to regard a subject area such as theoretical physics or mathematics as the equivalent of a 'genre' and to argue that philosophers' difficulties in grasping how imagination might apply or be appropriate in such cases has more to do with their/our lack of facility with the 'genre' whereas those within it are more comfortable about regarding such conceptual moves as 'imaginative game playing'. Obviously this would require some empirical research in support (but see for example Kasner and Newman 2001) and it would not block the further concern regarding Toon's account, specifically.

[26] It is significant, for example, that one of Toon's case studies—as illuminating and interesting as it is—is from chemistry and illustrates the way molecular models act as props. As an aside, he makes the further interesting point that in such cases, the tactile and manipulable properties of the physical properties become significant and thus what we have here is a form of non-propositional imagining (Toon 2012, p. 67).

appeal to the above distinction between sensory and conceptual imagination and again lump all of the above under the latter, but then one loses what appears to be distinctive about the Walton schema in this context—again, how is this 'conceptual imagining' any different from what the other accounts of modeling describe? Furthermore, '[o]n this type of view [the Waltonian], scientists would actually have to be aware of the fictions they are entertaining because they are constructing the fictions by entertaining them' (Weisberg 2013, p. 66)—and when it comes to the examples given above, they clearly are not.

Conclusion: Losing the Ontological Lightness of Being

In response to these concerns, the fictionalist might argue that she never intended her approach to apply to the above kinds of models, but only to those we can easily visualize, say, thus moving towards a form of pluralism. But once she concedes this, it seems hard to resist the pressure from the alternatives, which can also be pluralistic but with regard to scientists' cognitive abilities. Of course, imagination can be accommodated by these alternatives, but in its role as heuristic and developmental support (Weisberg 2013, pp. 67ff.). More importantly, as far as I am concerned, she loses the core advantage of her approach, which is the ontological lightness of being that comes with taking models to be imaginary. Indeed, as I have emphasized, in Toon's case we achieve a kind of ontological minimalism according to which modeling is nothing but imagining things about actual systems with no other entities involved or invoked. That advantage is clearly thrown out the window once one concedes that only certain kinds of models can be handled, while others are reified in various alternate ways.

Can we retain that ontological minimalism while accommodating the kinds of practices and attitudes that fictionalism appears to have such a hard time with? I think we can and showing how will be the topic of the next chapter.

7
Theories Eliminated!

Introduction

Let us consider again the question with which we began: What are scientific theories and models?[1] Or, to phrase it more explicitly, what is their ontological status?

As we've seen, within the framework of either the Syntactic or Semantic Approach we can articulate accounts that take theories to be either abstract or fictional. Fleshing these views out by importing various suggestions regarding the ontology of artworks carries some costly ontological baggage, as we've also discussed, and neither seems to possess any distinct advantage. How, then, do we adjudicate between them? Here again we can note that similar concerns arise in the philosophy of art.

Thus, in considering the ontological status of artworks, Thomasson asks,

> In considering how we might select from among the options in the philosophy of art, what exactly do 'we'—that is, philosophers of art—think we are doing?
> (Thomasson 2006, p. 248)

If we are trying to determine the ontological sort of thing that the terms 'painting' or 'music' pick out, then, she argues, we face a form of the 'qua problem' in the philosophy of language: in order to fix the reference of a name, we must know what *kind* of thing it is that we are naming. To know that, we must know what sortal term it falls under—that is, we must know it 'qua' such-and-such.[2] When it comes to artworks, this issue bites even harder:

> To establish the work of art as the referent of a name such as 'Guernica' or 'David', it seems, would-be grounders of the name of this or any work of art must have an idea of what sort of thing they are trying to name (a work of painting or sculpture), and what sort of thing that is: how it relates to those physical objects, in what way its identity, individuation and survival conditions differ

[1] This chapter draws on French and Vickers 2012, although with additions, clarifications, digressions, and sundry other changes. Once again I'd like to thank Pete for letting me use our joint work for my own nefarious purposes.

[2] In the context of understanding reference in the philosophy of language, this means that some prior descriptive element is required for such acts of naming to be successful (see Devitt 1981).

There Are No Such Things as Theories. Steven French, Oxford University Press (2020). © Steven French.
DOI: 10.1093/oso/9780198848158.001.0001

> from those of the physical bases, and so on (i.e. that *Guernica* would be destroyed by dissolving the pigmented surface of the canvas, though the canvas would not be). (Thomasson 2004, p. 86; see also 2003, pp. 144–5)

And likewise when it comes to works of music or literature: to be successful in establishing the referent of Beethoven's *Fifth Symphony* or *The Lord of the Rings*, we need more than just the presence of the relevant sound waves or words on a back-lit screen[3]—we need some knowledge, even if only tacit, of the ontological status of these sorts of artworks and how they are related, as entities of that sort, to the relevant performances or copies. And further, if we go up a level as it were, to consider what the terms 'painting', 'musical composition', or 'novel' refer to, we need some prior criteria that will enable us to pick out the relevant kind of artwork, rather than the canvas, sound wave, set of words on a screen, and so on. This, in turn, 'requires at least a nascent concept of the ontology of works of art of that kind, and of what distinguishes them from the physical entities in the immediate vicinity' (Thomasson 2004, p. 87).

In other words, what we need is a way of disambiguating the ontology. Thomasson's suggestion, already touched on in Chapter 5, is that '[t]he needed disambiguation...is provided by the beliefs and practices of those who ground and reground the reference of the terms in question ("painting", "novel" and the like), provided these are used as genuine sortal terms' (Thomasson 2006, p. 248). And sortal terms come with application and identity criteria; the first determining when the term may be successfully applied, and the second determining when it may be re-applied to one and the same thing. Together these yield persistence conditions under which we may continue to refer to a piece of music as this very piece of music, say.

Following her suggestion in (Thomasson, forthcoming) and translating this into theory talk, what we would be doing can be understood as taking the term 'theory' and asking what it is that this term picks out; that is, what is its referent? According to Thomasson, it is the *beliefs and practices of scientists* that provide the needed disambiguation, where 'theory' is regarded as a sortal term. However, it is, at best, unclear whether these beliefs and practices can determine the relevant identity criteria in the way that Thomasson supposes. Consider for example the various practices associated with quantum mechanics, covering the wide-ranging and significantly different presentations and interpretations of 'the' theory—can these actually provide the needed disambiguation and pin down 'that' theory? We shall return to this question in Chapter 8.

Indeed, we can view the shift within the history of the philosophy of science from the focus on theories to that of paradigms and research programmes and

[3] These being that which we are in causal contact with, of course.

the like... as reflecting meta-level dissatisfaction with 'theory' understood as such a sortal term. However, even following such a shift, issues of identity intrude: consider the questions of what constitutes a paradigm, or how one is supposed to delineate the hard core of a research programme, for example. The resurgence of the Semantic Approach can then be seen as a case of that particular pendulum swinging back, driven by further dissatisfaction with paradigms and programmes and such. However, one of the core advantages of this approach—namely its provision of appropriate mathematical resources for describing the inter-relationships between theories, theoretical models, data models, and so forth, or between theories and theories—undercuts its ability to resolve the identity issue: if theories are represented as partial structures, inter-linked horizontally, as it were, by partial isomorphisms, and also vertically, also as it were, up to mathematics and down to data models, by homomorphisms (Bueno and French 2018), then in precisely what terms is one able to delineate one such theory from another? The structuralists' divisions into 'theory-elements', 'theory-nets', and 'theory-holons' is a tacit recognition of the problems here and, clearly, if 'links' with other theories are taken to form the constitutive 'core' of a given theory, then the identity of the latter becomes constitutively dependent on that of others, and hence blurred in precisely the manner that is worrisome.

Thomasson's response to the concern about artworks is to insist that,

> to determine the ontological status of works of art of these kinds, we must analyze the practices involved in talking about and dealing with works of these kinds and see what ontological kind(s) they establish as the proper referents for the terms (assuming the terms refer). (Thomasson 2006, p. 249)

However, it then follows that certain questions regarding the ontology of art may be simply unanswerable. If identity and persistence conditions are determined by certain practices, and these draw no sharp line which could tell us how much of a novel, say, may be altered for it to remain the same novel, then there is no sharp fact of the matter as to when the novel is the same: 'when our practices are not determinative we shouldn't expect determinative answers' (ibid., p. 250). And just as some questions, such as the above, are too specific, others are too broad. Thus, the question 'What is the ontological status of works of art?' does not have a single answer, in part because 'works of art' covers too broad a spectrum of entity.

Here again we see a parallel with the case of theories and models: if one were to take their function (and hence, perhaps, nature) as scientific objects to be different in the way that some philosophers of science do (not me, I hasten to add), then one might be inclined to take the answers to the above question to be quite different. Someone who follows Cartwright (1983) for example and regards models as both constructed in a very different manner from theories (that is, from the 'bottom' up in some sense) and as serving a very different function (representation for example) could well take them to 'be' very different things.[4]

More significantly, Thomasson suggests that 'not all questions about the ontological status of Gs, where G is a general noun, are answerable, since not all general terms are sortal terms' (Thomasson 2006, p. 250). Appropriating this suggestion, if we were to agree that 'theory' is *not* a sortal term—and the kinds of considerations I have alluded to above suggest that we should not—then we should conclude, on Thomasson's kind of analysis, that our question regarding the ontological status of theories is not answerable either.

Now, that may be too quick. Thomasson also suggests that in the case of works of art, the relevant ontological 'kinds' may vary not just from field to field, which seems plausible, but also from place to place and even time to time! Could this be the case for theories and models in science? Well, we might draw on the diversity of models and modelling in biology (see for example Laubichler and Müller, 2007) to insist that these are very different kinds of things and practices from what we find in physics, for example. This may seem to run contrary to the kind of unitary framework afforded by the Semantic Approach for example but note: that framework merely represents, at best, what we call theories and models; what they are, ontologically, should not be read off the framework, lest we end up regarding theories as set-theoretic entities. Claims as to geographical diversity are harder to pull off, at least at the level of scientific practice, although of course at the level of the philosophy of science one might be able to make a case for distinguishing the 'London' account of models from the 'Leeds', or the 'Munich', or the 'Princeton' ones, but again, these should be regarded as different meta-level representational stances, rather than different views of what sorts of things theories are.

The suggestion that what a theory is, or is taken to be, ontologically, may vary from one time or scientific epoch to another looks more plausible, at least on the face of it. So one might argue that what theories were taken to be post-Galileo, say, were very different kinds of things from how they were viewed previously, given the emphasis on their being written in 'the language of mathematics'. Of course, someone might object that this is to again confuse the theory with its representation via some device, in this case mathematics. But then, granted the difference between Newton's geometrized formulation of his mechanics with Euler's more familiar re-presentation, can we say, in a non-question-begging fashion, that Newton's theory is something over and above these different formulations, such that it can be expressed without mathematics at all? One obvious answer to this question takes us back towards the Semantic Approach but, again, this is just another representational device (and I shall come back to what it

[4] Of course, that ontological difference might then serve to precisely exemplify the problems inherent on Cartwright's account, particularly for those of us who see the function of theories and models as itself laying on a spectrum.

represents in due course). The point is, it is not clear, at the very least, that even liberalized as Thomasson suggests for artworks, we can take 'theory' to be a sortal term in the sense required and as distinct from how we understand it within a particular philosophical representational framework.

One might try to relativize the ontological status of theories and argue that within a particular field (say, physics) 'being a theory/model' *can* be taken as a sortal term, with reference to a particular ontological kind, and that within another field (say, biology) 'theory/model' refers to another kind. But even if one rejects or downplays the differences between theories and models indicated above, it is not clear to me that the apparent diversity within a field—from overarching theories like quantum physics, to more particular examples like the London–London model of superconductivity—is straightforwardly reconcilable with such a view. Certainly the onus is on the proponents of such a claim to show that theory and/or model can function as sortal terms in this context.

We can extend this line of thought even further. One of the things we have learned over the past several decades concerns the heterogeneity of scientific practice. And this heterogeneity extends, of course, beyond inter-theory, theory-model, and theory-data relationships to embrace the roles of material models, apparatus, and objects in general within that practice. We do not have to go as far as Baird for example and the claim that material models carry representational meaning (Baird 2004), in order to accept that such objects should be accommodated within our description of such practice, as I have already noted. So for example the Leeds Museum's collection of anatomical wax models from the nineteenth century has been identified as of a kind that played a crucial role in embryological theory choice (Hopwood 2002). Likewise, Griesemer has argued that manipulable systems of material objects function as theoretical models and, in particular, has highlighted the presentative role of taxonomic models in theory development (Griesemer 1990). And we recall that the Semantic Approach has been criticized for its supposed inability to accommodate such material models (Downes 1992), precisely because of the aforementioned ontological problem of explaining how families of mathematical structures could possibly be related via isomorphism to physical objects. The response (touched on already) both foreshadows and underpins the position adopted here: material models can be *represented* by set-theoretical (and, in particular, partial) structures (French and Ladyman 1999). Adopting such a representational stance liberates us from having to find some gerrymandered way of capturing these material models and their relationships with other elements, *within a framework based on a particular ontology.*

The general conclusion, then, and exporting Thomasson's concerns from the philosophy of art into the philosophy of science, is that, first of all, there may be no one answer to our ontological question covering all kinds of the entity in question, and, secondly, where there might be an answer to the question as formulated with regard to a specific kind, it is to be sought in the relevant *practices*. I'm going to take this conclusion as prompting an *eliminativist* stance towards such an ontology, summed up in the claim, or slogan, *there are no such things as theories*. Like many such claims, it seems to fly in the face of the obvious. Surely there are theories, you exclaim! We can point to them, present them, talk about them, and so on. The challenge, then, is to account for that pointing, presenting, and talking or more generally, for the truth of various statements that can be made about theories, in our scientific and philosophical practices, while keeping our ontology as deflationary as possible.

In addressing this challenge, I shall again suggest that a certain move may be imported into the philosophy of science from the philosophy of art, a move that was itself originally imported from metaphysics and that offers a way of cashing out the truth of statements about theories, without being committed to their existence, in whatever sense. At its core, this move involves the introduction of 'truth-makers' that, as their name suggests, *make true* these statements, where I shall take these truth-makers to be the diverse and heterogenous *practices* of science.

Eliminative Fictionalism

Let's begin by recalling the following three statements:

M1: Musical works are abstract objects.
M2: Musical works are created.
M3: Abstract objects cannot be created.

As we also recall, Collingwood resolves the tension between this triad by dropping M1, whereas Dodd rejects M2, and Thomasson disputes M3. All of these moves come at a cost, crucially involving either ontological inflation or lack of clarity regarding the relationship between the artist and the artwork.

We might try to avoid paying this cost by accepting that 'the whole move to ontology in thinking about music is a mistake' (Ridley 2003). In part this suggestion is driven by the sense that there is such a diverse range of views about what a musical work is, with such different and competing reasons for and against, that the best response is simply to back away slowly, without making eye contact with any of them! We've already encountered this motivation in Thomasson's approach but in part this move is driven by reflection on the actual practices of performing and listening: when we listen to Beethoven's *Fifth*, or 65daysofstatic's *The Fall of Math*, we are not concerned with establishing the identity of the piece, or what its ontological status is, but whether it was a good or bad performance, whether, if live, it matched what we heard on a CD or piece of vinyl, or matched an earlier performance we heard, and so on (ibid., p. 207). Instead of worrying about the ontology, what we should be doing, as philosophers of music, is paying (more) attention to such evaluative issues and to the practices of performance and listening in general.

Of course, this looks like we're putting the cart before the musical horse—we can only determine the value of something in terms of what that something is; different kinds of things will merit different evaluative frameworks (Kania 2014). However, one person's dragging the poor horse back in front of some cart is another's question-begging move: why should we insist on fixing the ontology before determining whether a performance of Van Morrison's 'Into the Mystic' was a good one or not? To demand that we refrain from any such judgement until it has been established 'whether the prevailing ontological conditions are such as to warrant one's disapproval' (Ridley 2003, p. 210) seems utterly bizarre. Even to suppose that clarity on such conditions will help us better frame and judge such evaluations has been rejected as unpersuasive. In particular, it has been argued that the kinds of conditions that are typically laid down do nothing to help us in such judgements, and may even mislead us (ibid., pp. 211–12).

Even worse: such ontological conditions, as expressed for example via identity conditions for a musical work to be *that* musical work, may be not only impotent when it comes to evaluative judgements regarding particular performances, but may indeed be dependent upon the latter. Consider: the pianist Charles Rosen's performances of some of Debussy's works have been described as 'revelatory' (http://www.musicalamerica.com/mablogs/?p=9036) and we often talk of some performance opening us to a new understanding of the work concerned, or as showing us aspects we had previously missed. But then at best the relevant conditions have not helped us nail down the nature of the work concerned, or, at worst, it is only via such performances that we can say what it is. The upshot is that such purported ontological conditions should be viewed as, again at best, merely a set of expectations or defeasible criteria (ibid., p. 214). But in that case,

> if some—of perhaps even all—of a work's allegedly identity-confirming properties can fail to be reflected in a faithful performance *of* it, then those properties can have nothing to do with the work's being what, essentially, it is. Rather... one finds out what a work is, what properties it has, by experiencing performances of it, or by giving performances of it, and that is a process of discovery that may well have no determinate end. (ibid.)

The worry now is that all that this move is doing is shifting around the relationship between the ontological issues and the evaluative ones. Even if the former are now dependent on the latter, we are still concerned with them. The use of 'discovery' above is significant, as if there is still some*thing* to be discovered.

Better to drop this insistence that there are such things entirely and follow the suggestion that we should focus on the *practices*. The kinds of questions that we have considered here and in previous chapters can then be cut off at the knees by denying that there *are* musical works to begin with.

This is Cameron's nihilistic proposal (2008a; 2008b) but whether it should be labelled as 'fictionalism' (Kania 2008), 'eliminativism', or a form of 'nominalism' is open to debate. I'm going to avoid fictionalism due to its obvious connotations,[5] and also nominalism, because that is so broad a category, and go with 'eliminativism' since what is at issue here is whether musical works can be eliminated from our metaphysical pantheon.[6]

Cameronian Eliminativism with Added Truth-Makers

Cameron's (2008a) response to the above tense conjunction about musical works allows us to retain M1, M2, *and* M3 (in a sense), but also imposes a metaphysical framework that urges a (drastic) reduction in our ontological commitments. It is able to reconcile these features and relieve the tension by importing into the philosophy of music two well-known metaphysical manoeuvres.

The first is a version of 'truth-maker' theory. The core idea here is that true statements stand in need of truth-makers; i.e. that which makes the statements true. The underlying intuition is something like this: true statements, or propositions, are 'made' true, in some sense, by certain features of reality (facts, or whatever). Opinions differ as to what this 'making' consists of—whether it has to do with entailment, or primitive necessitation or some kind of 'grounding' (see MacBride 2014)—and both what the truth-makers are (facts, features, whatever) and what the truth-bearers are (statements, judgements, propositions, etc.). Here I'm going to try to sidestep as much of this debate as I can. I shall adopt the minimal assumption that the truth-bearers are *representational* (ibid.), for obvious reasons, and that the truth-makers are *fundamental features* of the world. So, we may say that the statement 'the electron has spin ½' is made true by those

[5] As one of the readers has suggested, if fictionalism is understood à la Walton and theories conceived of in terms of games of make believe and the latter further articulated in terms of certain practices then there may be little difference between this and the kind of theory eliminativism that I favour. However, here we might bring to bear the sorts of considerations presented in Chapter 6. In particular, such games of believe may only correspond to a restricted set of practices.

[6] Thus, Kania takes eliminativism to involve the removal of *x*s in favour of *y*s, whereas he thinks there should be neither (Kania 2008, p. 440). However, as applied to what I'm about to set out, this would mean no works and no *practices*, which would clearly be a step too far! However, insofar as nominalism is often taken to *identify* the relevant entity with concrete things—in this case, scores or performances—I agree that this is also an inappropriate term.

fundamental features of the world having to do with spin.[7] As for the relation between truth-makers and truth-bearers, I shall adopt Schaffer's suggestion that we think of it in terms of 'grounding', where this is understood to be an asymmetric, irreflexive, and transitive relation that thereby induces a partial ordering, with the relevant fundamental features as the minimal elements involved (see Schaffer 2008). Of course, cashing out these minimal formalities requires getting into the details of the relevant statements and corresponding elements (ibid.) and I'll come to that shortly. Let me return to Cameron and the new aspect that he adds to this framework.

Consider the statement 'electrons exist'. What makes this true? Standardly, we would take the relevant truth-maker to be (surprise, surprise) the set of elementary particles that we label 'electrons'. In general, on the standard understanding of truth-maker theory, the truth-maker for the claim '*x* exists' is always *x* (see for example Armstrong 2004). However, Cameron offers a new twist, by allowing for the truth-maker of the sentence to be something other than *x*:

> I think one of the benefits of truthmaker theory is to allow that '*x* exists' might be made true by something other than *x*, and hence that '*a* exists' might be true according to some theory without *a* being an ontological commitment of that theory. (Cameron 2008b, p. 4)

This then leads into the second feature of the account, which is a commitment to a form of *fundamentality*: our ontology should characterize how the world is at its most fundamental level. Elements of this ontology will be the truth-makers for sentences that mention both these elements and other, non-fundamental, features of the world. Now, it should be noted at this point that this commitment to fundamental elements—generally known in the trade as metaphysical 'simples'—has attracted considerable debate in the relevant literature (see for example Korman 2014 for an introduction). Again, I am going to be relaxed about this commitment. Standardly, it is associated with a rather crude atomistic ontology: the fundamental metaphysical simples out of which ordinary objects are composed are taken to be elementary particles, such as the aforementioned electrons. Elsewhere I have suggested that we need to move to a more sophisticated view according to which the relevant physical 'simples' are features of structure, which of course are not so 'simple' after all (French 2014; French forthcoming; see also Tahko 2018)!

In order to keep it clear which kinds of elements we are referring to—fundamental or non-fundamental—a distinction is also typically drawn between two different

[7] As a structural realist I would argue that those features also include the relevant symmetry structure and, relatedly, the spin-statistics theorem that dictates that fermions have half-integral spin.

types of statement: those of everyday English (or whatever language one speaks), and those that describe the world at its most fundamental level (Cameron 2008a pp. 300–1).[8] Statements in the latter look like statements of English,[9] so to distinguish them I shall follow standard form and use bold type.

Consider that much used and abused example of the statue (since we're talking about artworks, let us think of Michelangelo's *David*). Now, take the statement, 'there are statues'. In terms of what there is at the most fundamental level of the world, the corresponding statement, namely, '**there are statues**', is false, because *at that fundamental ontological level*, there are no statues, only elementary particles, arranged in a certain way.[10] However, we may still accept that the statement, in everyday English, 'there are statues' is true, *but* not in virtue of the fact that **there are statues**, rather in virtue of the fact that there are elements of our fundamental ontology that are arranged 'statue shaped' (Cameron 2008a, p. 301). Note how neat this device is: it allows us to be minimalists with regard to our fundamental ontology without having to back away from the truth of all the claims and statements we typically make in everyday English. We can continue to maintain that 'There is a table' is true, whilst denying that there are fundamentally, tables, because what makes that statement true is not some element of our ontology called a table, but rather certain features of the world—whether elementary particles arranged 'table-wise' or whatever (for further on how this approach can be extended to the objects of both the 'everyday' and of physics in a structuralist context, see French 2014).[11]

This distinction between statements at the everyday level and those at that of the fundamental, together with Cameronian truth-maker theory, then nicely relieves the tension in our conjunction as follows: M1 and M2 can be held as true,

[8] The language of the latter has been labelled 'Ontologese', which appears to agitate some folk (see the spat between Hirsch 2008 and Sider 2014). But really, nothing here much hangs on this label—if you are unhappy with the distinction between English and Ontologese, well, just calm down and dispense with it (Cameron 2010). All that I need for my purposes is the distinction between statements about those things (where this is to be taken broadly) that we are ontologically committed to and those we are not. Thus, it is a simple matter to translate the analysis to come so that the English/Ontologese distinction disappears, and is replaced by a distinction between truths and truth-makers, if that's what you prefer.

[9] They don't have to be—I could choose another language, such as Portuguese, or make up my own but, leaving aside whether something other than our everyday language would better describe our metaphysical commitments, that would be hugely unhelpful, as the aim is to describe those fundamental features which make true statements in English.

[10] Again, if you are uncomfortable with this talk of 'fundamental levels' then just think of the distinction in terms of where the relevant ontological commitments lie (see also Dorr 2007).

[11] Thus, the (eliminativist) ontic structural realist insists that '**there are particles**' is false, because at the most fundamental level there are no particles qua objects, only structures, whereas 'there are particles' is true, in virtue of the fact that there are structures 'arranged particle shaped' (that is, 'shaped' via the relevant symmetry groups for example; see French 2014; French and Ladyman 2011).

but we can take M3 to be false.[12] And we can do this because 'the way the world is fundamentally results in there being truths of English that proclaim the coming to be of certain abstract objects' (ibid.).[13] To put it another way, we can acknowledge that it is entrenched in our linguistic community to say that certain abstract objects (such as musical works) come into existence, and on those grounds we can assert that M3 is false. Thus, we can accommodate the claim that musical works are created, and not in a Doddsian sense of coming to be perceived from a certain position in music 'space', but in the more standard sense of coming into being.

However, what is *true* is the corresponding statement in the language of the fundamental, namely **M3: abstract objects cannot be created**, because *fundamentally* speaking abstract objects cannot be created (regardless of how native speakers of English are inclined to talk). And there is no paradox here either, because we needn't be motivated to believe the truth of the bold counterparts to M1 and M2. **M1** and **M2** are false because **there are no things that are musical works.** It is only M1 and M2 that are true. M2 is true by virtue of the fact that, according to Cameron, there are 'eternally existing abstract sound structures' which 'get indicated by composers, who lay down instructions for their performance' (ibid., p. 306; I shall address worries about the introduction of such structures below). And M1 is true by virtue of the fact that 'the objects that perform the role of musical works are abstract objects: they are sound structures' (ibid., p. 309). It is these abstract sound structures, of course, that function as the 'simples' that make true statements such as 'there are musical works'.[14]

Now, I am not going to adopt all aspects of Cameron's account. All I need is the distinction that allows us to say that '*a* exists' is true, in (colloquial) English, without having to be ontologically committed to *a*. What *makes* '*a* exists' true is the existence of other things that we do want to be ontologically committed to, *in some sense*. For Cameron these 'other things'—the truth-makers—must necessarily be at the fundamental level: they must be metaphysical 'simples' such as 'abstract sound structures', electrons, quantum fields, etc. Below, I shall extend this approach to incorporate those features that make true statements about theories.

[12] If one were tempted to adopt a reificationist form of the Semantic Approach and identify theories with mathematical structures, it is worth noting that Cameron applies his approach to mathematical ontology as well (2008b).

[13] Actually, there isn't such a motivation to resolve paradoxes in English, according to Cameron. In Cameron 2008b he argues that there can be true contradictions in English, even though fundamental reality is consistent. The logic of English should be paraconsistent, even if the logic of the language of the fundamental level (Ontologese) is classical. Azzouni (2007) takes a similar view. What really matters is that we don't have conflicts in our beliefs about how the world is, fundamentally.

[14] Likewise, and again, what makes the statement 'tables exist' true is *not* that '**tables exist**'. '**Tables exist**' is false because tables aren't part of the fundamental ontology on this view. For Cameron's defence of this particular consequence of his view, see Cameron 2008b, p. 6. Note that by introducing abstract sound structures, Cameron has already extended the set of features that count as truth-makers. In French 2014 it is extended still further and, as we shall see, I shall continue this trajectory below.

Cutting the Knot

Again we recall that a similar tension can be generated with regard to *scientific theories* and *scientific models*: Again, consider the following three statements:

S1: Scientific theories (and models) are abstract objects.
S2: Scientific theories (and models) are created.
S3: Abstract objects cannot be created.

As we have seen in some detail, we can run versions of the above positions in this context, with similar concerns that arise. Thus, we can deny S2 and take theories and models to 'exist' in some theory 'space' such as Popper's World 3. Leaving aside the problematic motivations for taking theories to be 'real', we face problems of access to such objects and of accommodating the heuristic features of scientific practice. Adopting Dodd's approach to artworks is of little help, as again it is unclear how these features effectively structure this theory space and the notion of having a 'clear conception' remains, ironically, obscure.

Or we can deny S1 as Collingwood does and adopt the radical view that what's going on when we read Einstein's paper or attend a scientific presentation is that we are reconstructing, more or less intelligently, the theory of special relativity. Alternatively, we can adopt the currently popular view of taking theories and models to be fictions, as discussed in Chapter 6 but again there is the problem of accommodating scientific practices. Finally we can back away from the tension in its entirety by insisting that it presumes stances that we should decline to engage with, an approach that we rejected as ill motivated and unhelpful in dealing with the kinds of questions we face as philosophers of science.

Instead we can dissolve the tension by eliminating the cause, namely the assumption that theories are 'things', whether abstract, fictional, or whatever. Obviously we don't want such elimination to force a revision of our language—we still want to be able to talk *about* theories and models.[15] Fortunately Cameron's truth-maker device allows us to do just that.[16]

Let us recall: All I need from Cameron's account is the distinction that allows us to say that '*a* exists' is true, in (colloquial) English, without having to be

[15] One of the readers queried why we would want to do that. But the point is, 'we' have no choice! Scientists, as well as layfolk and philosophers, talk 'about' theories all the time of course and if we want to maintain that some, at least, of that talk is true without being committed to theories themselves, then we need to adopt some such manoeuvre as Cameron's. We could, alternatively, demand that everyone drop such talk and use only the equivalent of Ontologese but that's a pretty severe line to take.

[16] Vickers (2014) also advocates a form of theory eliminativism as a methodological tool to resolve disagreements over whether for example a given theory (such as Bohr's theory of the atom or classical electromagnetism) should be regarded as inconsistent or not. Here my concern is less to do with methodology and more with *ontology*.

ontologically committed to *a*. The statement '*a* exists' is made true by the existence of other things that we do want to commit to. This will allow me to dissolve the tension inherent in S1–S3 and reduce our ontological commitments with regard to theories, models, etc., without sacrificing our 'talk' about those theories and models and their various features. Again, for clarity's sake, when it comes to that talk, I'll maintain the distinction between statements at the level of the 'everyday' and those at the fundamental level in this context, with statements in bold type taken to be about what (actually) exists when we come to consider the status of theories, models, and so forth.

So, to begin, S3 is the same as M3, hence we can draw directly on Cameron's argument that M3 is false. However, **M3** is true, because abstract objects, as standardly conceived, are causally inert and if creation requires causation, then they cannot be created (unless you adopt Popper's or Thomasson's views, in which case, you owe us an account of that creative relationship). What about the bold counterparts of S1, S2, and S3? If **S3** is true, then we need to be sure that **S1** and **S2** are not also true—this would just put us straight back face to face with the problem, only now expressed in the language of the fundamental level instead of English.

We can follow Cameron again here, saying that **S1** and **S2** are false because, fundamentally, **there are no things that are scientific theories**. In other words, **S1** and **S2** are false because they fail to refer. Since we are not ontologically committed to the existence of theories, it is false to say that theories are created or are abstract objects.[17] What *does* exist—what does the truth-maker work for theories—will be presented and discussed shortly.

Now, here is the absolutely crucial point: we *can* still talk about theories, their inter-relationships, empirical adequacy, simplicity, indeed their various qualities and so forth, *in English*. We can continue to discuss these qualities, write long essays (or even books) on them, argue (incessantly) about them, and so on. We can maintain all our usual talk about theories *but* what makes this talk true (or false) are not the entities directly referred to in it (namely the theories) but rather certain elements (still to be presented) that exist at the fundamental level, to which we *are* ontologically committed.[18] Thus, we retain all the advantages of being able to talk about theories in English, but we avoid having to inflate our ontology unnecessarily.[19]

[17] Recall: what I am insisting on here is the distinction between statements about those sorts of things to which we are ontologically committed and those to which we are not. However you want to label it, the language of '**there are no things that are scientific theories**' is that of the former.

[18] Another of the readers criticized this proposal on the grounds that it does what the logical positivists tried to do with regard to scientists' talk of unobservables, namely translate that talk into something else. That is not what is being proposed at all! The clue is in the phrase 'We can retain all our usual talk...' and the core idea is to ground the *truth* of such talk in the relevant practices.

[19] Of course, someone could still insist that we want to understand *why* claims about theories are true or false in English, in the sense that the truth-makers are describable in English, without having to introduce a language of the fundamental level ('Ontologese' or whatever), but this would be to beg the question. I am grateful to one of the referees of the original paper on which this chapter is based for pressing this point.

And so our dilemma is resolved. When stated in English it dissolves because S3 is false, and when stated in the appropriate language of ontological commitment it is resolved because **S1** and **S2** are false. Of course, this resolution comes with its own cost, namely accepting a distinction between statements in English and statements in whatever language one takes to refer directly to the elements to which one is ontologically committed. Everything comes at some cost, of course, but I think the costs here are lower than those associated with the alternatives. Furthermore, this distinction between the everyday and the fundamental is not exactly unfamiliar: many would maintain that what fundamentally exist are quantum fields, say, or strings, branes, or whatever, but that we can still utter true (and false) statements about tables, people, and so on. Nevertheless, I appreciate that this approach to the ontological status of scientific theories throws up some further questions that need to be addressed. First, however, I will set out what I take to be the truth-makers in this context, or the things that *do*, fundamentally, exist.

Bring on the Truth-Makers

As we saw above, Cameron himself suggests that what 'perform the role of musical works' on his account are 'eternally existing abstract sound structures' that are 'indicated' by composers (Cameron 2008a, p. 309). We could export this suggestion across to the philosophy of science and choose the kinds of abstract entities that are taken to exist in Popper–Thomasson land as the relevant truth-makers for the statements made in everyday English about *theories*. But we've already run through the costs associated with such a move, including, notably, ontological inflation and concerns about discovery/creation.

What else, then, would make S1 and S2 true without reintroducing abstract entities? A useful example here is that of the Supreme Court (whether US or UK). It looks like this is an abstract object (speaking in English), simply because there don't seem to be any good concrete candidates to identify it with. But, as Cameron insists, '[w]hat makes it true that there is a Supreme Court is... simply that there are various people performing a certain role' (ibid., p. 311). And people and their behaviour are, of course, *concrete* things: in particular, and obviously, they are *not* 'eternally existing', as abstract objects are. So here we have a clear case of something that might be regarded as abstract—namely the Supreme Court—(*speaking in English*),[20] but where the truth-makers for it are *not* abstract. And there doesn't seem to be anything to stop us applying this approach to scientific theories: scientific

[20] You could regard the Supreme Court as concrete (Cameron, personal communication) since for example you could assassinate all its members (heaven forfend) and state 'I just killed the Supreme Court'. However, it is unclear whether by eliminating all the members of the Supreme Court, one has eliminated *it*, as an entity.

theories can be regarded as abstract (speaking in English) but whatever is it that 'performs the role of theories' does *not* have to be abstract.

So let us ask the question again—'What makes S1 and S2 true?' Now, we don't have to feel that we are confined in our answer to *abstract* objects. Two candidates for the truth-makers have already been mentioned: the *thoughts* of individuals in the relevant scientific community and scientific *practices*.

Let us consider each of these options in turn.

Option 1: Thoughts

Here what matters vis-à-vis theory talk are the mental *representations* (about a certain domain of phenomena) common to individuals in the relevant scientific community.[21] However, the difficulties sketched previously now do not arise, since we are no longer *identifying* theories with these thoughts: fundamentally speaking, theories do not exist.[22] So we don't have to follow Collingwood in regarding theories, or artworks for that matter, as ideal, mental entities, in whatever sense; nor do we have to fret over whether in listening to a scientific presentation or reading a textbook we are reproducing or reconstructing the same theory or a kind of simulacrum.

But theory *talk* can be assessed by looking to what is going on in the minds of those in the relevant scientific community. One can then understand the truth of statements such as, 'Quantum mechanics is an elegant theory' as given in terms of some aesthetic element associated with the relevant mental phenomena. Likewise the statement, 'According to standard QM, undisturbed systems obey a linear dynamic' is made true by an understanding of the statement and its component elements that is shared by relevant scientists. Of course it may be that this statement turns out *not* to be accepted by everybody in the relevant scientific community, or even that the contrary claim is accepted.[23] Such cognitive facts may then in turn support claims to the effect that 'QM is not committed to a linear dynamics'. The issue as to whether QM, as a theory, should be considered broad enough to encompass linear and non-linear dynamics or whether the introduction of forms of the latter (for undisturbed systems) eventually contribute toward the introduction of an entirely new theory again takes us to the further issue what would be the identity conditions for QM, or theories in general. The

[21] I'm again relying on the input of Pete Vickers as presented in French and Vickers 2013.

[22] Of course, since the Time of Frege, at least, there have been arguments against the identification of abstract objects with collections of psychological objects. But again, it is important to appreciate that we are not *identifying* theories with such objects or collections thereof; rather the latter are being invoked as truth-makers for sentences about such theories.

[23] There is, indeed, an ongoing, if perhaps rather marginal, research programme in non-linear quantum mechanics; see for example, the papers in Aerts, Durt, and Czachor 2002.

approach being set out here allows us to sidestep this issue since, fundamentally, there are no theories, qua objects, and hence no identity conditions to worry about!

Of course, this is still a tricky line to take in some respects, not least because there are difficult questions to ask in the philosophy of mind. How exactly should we understand these 'mental representations' that do the truth-making work? What is the relationship between a mental representation in one mind and that in another? Can these ever truly be exactly the same? If not, how should we think about the *similarity* of one mental representation to another? These questions lead into variants of the concerns raised with regard to Collingwood's stance on artworks: if you and I have different relevant mental representations, can they both be taken as truth-makers of statements such as the above? Or should they be taken as exemplifications of the same truth-maker? In which case, how similar do they have to be for that to be the case? Or should they be taken as different truth-makers entirely, so that there may be as many truth-makers for the statement as there are minds thinking about it?

Perhaps it might seem plausible to say that in such cases there is only one truth-maker for a statement like 'According to standard QM, undisturbed systems obey a linear dynamic', namely a complicated network of overlapping mental representations in the minds of individuals in the relevant scientific (sub-) community. This might help assuage the concern that given how complicated theories can be, and how they involve abstract ideas, and, often, complicated mathematics, nobody can hold such a scientific theory in their own head.[24] Alternatively, we could follow Giere (2002), who argues that even multiplying together 456 and 789 usually means 'creating and manipulating *external representations*' (ibid., p. 288, original emphasis). In other words, we have to start writing things down to work through the problem. Similarly, it might be claimed, theories cannot be conceived unless we start writing things down to assist our minds.[25] Giere goes on to recommend that we think in terms of *distributed cognitive systems*, essentially a combination of the internal (whether in a single individual or in many) and the external. We could then adapt this to suggest that only those truth-makers that relate to fragments of a theory exist within any one mind at any one time.

But of course (and here's the kicker!), as truth-makers, these mental representations are still inter-subjectively inaccessible—or at least, they are as long as we keep the resultant speech, behaviour, and, crucially, *practices* out of the picture. But if we include those, we move to a very different picture, as we'll now see.

[24] It is sometimes said to be a truism of cognitive science that every mind differs from every other, because every brain differs from every other. If scientific theories are described in terms of 'configurations of synaptic weights' in the brain (Churchland 1989, ch. 9), then given the neurophysical composition of the human brain, it could hold $10^{100,000,000,000,000}$ different theories, or, in the terms outlined here, their truth-makers.

[25] And here one might draw on some of the arguments associated with 'extended cognition', for example (see for example, Clark and Chalmers 1998).

Option 2: Practices

The alternative option that I want to explore takes the relevant truth-makers—that is, all that *really* exists in this context—to be the complex set of practices of the scientific community: the writing and dissemination of articles, the performance of experiments, the kinds of heuristic moves already mentioned, and so on.[26] Now, there is of course a huge literature, particularly in sociology, on the nature of practices, which can be described quite generally as 'arrays of human activity' (Schatzki 2001, p. 11).[27] Such arrays are of course governed by certain norms and there is again a rich literature on such norms in scientific practice (the classic kick-off point is Merton 1942). One might worry[28] that, having eliminated an ontology of theories and models, I've left in place an ontology of norms. But however we regard such norms, whether as real entities 'that must manifest themselves in certain empirical facts or processes', or as ideal entities 'that have a propositional content expressing some normative aims or claims' (Koller 2014, p. 156), they will be present whatever stance we take towards theories. And my aim here is only to offer a minimalist account of the latter. Neither shall I pursue the nature or, in particular, ontology of the relevant practices any further, for a similar reason. All I need for my purposes is their ability to act as truth-makers for statements putatively about theories. So I shall take them to be either concrete entities or reducible to such and within this particular context ontological commitment to such practices can be taken at face value. These practices will then collectively act as the truth-makers for our theory talk.

But please note again: this is not to *identify* a theory with the set of actions that are taken with regard to it, or practices associated with it. That would be to say that a theory *just is* the writing of articles, the carrying out of experiments, and so on. However, if a theory were to be identified with *individual* practices, an obvious worry would be which practices, or which collection of practices, should we pick? We could adopt a relaxed stance on this issue and for example suggest that it is a matter of convention where we draw the line, as it were. Alternatively we could identify theories with certain *types* of practice, but then, of course, we are back with identifying theories as abstract objects.

Neither is the view I am proposing the same as that of Cartwright and Suárez for example who write that 'theorising *is constituted by* a heterogeneous mixture of journal articles, textbooks, lectures, PhD seminars, practices, techniques,

[26] We recall that according to Schiffer, we know about propositions by participating in the relevant linguistic and conceptual practices and it is these that make true our proposition talk. Thus, if one were keen to maintain the view that the best way of characterizing theories is via sets of propositions, one could still adopt the kind of deflationary view on offer here.

[27] One can characterize them further as embodied, materially mediated arrays of human activity (Schatzki 2001, p. 11) but I'll stick with the broader understanding here.

[28] As a reader did.

explicit explanations, implicit half-formed understandings and the like' (Suárez and Cartwright 2008, p. 79). Here it is 'theorizing' that is taken to be constituted by practice and if that is understood as distinct from the creation of theories, qua *objects*, then to some extent it bears comparison with what I am suggesting here. Cartwright and Suárez's arguments for such a position are entirely different, however. Furthermore, their view is associated with an instrumentalist view of theories that sees them as 'tools', whereas I do not view them 'as' anything. This is the crucial difference: there *are* no theories on my view—just theory talk whose truth-makers are the sorts of things indicated above. Recalling the statue example, one might sloganize this view as follows: there are no theories, just 'theory-shaped bits of practice'!

Here again we can discern a difference between this form of eliminativism and the kind of fictionalism that maintains that talk about numbers, say, really amounts to talk about concrete objects. On that sort of view we need 'bridge principles' that relate the two domains and that give us rules for determining why it is appropriate to say '2 + 2 = 4' for example but not '2 + 2 = 5', even if there are no numbers. But this is not my view. I am not saying that talk about theories really amounts to talk about practices—again, the claim is that talk about theories is *made true* by the practices. Thus, there is no need for bridge principles; or, better, the relevant rules that link theory talk to practice are simply those pertaining to the notion of truth! Of course, we might still wonder when, or indeed whether, it is appropriate to say that 'QM is a very elegant theory' and here opinions may differ, depending on how we regard the relevant practices (I shall return to such claims regarding the aesthetic features of theories shortly). However, when it comes to a statement such as 'According to standard QM, undisturbed systems obey a linear dynamic', the reasons why it is appropriate to say that and not 'According to standard QM, undisturbed systems obey a non-linear dynamic' have to do with the practices associated with using Schrödinger's equation and so forth. Those practices make the former statement true.

But what about the claim that 'QM is empirically adequate'? This last can be understood in terms of the embedding of empirical sub-structures into theoretical structures, according to the Semantic Approach. But it is not the embedding itself that makes true that claim, since that would presuppose there is an entity—the theory—into whose structure the empirical sub-structures are embedded. Rather, it is made true by a complex nexus of practices involving the obtaining of predictions or the identification of experimental consequences more generally, the testing of such consequences through experiments, and so forth. These practices result in datasets for example and underpin models of that data, and of the experiment, and so on, as Suppes famously described and all of these, or the nexus they constitute, can then be represented (by *us*, philosophers of science) set-theoretically as empirical sub-structures. However, when it comes to what functions as the relevant truth-maker, we should not confuse that representation with the practices themselves. I'll come back to this issue of representations and what we are doing as philosophers of science (or commentators on science, more generally) in the final chapter.

Now, consider again the statement 'According to standard QM, undisturbed systems obey a linear dynamic'. First of all, note that I am not asking what makes true the claim that 'undisturbed systems obey a linear dynamic'—that's straightforward for the realist: it is certain features of the world. (And, again, I'll come back to realism in Chapter 9.) What I am asking to be considered is the claim, '*According to QM*...'—or, in other words, the claim that it is a feature of the theory that it states that undisturbed systems obey a linear dynamic. And of course, it is not as if we can pick up the theory, like an object, scrutinize it and conclude 'Yup. There's that linear dynamics alright...'! Rather, again, to determine if the claim is true, we draw on certain practices. These can be quite minimal in this case, as in simply opening a (trusted) textbook, noting Schrödinger's equation, and concluding, 'Yup. That's linear right enough'. Less shallowly, we might reflect on what is meant by 'undisturbed' and 'linear', perhaps taking the appropriate definitions of these terms and demonstrate the truth of the claim ourselves, working through any mathematics involved, where such 'working through' itself constitutes a form of practice, of course. More likely, we would pick up the textbook again and look for the relevant statements supporting such a claim, these statements being the product of, and thus encoding, in a sense, the relevant practice.[29]

The Aesthetics of Theories

Finally, let us return to 'QM is an elegant theory'. Scientists themselves often make statements such as this that appear to assign aesthetic values to theories (for a useful selection of quotes from scientists extolling such values, see Ivanova 2017). Furthermore, it has been argued that such features as beauty and elegance play a role in not only the 'discovery' of a theory—that is, as a heuristic factor—but also with regard to the *justification* for choosing the theory (ibid.). Such claims are contentious, not least because they demand explication with regard to the purported relationship between beauty or elegance and truth (ibid.).[30] Here I am not

[29] We should acknowledge the 'epistemic dependence' emphasized by Hardwig (1985) for example. Indeed, this is exemplified in modern science itself, with its large research teams and highly specialized skill sets (see the discussion in da Costa and French 2003, p. 76). But that the relevant practices are undertaken or performed by others, specialists in their field, does not, of course, undermine the point that it is these that act as truth-makers.

[30] In this regard Ivanova (2017) makes an intriguing suggestion that exports certain results at the intersection of aesthetics and psychology into the philosophy of science: certain studies have shown that exposure to so-called 'bad' art is not correlated with an increase in the subjects' aesthetic appreciation, leading to the conclusion that some feature over and above such exposure must be responsible for the subject's aesthetic responses to artworks (Meskin, Phelan, Moore, and Kieran 2013). She

so interested in that issue, but rather in the implication that statements such as the above seem to refer directly to features of the theory *as an entity*.

To approach this issue, let us briefly consider how similar claims about artworks are treated in the philosophy of art. Setting things out in broad terms, there is a well-known split between those who argue that aesthetic qualities such as 'elegance' supervene on non-aesthetic properties of the artwork and those who deny this and argue for example that context plays an important role in the assignment of such qualities (for a useful overview of the debate, see Hick 2012; also Sibley 2001). Consider for example Turner's painting 'Thomson's Aeolian Harp' (http://www.tate.org.uk/art/artworks/turner-thomsons-aeolian-harp-tw0579), described as a 'truly elegant painting' (Ziff 1980). On the one hand, the property of elegance might be taken to supervene on various features of the painting, both structural and non-structural, such as the balance of shade and light between the land and the sky, the sweep of the river in the centre, which draws the eye to the figures on the right, and so on.

On the other hand, it could be argued that such an ascription can only be made in a certain context—in this case that which embraces our current aesthetic sensibilities and our understanding of Turner and *his* context. Present the painting to a devotee of the Bauhaus school and they might twist Constable's praise of Claude (who inspired Turner at this stage of his career) and dismiss it as entirely too 'amiable'. Still, we could fold such contextual or non-intrinsic features into the supervenient base and still maintain that the quality of elegance derives, at least in part, from features of the painting, as an art object.[31] In either case, it is the object itself—in this case a painting[32]—that is said to possess the relevant aesthetic quality.

And it would seem that we could reproduce these sorts of moves when it comes to scientific theories and models. Lets take an obvious example first, where the model is a physical object, namely the Crick and Watson model of DNA. Here the physical entity might also be said to possess a certain elegance, in exactly

suggests that if such results could be reproduced with regard to scientific theories (and models), we would have grounds for similarly concluding that exposure and habituation cannot solely be responsible for scientists' appreciation of such aesthetic features.

[31] Hick himself argues for the view that 'an object's aesthetic properties just *are* its realized powers to produce aesthetic effects of particular kinds in suitable perceivers under suitable conditions' (Hick 2012, p. 314). Leaving aside the issue of whether artworks can really be said to possess 'powers', in a metaphysically robust sense, as Hick notes this raises questions about which perceivers? And under what conditions? Interestingly, however, he acknowledges that scientific and mathematical formulae may be accredited with aesthetic properties and, as with everyday or natural objects, also so accredited, it would obviously be inappropriate to regard such properties as dependent on certain artistic categorizations or art-history contextualization.

[32] Aesthetic qualities are also attributed to musical works, of course. However, here we might want to pay attention to the distinction between attributing elegance, say, to the work itself—conceived as an abstract artefact or whatever—and the performance. Thus, one of Bach's toccatas might be described as 'having a tendency to ramble' yet as played elegantly by a certain pianist (see Distler 2015).

the same aesthetically informed sense as a sculpture, say. (It might also be said to be elegant in a scientifically informed sense, but I'll come to that shortly.) And we can obviously extend this to scientific drawings, pictures, diagrams, and so forth; consider the 'Beautiful Science' exhibition held at the British Library, for example (http://www.bl.uk/reshelp/experthelp/science/inspiringscience/2014/beautifulscience.html)—who could deny that NASA's visualization of the ocean currents is beautiful and aesthetically pleasing?

But what about non-physical models and theories? Take the example of Newtonian mechanics: it might well be argued that this is an elegant theory, where this quality can be attributed to the internal structure of the theory,[33] or its simplicity, or some combination of these and other features (just as with Turner's painting); or it might be insisted that such an attribution is contextual—so, some might argue that Newton's original geometric presentation of his theory in the *Principia* is not elegant at all (see http://cudl.lib.cam.ac.uk/view/PR-ADV-B-00039-00002/1) and that it only acquires this quality when formulated in the modern notation (due to Euler; see Stan, forthcoming). Relatedly, it is well known that expressing Newton's equations in polar coordinates, say, can be clumsy and tedious, and that the Lagrangian formulation, which is independent of the coordinates, offers a particularly elegant presentation of the theory.[34]

Exporting such moves from aesthetics to the philosophy of science might suggest that the truth-maker of our claim 'QM is a very elegant theory' is the quality of elegance as possessed by the theory qua object, just as in the case of the Turner painting. But this is to move too fast. First of all, note that in the (non-physical) examples above we are talking about different formulations and presentations; indeed, the attempt to construct a parallel form of contextualization makes this clear. So we can press this point and insist that whatever quality scientific elegance is or consists in, it is attributed not to 'the theory', as an object, whether abstract or not, but to a set of symbols laid down in a certain order, whether on paper or a whiteboard or whatever. And to infer from this that such qualities can also be attributed to a theory, because such formulations are *of* the latter, is to beg the question, of course! Thus, we can argue that all that the above examples show is that a set of symbols in a certain order possesses aesthetic 'elegance', say, and that insofar as this is what we mean when we utter the above assertion, then it should be taken as shorthand for 'QM as expressed via a certain set of symbols is a very elegant theory'. And it is no objection to my account to say that the truth-maker of 'That expression of Newton's theory is elegant' is a certain aesthetic quality possessed by a set of symbols set down in a particular order.

[33] There may be a distinction to be made between the attribution of aesthetic features to the relevant *mathematics* and to the theory 'as a whole' (for examples of aesthetic responses to mathematics, see again Ivanova 2017).

[34] For a discussion of the virtues of the Lagrangian formulation over the Hamiltonian formulation, see Curiel 2014. I'll come back to this example.

Alternatively, we might insist either that it is downright inappropriate to attribute aesthetic qualities to scientific theories and models or, at the very least, that what is typically meant by 'elegant' here is not what is meant in the case of paintings or musical works. So, a typical way of cashing out what we mean by elegance in the scientific context is via a combination of parsimony and power, for example (for an overview of such reductive moves, see Ivanova 2017).[35] With regard to the former, this can be explicated in terms of *qualitative* parsimony, in the sense that a theory with fewer *kinds* of entities might be preferred over its rivals and *quantitative* parsimony, in which preference is given to the theory that minimizes the *number* of entities postulated. Not only are such preferences evidenced in the history of science (classic examples that are used include the postulation of the neutrino in beta-decay, Avogadro's laws, and the positing of the existence of Neptune) but epistemic justification can be given for them: for non-parsimonious theories to appropriately entail the evidence, certain additional costly (and context dependent) assumptions must be introduced that their parsimonious competitors can avoid (Jansson and Tallant, 2017). Thus, the truth-maker of 'Theory *T* is more parsimonious than theory *T**' will be found in the practices we undertake in showing how *T* and *T** entail the evidence with those of the former being less 'costly' (less complex, having fewer additional assumptions, etc.) than the latter.

What about power? This can be understood in various ways but typically as unificatory or explanatory.[36] Thus, we could regard the theory that expresses or includes the conservation of momentum for all systems as more powerful and hence as more 'elegant' than a theory that incorporates conservation of momentum for elastic processes only (Post 1960, p. 35). In such cases, the truth-maker for 'Theory *T* is more powerful than theory *T**' can be associated with the practices we engage in when we (apparently) apply *T* and *T** to phenomena, where the range of phenomena will be wider in T's case and hence the number and kinds of practices (theoretical and experimental) will be different. Likewise when it comes to explanatory power—insofar as *T* offers a better explanation than *T** that will be manifested in the different practices involved in each case (involving fewer, simpler, etc. deductions for example on the Deductive-Nomological view of explanation, or featuring fewer, again, or different difference-makers according to the 'kairetic' account; see Strevens 2008).[37]

The upshot then is as follows: if what we mean by 'QM is a very elegant theory' is just that a certain expression 'of' the theory possesses certain aesthetic qualities,

[35] Cf. Post's distinction in a much-neglected paper, between 'linguistic' and 'semantic' simplicity (Post 1960), where the former refers to the kind of simplicity in or elegance of formulation already discussed, and the latter covers unificatory and explanatory power.

[36] And on the view that explanation fundamentally consists in unification then the latter will just collapse into the former.

[37] Balancing parsimony and power may be tricky of course (see Woodward 2014).

then insofar as that expression is not the theory, nor is it *of* the theory in any metaphysical serious sense, then this is straightforwardly accommodated. But if we mean something scientifically 'deeper' by elegance, then insofar as that can be understood in terms of some combination of parsimony and power, the statement is made true by the relevant practices, involving for example the ease of deduction of certain (written, typed, scrawled, . . .) statements from the axioms or fundamental claims of the theory, the way in which a wide variety of claims (both theoretical and empirical) can be obtained from these axioms and so on (the details of which aspects of practice will act as truth-makers will of course depend on precisely how one understands both parsimony and power, as the different accounts of explanation illustrate).[38]

Having set out how the truth of statements that appear to be about theories may be grounded in sets of practices, taken as the relevant truth-makers, let us run through some obvious concerns that might be raised.

Concerns

The first worry you might have is that this approach may not seem to accord with how we, or more pertinently, scientists, use, refer to, describe, and so on, theories. When quantum theory is referred to in the latest issue of *Nature* for example that referral does not appear to bring in train a whole complex set of practices but rather appears to denote a single entity. But of course, appearances can be deceptive! As we'll see in the next chapter, what we take to be 'quantum theory' and how we delineate quantum physics from classical physics are deeply problematic issues—indeed according to one analysis of the issue of when two 'formulations' of a theory are equivalent or not, there are as many different theories as there are interpretations, and as many of *those* as there are metaphysical positions on the nature of objects, properties, laws, and so on. I shall come back to this issue in Chapter 8.

The point is the name 'quantum theory' strictly speaking, on this view, does not denote anything, neither some abstract entity, whether eternal or artefactual, nor some mental representation. But the statements in which that name appears are made true (or not) by the whole set of practices, just as those regarding 'The Supreme Court' are by those of the set of five individuals (say). On this view, although it remains sensible to talk of theories as abstract, or artefactual, or as mental representations *in English*, all that really exist are concrete practices, and it is the failure to appropriately distinguish our talk between colloquial English and

[38] And again, with all these statements, I am not talking about the truth or falsity of claims the theory makes about the *world*, but rather the truth or falsity of claims made about the *theory*.

the more restricted language that expresses our ontological commitments to such practices[39] that generates the tension.[40]

The second concern is that even if the relevant scientific community were to stop writing articles, carrying out experiments, presenting their findings, etc., we might still want to say that a given scientific theory 'existed'. Now, we've been here before of course and the insertion of scare quotes indicates that we need to be a little careful with how we understand the term. Certainly, on the view being defended, theories don't exist, have never existed as such, and hence will not continue to exist (or stop existing) were science to come to an end. But it does seem right to say that it would be appropriate to *talk* of theories as 'existing', in English, even if all of the relevant practices were to be halted. But how can it be appropriate, if the truth-makers for theory talk *are* the practices?

An obvious response might be to say that we could still talk about Einstein's theory, say, but that what would make that talk true would be a set of past and no longer evolving practices. We could still write about and describe that theory and, in that sense, our talk, in English, of the theory as 'existing' or as having various qualities etc. would still be apt, because that talk would be made true or false by the relevant past practices and associated discussions etc. which can still be invoked if necessary. And of course it is not at all strange to say things about any theory *now*, which statements are made true because of what happened in the *past*, as historians do this all the time! Furthermore, one could insist that as long as the relevant books, papers, etc. continue to exist, then the reading of such books, the working through of examples, and so on constitute a set of practices sufficient to constitute truth-makers for the relevant claims.

But now, what if (trigger warning: doomsday scenario!) not only the human race but all sentience across the universe were to cease to exist, and all traces of our existence, our practices, etc. were wiped out, so that there was no longer even the possibility of ever invoking these sets of practices, and indeed, no traces of past practices to mediate that invocation? In that bleak context, there would be nothing to now make true statements about theories presented in the past but that seems to me to be a not particularly problematic consequence, for obvious reasons. Indeed, one might worry that there is some question-begging going on here in that the sense of appropriateness above depends on blurring the English/Ontologese distinction and taking 'general relativity' to be referring to something that exists over and above the relevant set of practices.

Nevertheless, a niggling concern might still remain and the main thrust of this objection could be pushed a little further. Suppose a scientist, faced with an

[39] That is, a form of 'Ontologese' appropriate in this context.

[40] And if one is uncomfortable with this idea of names like 'general relativity' being strictly denotationless, one could maintain that the term refers to the theory of general relativity but insist that having a singular term of English occur in a true sentence of English doesn't suffice to bring ontological commitment to the denotation of that singular term. Thanks to Ross Cameron for pointing this out.

unusual or unexpected phenomenon, were to suddenly think up a theory to explain it. At that precise moment it might seem perfectly natural to say that the theory exists, but there haven't *been* any relevant practices that could make this true. Taking *future* practices as the relevant truth-makers seems to be pushing the relationship between our language and the world a bit too far. Furthermore, suppose that the scientist unfortunately dies before she manages to pass on her theory. To say 'the theory existed' is true in virtue of the practices that *would* have happened if she had lived leaves it completely undetermined what those practices could be and hence seems unpalatable, as does the alternative claim, to say, in English, that the theory never in fact *did* exist.

Of course, one straightforward response is simply to deny that scientists 'suddenly' think up theories and to insist that there is always a heuristic context in which certain moves are made that in turn 'bleeds over' into what has been called theory 'pursuit'. In that case, the claim would be, there are always sufficient bits of practice to function as the truth-makers of the relevant claims, particularly as the latter will be of the form 'this theory is worth pursuing' or 'this theory shows promise', as they will be, since, by assumption, the theory has not yet been thrown to the justificatory wolves. And if it were to be insisted that our putative scientist was never involved in any such heuristic moves and that the theory really did just 'spring' into her mind (as has indeed been claimed about certain scientific theories), we might be tempted to insist back that in that case what we are dealing with is not actually a 'theory' but just a speculative thought.

The strength one attributes to the above objection may depend on what one takes to count as a 'theory'.[41] Perhaps some would want to insist that, in the previous example, the scientist really *did* have a 'theory', and not merely a 'speculative thought', and that grounding the truth of claims about theories in practices thus comes at too high a cost. But notice: we are envisaging a circumstance in which she comes up with a theory, 'in her head' as it were, and makes no claims about it, puts forward no statements, presents no findings, etc. In that case, there are indeed no practices to act as truth-makers but there are no statements to act as truth-bearers either!

However, you might insist, our scientist friend still 'has' the theory 'in' her head, so doesn't this demonstrate that theories can still be said to exist in the absence of the relevant practices? Setting aside any question-begging going on here, the obvious move is to suggest that insofar as the scientist is thinking about

[41] There is more to say about evaluating the counterfactual claims here and the truth of the assertion 'had there been no human practices and thoughts, there would have been no scientific theories' may depend on our views of such evaluation. Thus, if it is accepted that judgements about such possible worlds are made from this, the actual, world, then perhaps we can maintain that we have actual truth-makers for the counterfactual claim that there would still have been theories, even if thinking about that possible world *as actual* yields the conclusion that there are no theories there. Again, I'd like to thank Ross Cameron for suggesting this move.

the theory, or could be thinking things about it, these 'internally voiced' statements could be regarded as the relevant truth-bearers, with the mental thought processes that lead to the construction and justification of the theory functioning as the associated truth-makers. Again, of course, we have the problem of the intersubjective accessibility of such practices but honestly, insofar as she never publicly talks about her theory, never even mentions it and takes it to her grave, I don't think this is a particularly worrisome scenario.

Finally, what about Popper's argument that theories can sometimes surprise us, just as physical objects can, and hence should also be taken as real? Let's consider his example: 'the deduction of $E=mc^2$ from special relativity was surprising'. How can the relevant truth-makers, whether our own thoughts or practices, surprise us?

Certainly our own *thoughts* can surprise us if we accept that those thoughts (or their propositional representations) can have consequences we haven't deduced yet. We might reject the claim that consequences are somehow contained within the relevant premises and argue that the element of surprise comes because we have certain beliefs plus certain rules for generating new beliefs therefrom, not because the new beliefs actually already exist and we *discover* them as we make our inferences. We talk *as if* the latter is true, but that doesn't have to mean that it *is* true, ontologically speaking. As I have emphasized, the distinction between English and the language of the fundamental level (Ontologese or whatever) allows us to speak in one way but to have quite different ontological commitments.

What about our practices? Wherein might lie the surprise there? Well, imagine that you are reading, first, Einstein's paper 'On the Electrodynamics of Moving Bodies' from 1905 and then 'Does the Inertia of a Body Depend Upon Its Energy Content?' published later that year. Where does the surprise associated with $E=mc^2$ come from? Here again we might usefully draw on work in the philosophy of art and specifically work on surprise in music, where the standard account holds that musical surprises are due to the thwarting of prior expectations (see for example Huron 2006). Could this account for the surprise felt when your eye lights upon Einstein's famous result? It is, at the very least, unclear that you would have had any prior expectation regarding this result before coming across it.

Judge suggests that there is no need to invoke any future-directed state involving expectation and argues instead that 'Many musical surprises can be explained by the falsification of assessments of the present, rendering the appeal to expectation unnecessary' (Judge 2018, p. 226). The example she gives is of Led Zeppelin's 'Black Dog', in which the beat suddenly changes eleven seconds in and as a result, she writes, 'You are forced to revise your model of what you're presently hearing, and this forced revision is felt as surprise' (ibid., p. 227). Now, talk of models and falsification might lead you to think that this view would mesh nicely with a Popperian philosophy of science! However, the implication that $E=mc^2$ does not, of course, falsify special relativity—on the contrary. More acutely, there does not appear to be anything equivalent to an 'assessment of the present' that is falsified by this result—of course, when reading the 1905 paper one forms a certain assessment but without going into the phenomenology of individual readers' experiences, there still does not seem to be anything to this that could be considered falsified. Even casting things in the more appropriate terms of a model that is revisable does not seem to capture the sense of surprise in this case.

Nor can it be due to the discovery, taken on its own, that a change in energy is related to a change in mass. This was already well known, although it had been established only with regard to electrostatic fields (which contribute to the mass of a charged object) and not in the above form (see Janssen et al. 2007). Granted that Einstein himself explicitly acknowledged that he obtained the result from Maxwell's electrodynamics and his own principle of relativity, it is of course much more general than the aforementioned previous expressions. Having said that, the current (anachronistic) understanding that he showed how it followed from the symmetries of space-time obviously cannot be the source of any surprise he might have felt, pre-Minkowski, although it might have been the grounds for Popper's.

Nevertheless, this seems to point in the right direction: what provoked Einstein's interest and occasioned Popper's surprise was the establishment of a *relationship* between the core principles of his theory and the further relationship between energy and mass, where the element of surprise is perhaps enhanced by the form of that latter relationship being different from that previously obtained and by that relationship being generalizable beyond the electromagnetic context.[42] This looks set to collapse into the Wittgensteinian account, mentioned in Chapter 5: surprise is due to our not being logically omniscient. If Einstein had been more gifted logically he would have immediately 'seen' the result and not been surprised by it. However, it is not quite correct to say that the result can be straightforwardly deduced and there has been some discussion as to whether Einstein's derivation is correct or not (see Hecht 2011).

That granted, we can easily see how this can be accommodated within a practices-based account, where these practices include such derivations, and these involve either the mental or physical setting out of the steps involved. That these *practices* may yield something unexpected, given our cognitive limitations, does not demonstrate that behind them, as it were, lies something—the theory—that exists in some sense, akin to a physical object itself. Instead the appropriate *truth-makers* for claims that theories surprise us lies with those practices of setting out the steps of the derivation and ultimately, perhaps, with the associated phenomenology of understanding.[43]

[42] Einstein himself suggested that it might by tested via radioactive decay.

[43] One of the readers asked whether I would extend this account to thought experiments—the short answer is 'yes'! Setting aside the very interesting question whether thought experiments are like models or simulations, they raise the same ontological issues regarding identity as theories do and can likewise be fed into the eliminativist maw. Statements 'about' them, including aesthetic claims (Murphy, forthcoming) are made true by the relevant practices, whether physical or mental.

Conclusion: Shifting from Ontology to Practices

I have argued that the tension associated with musical works retains its force when it is applied to scientific theories: *prima facie* it isn't at all obvious what statement to reject. However, Cameron's truth-maker approach provides a natural solution. The tension divides across two contexts: one covering colloquial English and the other associated with the language that expresses our ontological commitments, whether that be 'Ontologese' or whatever. In English it is resolved because it is natural to say that abstract objects *can* be created. In the language of ontological commitments it is resolved because, fundamentally speaking, scientific theories do not exist. Instead what exists—what makes talk of theories and their properties true or false—are the relevant scientific practices.

It might be claimed that this approach is at best only as convincing as Cameron's original argument, which is itself contentious (for, among other things, presupposing that abstract objects cannot be created). However, the above conclusions do not depend on how good Cameron's account is *in the first place*. Rather, what I am doing here is adapting Cameron's account and then *showing* that it is a good approach in the current context because it is able to dissipate the tension without necessitating ontological inflationism. Of course, there are alternative ways out by for example claiming that abstract objects *can* be created, but as noted above, repeatedly, I don't think this approach holds much appeal when it comes to scientific theories. More importantly, I insist that we can resolve the tension without the ontological costs of these alternatives.

This shift away from the ontological status of theories and models and towards the nature of practices will be appealing to some. But it raises further issues: in particular, if, as has been discussed in Chapter 3, theories *represent* aspects of the world, how can we understand the nature of that representational relationship when one of the relata simply isn't there?! More generally, what are the implications for realism? And if theories do not exist, what are we, as philosophers of science, supposed to be doing when we characterize them in terms of either the Syntactic or Semantic Approaches? I will try to answer these questions in the final chapter. Before we get there, however, let me address the concern that the history of science itself presents us with theory after theory, indeed, a whole parade of them and hence to insist there are no theories is to fly in the face of the historical 'evidence'.

8

Theories in History and Practice

Introduction

An obvious reaction to my claim that there are no such things as theories is to look to the history of science, or the opinions of practising scientists, or the practices of such practising scientists and point out the various theories that appear to be so clearly on display. However, I think this parade of putative theories is in fact a construction, whether by historians or scientists themselves. Let's examine some examples, beginning with quantum theory but also embracing what might seem to be a less contentious case, namely classical physics.

A Little History of Quantum Physics

As is very well-known, during the mid- to late 1920s there were various alternative theoretical constructions in play, including not only Schrödinger's wave mechanics and Heisenberg's matrix mechanics (see Janssen forthcoming), of course, but also Dirac's 'general science of non-commuting quantities' and Weyl's group-theoretic approach (see Bueno and French 2018).[1] However, as it turned out, despite Dirac's aversion to the latter, his 'transformational' approach is mathematically the same as Weyl's. And as Schrödinger indicated and von Neumann subsequently demonstrated, the former's mechanics and Heisenberg's are also equivalent (Muller 1997; I'll come back to this example). Now, for many commentators, including physicists as well as philosophers of physics, it is the von Neumann formulation, with its representation of states as vectors (or more generally, rays) in Hilbert space, and observables as operators, that provides *the* theoretical framework 'of' quantum mechanics—question begging alert!—although many, especially physicists themselves, would agree that Dirac's approach, with its 'bra' and 'ket' formalism, offers certain pragmatic advantages (even though early reviews of his famous *Principles of Quantum Mechanics* in which he presented this formalism decried the lack of examples and exercises; see Brown 2006 p. 389–91).[2]

[1] Wallace suggests that quantum mechanics should be considered more of an overarching theoretical framework into which theories of, for example, electromagnetic phenomena, can be slotted, rather than a theory itself (Wallace 2020).

[2] Einstein, for example, said that it was to Dirac that 'we owe the most logically perfect presentation of this theory' (in Kragh 1990, p. 78). However, many of the reviewers of Dirac's book complained that it was too 'abstract' and 'symbolic'; see Kronz and Lupher 2012).

There Are No Such Things as Theories. Steven French, Oxford University Press (2020). © Steven French.
DOI: 10.1093/oso/9780198848158.001.0001

von Neumann himself was dismissive of Dirac's framework incorporating as it did the infamous 'delta function' which von Neumann regarded as mathematically self-contradictory (see Bueno and French 2018, ch. 7). But then he also became dissatisfied with his own Hilbert space formulation, and attempted to delineate an entirely new framework based on a mathematical structure known as continuous geometry (Kronz and Lupher 2012; see also Redei 1996). This was ultimately unsuccessful but did indicate some useful ways forward (Kronz and Lupher 2012; Bueno and French 2018, ch. 6). As for Dirac, again as is well known, the delta function was rendered mathematically 'kosher' in hindsight through the work of Schwartz[3] and the former's approach in general was put on a sound mathematical footing via what is known as 'rigged' Hilbert space (again see Kronz and Lupher 2012). Interestingly it is claimed that this 'formulation' ('of quantum mechanics') can handle a broader range of phenomena than that of separable Hilbert space (ibid.; see also Antoine et al. 2009). And just as von Neumann criticized Dirac for his lack of rigour, so Weyl admonished advocates of Heisenberg's matrix mechanics as introducing treatments of variables that were 'mathematically unsatisfactory and physically unfeasible' (Scholz 2008), offering his group theoretic approach as a way of yielding 'deeper insight into the true state of affairs' (ibid.). In particular, Weyl was able to obtain Schrödinger's formulation on the basis of group-theoretic considerations[4] and, more generally, what he regarded as an appropriate structural characterization of quantum kinematics. This was further extended by Wigner's famous work on the irreducible representations of the Poincaré group and together with Wigner's own earlier work, developed from a more problem solving perspective, group theory provided another rich set of formal resources (see French 1999; Bueno and French 2018). Within this framework, the irreducible representations of the relevant groups yield the transformation formulae of the vectors of Hilbert space and thus group theory is able to describe the quantum mechanical relations and in a way that is not dependent on the form of the dynamical laws or on assumptions regarding the forces involved (Weyl 1931, p. xxi).

So, although these different mathematical frameworks can be shown to be inter-related—wave and matrix mechanics are just different representations on Hilbert space; Dirac's transformational account was equivalent to the group-theoretic (see Coleman 1997, p. 13); the latter yields the Hilbert space formulation via its representations—they embodied different motivations and offered different advantages. In particular, we all know (don't we?!) that Schrödinger was a 'naïve' realist, defending (hopelessly, or so it is typically claimed) a wave-based conception, whereas Heisenberg was—to put it crudely—an equally naïve

[3] The delta function was reformulated in terms of—but should not be identified with—a distribution in Schwartz's theory (Bueno and French 2018, pp. 136–7).

[4] 'In the end, the Schrödinger characterization of a free particle turned out to be nothing but a well-chosen basis description of the irreducible ray representation of the non-relativistic kinematical group R^{2n}' (Scholz 2008, p. 257).

positivist, focusing on the representation of observable quantities. Of course, their attitudes and those of Dirac's were more complex than that (see Bitbol 2011; Camilleri 2009a; Kragh 1990) but even if one felt that such attitudes have more to do with the stance one should take with regard to 'the' theory, rather than how one delineates the latter, the crucial point remains that *the quantum 'revolutionaries' differed with regard to what they took 'the' theory to be and what principles they felt sat at the heart of it.* Thus, for Heisenberg it was wave-particle duality, understood, at least early on, in the context of Bohrian complementarity (see Heisenberg 1930). Yet, this did not feature at all in Dirac's (1930) book, *The Principles of Quantum Mechanics*; rather he emphasized the analogy with classical mechanics afforded by the relationship between Heisenberg's non-commuting products and the Poisson brackets of classical dynamics (Kragh 1990). And as we have noted, von Neumann, who in his *Mathematische Grundlagen der Quantenmechanik* of 1932 undertook to provide quantum mechanics with a secure mathematical foundation, rejected Dirac's framework as insufficiently rigorous.

All of these authors were obviously seeking to disseminate what each thought were the basic precepts of the new theory. As Kragh (1990) puts it, with regard to Dirac's *Principles*: 'He wanted to *shape* a theory which had not yet found its final shape'. Which raises the obvious question(s): how and when does a theory get its final 'shape'?

As to the how, an important role is played by books such as those already indicated. Of course, the nature of this role and the impact of the books varied across time. Although Dirac's monograph had a significant impact on both students and colleagues (Brown 2006; Kragh 2013), much of Weyl's and Wigner's work in this area lay comparatively neglected for many years, only to be revived when the foundations of what is now known as the 'Standard Model' were laid down in the 1950s. And von Neumann's (1955) *The Mathematical Foundations of Quantum Mechanics*, although published in German in 1932, did not appear in English until 1955 (as the citation makes clear).

In addition to these 'quantum revolutionaries', we might also consider those scientists who followed the vanguard, who also had an important influence on shaping the field and hence what we now view as 'the' theory. Thus in the UK, Darwin's (1931) book, *The New Conceptions of Matter* had a significant impact on those, both scientists and lay people, who were not directly engaged in these developments (Navarro 2009). Here, both Heisenberg's matrix mechanics and Dirac's approach were deemed to be too formal and, in particular, viewed as incapable of supporting a *visual representation* which Darwin saw as crucial for understanding the theory. Thus he regarded these formal frameworks as just so much mathematical 'scaffolding' (his term) from which 'the structure of the atom' should be liberated, setting de Broglie's principle and Schrödinger's equation at the centre of this new physics. In the USA, on the other hand, Ruark and Urey's *Atoms, Molecules and Quanta* (Ruark and Urey 1930; see Cartwright 1990) first introduced wave and matrix mechanics as independent theories, before presenting the connections between them.[5]

The reason for emphasizing these different approaches as embodied in the textbooks of the day is because they illuminate and further drive home the point that quantum mechanics, as a theory, did not emerge, blinking, into the light after the 'revolution' of 1925–7; rather, what came to be accepted as the core of 'the' theory crystallized over time as a result of numerous contingent and specific factors. As Kragh puts it:

> In the early stages of a new science, discipline, or research field, textbooks play an important role by legitimating the field and formulating the principles on which it builds. Whether explicitly or implicitly, the first generation of textbooks articulates the constitutive features of the new research field, which is particularly important in changes of a more revolutionary nature, such as quantum mechanics. Because the field is not yet fully consolidated, early textbooks may differ considerably in their understanding of the field, both as to content and methodology. It is almost inevitable that what an author presents has the character of a partisan text, at least in the sense that the book reflects the author's view of the new field of science. (Kragh 2013)

We can substitute 'theory' for 'field' here and the point remains. Kaiser likens such early textbooks to test bodies used to mark out an invisible (physical) field (like iron filings around a magnet, or a charged sphere), enabling historians to chart the 'conceptual trajectory' of a (disciplinary) field (Kaiser 2013). But then he asks, 'why should we assume that a research-oriented conceptual trajectory existed prior to or independent from all these pedagogical exertions?' Indeed, he continues,

> [r]ecent scholarship has highlighted the striking heterogeneity—even cacophony—of competing assumptions, approaches, and interpretations during the early years of quantum theory, even among physicists who worked closely together and whose views had earlier been considered synonymous.... The wide array of textbooks sampled in this volume only reinforces the point. Indeed, we might well wonder whether any coherent conceptual trajectory connected, say, Planck's publications in 1900 with Heisenberg's, Born's, Jordan's, Schrödinger's, or Dirac's papers in the mid-1920s. (ibid.)

Thus, instead of assuming there is a natural or an already given 'conceptual trajectory' to be mapped, or reflected by such books, we might think of such

[5] Geographical context is obviously important but so far as I know there have been no specific studies of this. However a survey of early 'statistical' interpretations of quantum mechanics in the USA and USSR can be found in Pechenkin 2012.

works as helping 'to reduce the ever-multiplying possibilities, producing what would later appear to be a recognizable conceptual path' (ibid.). Again, substituting 'theory' here, as appropriate, we might see such books as actually contributing to what came to be seen as the production or construction of quantum mechanics, qua *theory*, in this era.

Furthermore, although one might be inclined to dismiss these contrasts as a more or less natural result of the contestation that always follows a major scientific advance, with different parties pushing their different agendas, the issue of how we should delineate 'the' theory has continued to resonate. Certainly these quick remarks[6] do, at least, indicate that what was taken to be the theoretical content of 'the' theory, or even the extent to which it could be taken to 'have' such content, was disputed from the very beginning of the quantum revolution. And of course this point is sharpened further by the well-known divergences between the different 'interpretations' (so called) of quantum mechanics, mentioned in the opening pages of this book: if part, at least, of the theoretical content of 'the' theory is expected to be cashed out in stating how the world is, or could be, according to that theory, then these interpretations offer alternative contents and the continuing debate demonstrates that this issue is not confined to the quantum revolution itself, nor its immediate aftermath.

Now, one might balk at such a claim, of course. Take the Everett or 'many worlds' interpretation which in its original guise, at least, was presented as a kind of minimalist interpretation that took the theory 'as is', without any postulate of wave function collapse or any transformation of the core equations (for the best account of this interpretation, see Wallace 2012). But as standardly understood, this renders quantum mechanics as a theory, not of this world alone, but of 'many worlds' (the clue is in the name), or of 'the multiverse'. How does that not change our understanding of what the theory is, or is about?!

Or consider the Ghirardi–Rimini–Weber account, which explains definite measurement results by postulating the spontaneous collapse of the wave-function, where the probability of collapse is dependent on the number of particles in the system (for a single particle we might expect one such collapse in a hundred million years; Ghirardi et al. 1986). This postulate is then incorporated within 'the' theory as one of its fundamental laws—again, in what sense can this then be said to be 'the same' theory as that envisaged by Dirac, Heisenberg, Schrödinger, et al.?

Finally, there is the Bohm 'interpretation', as previously mentioned, sometimes (particularly, more recently) called the de Broglie–Bohm *theory* (see https://en.wikipedia.org/wiki/De_Broglie–Bohm_theory; Goldstein 2013). Originally, as presented by Bohm himself, the core equations were laid down as obtainable

[6] And there is a crying need for a serious, coordinated study of the reception of quantum physics post-1927 and across various disciplinary and geographical boundaries.

from Schrödinger's equation via a straightforward mathematical transformation.[7] Here we again obtain a different ontology, with well-defined particle trajectories on the one hand, but the infamous pilot wave or quantum potential on the other. More recently, Bohmian *mechanics* (for want of a better term and in an effort not to beg any questions) has been expressed as a distinct theory, 'defined' (ibid.) by Schrödinger's equation and the so-called 'guiding equation', understood as 'the simplest first-order evolution equation for the positions of the particles that is compatible with the Galilean (and time-reversal) covariance of the Schrödinger evolution' (ibid.). The emphasis on particle positions here is significant, not least because the 'primary', sometimes called 'primitive', ontology of the theory is that of particles, with the wave function regarded as secondary and, notably, the problematic quantum potential not even mentioned, or, at best, regarded as merely reflecting the wave function (ibid.). Thus, we recover (more or less) an apparently metaphysically unproblematic ontology but at the expense of an additional equation, yielding what appears to be a clearly different theory, albeit still empirically equivalent to 'standard' quantum mechanics.[8]

The Bohm case also offers a useful example of that trope so beloved of the science studies community, namely the 'contingency' or historical 'situatedness' of these accounts: Cushing famously argued that the Bohm theory/interpretation was effectively removed from play as a viable alternative to what became the mainstream 'Copenhagen Interpretation' due to an argument presented by Pauli at the 1927 Solvay conference, an argument that was subsequently shown to be flawed (Cushing 1994). Beller argued that the Copenhagen Interpretation itself, as it has come to be understood, was constructed via a 'dialogical' process in which different principles and theoretical features were woven together in a manner that was driven by the contingent forces powering the debates at the time (Beller 1999).[9] And Camilleri has insisted that the Interpretation as such—that is, as a more-or-less unified interpretation 'of' quantum mechanics—only came into focus via the opposition of Soviet scientists (Camilleri 2009b).

The point, then, is that we need to abandon the idea that the history of the field, or the relevant practices of the scientists in general, supports the claim that there is 'a', or 'the' theory of quantum mechanics, as a unitary and well-delineated entity, with definite identity conditions. This was clearly not the case at the

[7] Essentially Bohm expressed the wave function in polar form and rewrote Schrödinger's equation, yielding a pair of coupled evolution equations.

[8] The waters are further muddied by, first of all, the fact that Bohm's original formulation yields the guiding equation for particles without spin; secondly, it is claimed that 'it should require no imagination whatsoever to guess the guiding equation from Schrödinger's equation' (Goldstein 2013); but thirdly, it has been suggested that the wave-function itself should be regarded as law-like in this context (Esfeld et al. 2017).

[9] Indeed, Beller argues that these principles and features themselves became established as such—that is, as features of the emerging theory—via a process of dialogue between the scientists concerned; an argument that brings to the fore the kinds of practices I am keen to emphasize here.

time of the so-called quantum revolution, nor in the immediate aftermath, nor subsequently, if we understand a theory, qua entity, as incorporating some claim as to how the world is, or could be (I'll come back to this last point below).

Concern: Scientists' History

Two immediate objections can be raised. The first is that physicists themselves talk as if there is 'a' theory of quantum mechanics and also as if there are different formulations of that theory and further that the different interpretations, again 'of' that theory, can be regarded as distinct from the theory itself. The obvious and blunt response is that this is just talk. In their practices they display a different attitude, using whatever tools are to hand to crack whatever problem or issue they are dealing with, without thought as to whether in each case they are bringing to bear just another facet of 'the same' theory or not. But of course, given the 'high' metaphysics involved—of objecthood, abstractness, and so forth—how could it be otherwise?! A slightly more nuanced response is to suggest that insofar as it can be taken seriously, such talk arises from physicists'—and perhaps scientists' in general—own construction of their history. It is well-known that the kinds of histories one finds in science textbooks, or (purported) histories of science written by scientists (stereotyped as either retired or coming to the end of their working life) differ, sometimes wildly, from that offered by historians of science. In such cases, Carlyle's dictum that 'The history of the world is but the biography of great [white] men' (Carlyle 1840, p. 34) still seems to hold sway and the history of science as the history of great (white) men remains alive and well. Interestingly, this is so not only when it comes to scientists reflecting on the achievements of their peers, but when they turn to their own achievements, whether that be in Nobel acceptance speeches or autobiographical tracts.[10]

Since I've invoked the example of quantum mechanics, consider Planck and the standard account of his discovery: very briefly, by considering experimental results regarding the distribution of energy over frequencies in black body radiation, Planck came up with a 'phenomenological' formula in 1900 that attributed a certain 'quantum' of energy to each of the various modes of vibration of the radiation. Einstein then deployed this notion of quanta to explain the photoelectric effect—whereby light above a certain frequency causes electrons to be emitted from a material—and the rest, as they say, is history (!). This is the account that Planck himself gives in his Nobel Prize acceptance speech (Planck 1920). However, as Kuhn has shown in great, some might say pitiless, detail, Planck did not consider the radiation itself to be quantized, only the resonators of the black body

[10] The role of such autobiographies in constructing certain historical narratives could also use closer examination.

cavity, nor did he use the term 'quantum', preferring 'element' (Kuhn 1978). It was Einstein who set the quanta free as it were, and thus together with Ehrenfest,[11] ushered in quantum theory, and it was only after 1906 that Planck himself started referring to the quantum of action or of energy. As Kuhn notes, Planck, in his Nobel speech and after, could not resist applying to the 1900 developments the terms that were only coined subsequently, thus reconstructing his own history to suit the later narrative (ibid., p. 363).[12] Jon Dorling, in his history of science lectures in the HPS Department at Chelsea College in the 1970s noted that Planck records exclaiming over his great discovery to his son at the time, but, Dorling argued, he was, in fact, excited over the fact that his formula allowed much more precise calculations of certain fundamental constants, such as Avogradro's number.

As Kuhn says,

> Later accounts of a discovery typically redescribe it, but again in the conceptual vocabulary of the period in which they are prepared. The result is the linearized or cumulative histories familiar from science textbooks and from the introductory chapters of specialised monographs. These accounts, however, almost never withstand detailed comparison with documents from the period of discovery.
>
> (ibid., p. 366)

And, as he goes on to note, when the contrast is pointed out, to scientists themselves and others, the response is often to claim that the discoverer must have been confused, understandably so. But as Kuhn insists, this way of talking 'suggests that the discovery, of which its author is said to have had only a confused view, had already been made, was somehow already there, in the author's mind' (ibid.). The discovery, on this view, is somehow clouded and the discoverer, a kind of 'sleepwalker' groping his way towards it (again, we might recall Dodd's approach and his idea of eventually arriving at a 'clear conception' of the theory).[13] But again, as Kuhn states, this is an incoherent notion of discovery, since it renders it dependent on a prior grasp, if only a feeble one, perhaps, of what is to be discovered. And from the perspective outlined here, it assumes the very thing I have tried to argue against, that the relevant theory is 'out there', wreathed in World 3 mists, waiting to be discovered and brought into the light![14]

[11] Ehrenfest played a significant role because he realized that the form of counting of these elements/quanta that Planck used could not be classical (see also French and Krause 2006).

[12] Einstein himself likewise indulged in such reconstruction: as we have already noted, he insisted in later life that the Michelson–Morley experiment played no role in the development of special relativity, despite earlier admissions to the contrary. Such reconstructions of personal history may also then be entwined with the construction and reconstruction of the scientist's professional 'persona' (see van Dongen 2017).

[13] The reference here, of course, is to Koestler's claim that scientists are generally unaware both of what guides their research and what its full implications are (Koestler 1959). Toulmin famously dismissed the argument for this claim as 'paltry' (Toulmin 1962, p. 503).

[14] Kuhn himself was guilty of relying on scientists' own testimony in his (1962).

As Wilson has recently noted, Kuhn, in emphasizing the way that scientists, and science textbooks in particular, suppress the existence of revolutions, himself ignores the opposite phenomenon, namely the way that change and discontinuity can be exaggerated (Wilson 2017a). An obvious and ironic example has just been sketched, with Kuhn's own book (1978) illustrating nicely how techniques, devices (mathematical and physical), principles, etc. were carried over from classical mechanics and deployed within the emerging practices of quantum physics. This is not to say that there aren't differences, of course—one of the most striking being the very feature emphasized by Kuhn, namely that although Boltzmann's combinatorial technique was adopted by Planck, Einstein, and, not surprisingly, Ehrenfest, the counting involved was different (in Planck's case at least), in the sense that, to put it rather crudely, permutations of particles were excluded from the relevant mathematical expressions (see French and Krause 2006)—but as Saunders has emphasized, these should not overshadow the commonalities, particularly as represented by certain mathematical forms that show a remarkable degree of 'heuristic plasticity' (Saunders 1993).

The Darwinian 'revolution' offers another example, as Wilson emphasizes recent work that rejects the view that Darwin overturned '2000 years of Platonic-Aristotelian consensus' (Hodge and Radick 2009, p. 258). These cases feed into a general thesis about scientists' reconstruction of their own past: 'science (or rather, each of the individual sciences) necessarily constructs—as an integral aspect of its endeavours—its own imagined past, a fanciful genealogy of itself' (Wilson 2017a, p. 814). This is an 'interior' form of this imaginary past, one that is created by scientists for scientists themselves (that is, for their students, peers, and so on).

Likewise, Brannigan (1981) has argued that scientific discoveries are in fact retrospective constructs, framed and presented in the way that they are in order to serve the needs of current scientific practice. The example he gives is that of Mendel's work, often, perhaps typically, portrayed as having been ignored and then 're-discovered' as a foundation stone of genetics. But in fact, this work was *not* forgotten—it was widely cited in other work on hybridization—it was not originally on or about genetics, and the supposed re-discovery was actually a form of appropriation to the cause of evolutionary theory. Thus, what was actually—from a Kuhnian perspective—a piece of normal science within the hybridization tradition is effectively rebranded as a 'revolutionary' achievement by early twentieth-century geneticists. And, as Wilson notes, this very Kuhnian distinction becomes relativized to the appropriate context. As for Mendel, so for Planck.[15]

[15] As Wilson notes, this theme was then taken up by Schaffer (1986) who argues that this emphasis on and construction of 'discovery' is what distinguishes 'science' from 'natural philosophy' and also that, in the context of a Whewellian ethos, such discovery was seen as individualistic, powered by the 'flash of genius' and practice fixing (Wilson 2017a).

Wilson draws these strands into his bigger theme, which is that whether science is inventing discontinuity or suppressing it, constructing or reframing discoveries, it is inventing a past for itself.[16] And that this phenomenon deserves greater attention than it has received so far from historians and philosophers of science. He concludes by urging historians to consider in more detail the different ways in which this past may be invented, depending on discipline, era, and other contextual factors, and also to examine in particular the notion of discovery and how it has been viewed and represented across the years (Wilson 2017a).[17]

All of this is grist to my grinding mill, of course, and the crucial point is that we should not take what scientists say about their own history as sacrosanct or otherwise privileged, epistemically or otherwise—these histories are (re-)constructed depending on the interests and concerns that are currently in play.

Concern: Quantum Physics Is Special

The second concern that might be put forward is that quantum mechanics is somehow a special case and that this 'constructivist' line could not possibly be applied elsewhere. Well, we can easily go forward from the quantum revolution and ask the question: what is quantum field theory (QFT)? Is it the axiomatized construction beloved by those who are members of the so-called 'Algebraic QFT' camp? Or is it that which physicists themselves actually use? These are significant questions because QFT is widely lauded as yielding some of the most precise predictions ever made in science[18] and therefore as clearly being worthy, and perhaps more so than other theories, of realist acceptance and also because these questions bear on concerns that are important for philosophers of physics and realists alike (Ruetsche 2011). Advocates of Algebraic QFT insist that the only way one can avoid the infamous infinities that plague 'the' theory is by reformulating QFT on an axiomatic basis. The problem is that, as typically formulated, these axioms do not cover or accommodate interactions and hence if Algebraic QFT is taken to be 'the' theory, it is strictly empirically inadequate. In particular, the axioms do not cover the well-known Standard Model of high-energy physics, recently given a further epistemic boost by the discovery of the Higgs boson. Those who urge philosophers to shift their focus to what physicists actually use in

[16] He also notes a corresponding phenomenon in philosophy, as observed by Watson who coined the phrase 'shadow history' (Watson 1993).

[17] Shifting to the meta-level, Thomas (2017) suggests that historians like to see and portray both scientists and lay-people as labouring under various misconceptions about the history of science and themselves as capable of dispelling such misconceptions, a perception that is in fact problematic; for a response, see Wilson 2017b.

[18] So, the Lamb shift is typically cited in this context, where this refers to the shift in certain energy levels in the hydrogen atom and allows a measurement of the fine structure constant to better than one part in a million.

their practice(s) have argued that this—termed 'Lagrangian' or 'naïve' QFT—can be rendered perfectly well-defined and kosher, subject to certain caveats,[19] and, furthermore, that it is, in effect, 'the' QFT that underpins the Standard Model (Wallace 2001).

Despite some conciliatory moves early in this particular debate, these two options have been presented as opposing horns of a form of underdetermination, with grounds given for favouring one over the other (Fraser 2009). What we have here is a dispute over what should be 'the' theory of QFT with (at least) two constructions in play and various reasons given for adopting one or the other. To what extent those reasons are decisive is dependent on the weight given to such virtues as consistency,[20] for example, or to the promise of a research programme but the point I want to emphasize here is that both options can be seen as constructions or, better perhaps, representations and that we should not blithely accept that there is a real issue as to which one should count as 'the' theory (for further on this and related issues, see Fraser 2016 and forthcoming).

And the same point about physicists' reconstruction of their (favoured) past can be made here too: see for example Brown's essay review of Schweber's book *QED and the Men Who Made It* (Schweber 1994; Brown 1996), where he takes Schweber to task for attributing the discovery or, significantly, 'making' of quantum electrodynamics to just four people, downplaying or ignoring altogether the contributions of many others, for over-emphasizing the American context of this discovery/construction, for, relatedly, over-emphasizing the pragmatism or practice-based approach to physics that was supposedly an 'American trait', and so on.

Still, one might continue to be unimpressed, maintaining that even more so than QM, QFT is still in an indeterminate conceptual state and further work is needed before we can delineate the outlines of the theory itself. So, let's shift 'backwards', in a sense, and ask, 'what is *classical* physics?'

Of Lifts and Quilts and Facades

This is the question with which Gooday and Mitchell kick off their historical analysis of the distinction between 'classical' and 'modern' physics (Gooday

[19] Notably that we should take seriously the high-energy cut-off introduced to deal with the infinities and should accept that the justifications for this cut-off imply that Lagrangian QFT is only an approximation to some 'deeper' or more fundamental theory. (Of course, that is the situation in which many theories find themselves! More significant, perhaps, is the claim that 'QFT only makes sense if we include in it some vestigial aspects of the very theory which we expect to replace it' (Wallace 2001, p. 75). We recall that one of the heuristic moves noted by Post (1971) is to identify the 'footprint' in one's current theory of its successor and here we have the claim that to *be* regarded as a theory, QFT must incorporate that heuristic manoeuvre.)

[20] Lest this be thought of as a no-brainer, it is of course not the case that the only way of dealing with apparent inconsistency is to set up a clearly consistent set of axioms, as we noted in Chapter 3, following Vickers 2013.

and Mitchell 2013). Beginning with Staley's claim that the pair were actually 'co-created', or constructed simultaneously by Planck at the Solvay Conference of 1911 (Staley 2008), they argue that in fact the distinction emerged over a long period of time, extending into the 1930s, and depending on the geographical location considered. Now, it is not perhaps surprising that the distinction between classical and non-classical physics should only be made after the latter became reasonably clearly articulated as the appropriate contrast to the former, but what's interesting is Gooday and Mitchell's claim that one needs to be clear whether the physicists at the time were talking about classical physics in its entirety or particular classical branches of physical theory. Staley takes Planck to be concerned with the former, but they insist his focus was on the latter and on the difficulty in accommodating quantum concepts within the frameworks of distinct classical theories such as classical mechanics and electrodynamics.[21] Furthermore, they argue, '[t]he specific and systematic application of the term "modern physics" to quantum and relativistic phenomena did not take place concurrently with the articulation of "classical physics", but some time later' (Gooday and Mitchell 2013, p. 726).

Furthermore, many physics textbooks tend to emphasize the classical/modern distinction as representing a distinctive conceptual break, or revolutionary moment, while others, and sometimes the same books, note the continuities in theoretical practice (we recall the remarks made above). Again, the distinction gets applied in different ways to emphasize either continuity or change, depending on the pedagogical or more broadly cultural aims and interests involved,[22] yielding different versions of 'classical' and 'modern' physics, each with their own strengths and weaknesses. As Gooday and Mitchell again note, classical mechanics plays a central role in this articulation, as classicality is taken to be a function of deploying the kinds of devices and techniques used in the former.[23]

The conclusion, then, is that classical physics only ever existed in the limited sense that the label was developed and attributed by theoreticians in the early twentieth century 'who sought to preserve a restricted role for established theory and techniques whilst setting forth a future research programme based on new forms of theorizing' (ibid., p. 751). Any reference to 'it' prior to 1900 implicitly adopts 'an anachronistic perspective that was created to legitimize the new foundations

[21] A further interesting aspect is the normative element in the sense that these issues and contrasts were taken to have profound implications for the direction of future physics.

[22] 'First, to reveal clearly the (familiar) techniques upon which relativity theory and quantum theory are built; second, to focus attention on the rejection of Newtonian absolute spacetime and Laplacian determinism; third, to isolate the physics of everyday objects from the physics of more extreme realms (the 'tiny', 'huge', and 'very fast'), and to prove its adequacy for many experiments and practical applications; and fourth, to provide an antonym for twentieth- and twenty-first-century fields of research in physics that define themselves in terms of their rejection of various common-sense assumptions' (ibid., p. 727).

[23] Thus, they record Eddington's claim that 'the general idea is that the scheme of natural law developed by Newton in the *Principia* provided a pattern which all subsequent developments might be expected to follow' (Eddington 1928, ch. 1).

for physics proposed within relativity and quantum theory' (ibid.). As an antidote to such anachronisms, studies of the relevant continuities can be deployed and presented as 'ironic' rejoinders to Kuhn's claim about the rendering invisible of revolutions by the adherents of the new paradigm: rather than committing a form of patricide, physicists constructed a 'classical' identity for their forebears in order to serve their own interests. Thus, 'the apparent unity of "classical physics" [should be seen] as the *post hoc* creation of twentieth-century theoretical physicists seeking to consolidate new departures within their discipline' (ibid., p. 722).

But moving to the particular, what about classical mechanics itself? Surely, one might say, there is no doubt about its identity—we simply have to recall and write down Newton's laws and we're done! Setting aside for the moment the whole issue of whether Hamilton's 'formulation' counts as such (and we'll return to this shortly), the form in which these laws were given in the *Principia* is, of course, very different from how we would write them today, as we have already noted. Furthermore, they have been subject to different interpretations that in some cases undermine their status as laws, at least as standardly conceived: Poincaré, for example, argued that the first law is a *convention*; the second has often (perhaps erroneously) been taken to provide the *definition* of 'force', and the status of the third has been described as 'hazy'. Indeed, as (a different) Wilson warns us,

> Classical mechanics is frequently characterized as 'billiard ball mechanics' or 'the theory of mechanism' on the grounds that the science treats its materials in the manner of colliding particles or clockwork. The reader should approach such stereotypes with caution because the basic framework of classical mechanics has long been subject to divergent interpretations that unpack the content of Newton's 'three laws' in remarkably different ways. These differing interpretations provide incompatable catalogs of the basic objects that are supposed to comprise the 'classical world'—should they be point masses, rigid bodies or truly flexible substances? (Wilson 1996)

Taking up that last question, as originally stated these laws could not be applied to rigid or deformable bodies, and it was Euler (again) who generalized them, although Euler's laws can also be taken as a distinct set of axioms for the behaviour of such bodies. But the basis of this generalization is not conceptually straightforward because thinking of rigid bodies or continua as merely 'swarms' of point masses held together by short-scale cohesive bonding cannot serve to underpin the empirical success involved, nor will it help illuminate the various conceptual issues in play (Wilson 2014).[24] In particular, what might be seen as the 'triumphant

[24] Indeed he notes, nicely meshing with Gooday and Mitchell's thesis above, that 'a considerable portion of our modern appreciation of the subtleties of mechanical doctrine was prompted in the 1920s by the modelling requirements of the paint and rubber industries as their technicians sought reliable guidance in modelling such oddly behaved materials in a satisfactory manner' (Wilson 2014, p. 5).

hegemony' (ibid., p. 103) of classical mechanics owes a great deal to the often hidden contribution of what Wilson terms 'lifts', which are basically devices and manoeuvres that take one between different levels of description, demarcated by different characteristic scale lengths and typically different ontologies—so consider the example of a steel beam and the shifts involved as we move from the level of the 'bulk' steel to that of the crystalline grain, and then to that of the molecular lattice, and finally to that of point mass atoms bound together. And these lifts may be infected with various dubious presuppositions, such as, and typically, that certain rules and principles applicable at one scale can be exported unproblematically to another.

Thus, the very notion of 'force' alters its significance via such lifts: consider friction, for example, regarded at one level as a straightforward Newtonian force opposing forward motion, but from the perspective of another, this 'force' incorporates the stretching effects that the mass of the object has upon the material, causing it to travel further than is apparent. Another classic example is that of the viscosity of a fluid, typically analysed in terms of the shear 'forces' on units or blocks of fluid that from a foundational perspective are, at best, ontologically ephemeral but essential for the relevant description at the level of fluid mechanics. Contentiously, perhaps, Wilson claims that axiomatic presentations simply do not accommodate these shifts in ontological perspective and we are left with 'doctrinal holes', the filling in of which raises deep conceptual issues. But more significantly from our point of view, these lifts and strategies, devices and moves of various kinds, form a crucial part of the practice of modelling, generating a 'compendium of descriptive lore' in terms of which classical mechanics is best viewed as a series of descriptive patches, linked together by these very manoeuvres (ibid., p. 19).[25]

This feeds into Wilson's own more general account of the nature of theories, according to which we should abandon the attempt to impose 'internal conceptual closure' in such cases and instead replace theories, as our unit of philosophical interest and as standardly conceived, with 'theory facades', which are quilt-like assemblages that 'look kinda like theories if you don't look at them too closely' (2014, p. 20; 2006). From this perspective, one can better appreciate and understand the kinds of moves one finds in textbooks of classical mechanics, for example, as one moves up (or down) from one descriptive level to another and

[25] According to Wilson (2014, p. 103), the factors that lay behind certain of these lifts had the effect of encouraging core themes within the philosophy of science, notably with regard to the role of idealization: in order to model materials as a continuum, physicists had to attribute to the lower-length scales characteristics that they knew were not actually possessed, thus generating the preliminary essential idealizations that are now taken to be required by philosophers of science (see our earlier discussion in Chapter 2). Such idealizations are regarded as 'fictive scaffolding' or ladders (see also Janssen forthcoming) that may ultimately be thrown away, a move that is made time and again in the history of science and in reflections upon it, and which elsewhere I've called Poincaré's Manoeuvre (French 2014, p. 67).

also the kinds of conceptual shifts associated with such moves. A useful comparison might be made with the kind of 'collage' account offered by Cartwright and her co-workers, as mentioned earlier (Cartwright, Shomar, and Suárez 1995), although Wilson is careful to distance himself from the practices-focused 'disunity of science' movement on two grounds: first, because he still recognizes the value of formal representations and axiomatizations for describing the 'concrete structures' within a facade (while acknowledging that these representations do not contain the resources either for preventing the various patches from effectively running into and interfering with one another, or for describing the ladders and lifts one must use when local resources fail); and second, because we cannot deny that in the practice of physics, pedagogical and otherwise, there is a 'modelling recipe' built around $F = ma$, leading to a kind of family resemblance between the various and different manifestations of the term 'force' even though, according to Wilson, 'that structural "commonality" is too weak to prevent the term "force" from experiencing localized "property dragging" that eventually blocks any effort to bind all of "classical physics" into a plausibly axiomatized whole' (Wilson 2014; [preprint] p. 23).

Here we have further grist to grind, not least because with a description of the relevant practices in terms of Frankensteinian 'theory facades', consisting of patches and lifts, holes and ladders, moves and manoeuvres of various kinds, all cobbled and bolted together, it is even more difficult to see how any abstract object, whether in World 3 or wherever, whether artefactual or whatever, could reasonably be taken to be referred to by such a description. And this is good ol' classical mechanics we're talking about! What Wilson's detailed analysis gives us, then, is a further set of descriptive resources at the meta-level of the philosophy of science, in terms of which we can represent, for the kinds of purposes he himself sets out (2006; 2014), the actual practices that serve as the truth-makers for statements about what we take to be 'classical mechanics'.

Even if we were to restrict our discussion to classical point mechanics, there is the further point that it is often presented using other forms of variational technique, such as, again, Hamilton's principle and other varieties of 'least action'. Indeed, if we were to open a standard undergraduate textbook in classical mechanics, we would typically be presented with not just, or perhaps not even, objects and forces, but potentials, 'action', and so forth, as we are presented with the Hamiltonian and Lagrangian 'formulations', mentioned above. This then forms the basis of a powerful critique of standard scientific realism, if this is seen as 'populating the world with a clearly defined and described set of objects, properties...' (Jones 1991, p. 186). Since these different formulations of classical mechanics all proffer different sets of such objects, properties, and processes, this realist vision, Jones argues, cannot be achieved. There are, of course, ways we might respond to this challenge but setting that aside, we can recast Jones's point as follows: if a theory is 'about' some definite set of such objects, properties, and processes, then shouldn't we take these 'formulations', which appear to be clearly 'about' different things as actually different 'theories'; or at the very least, as undermining the claim that there is just one theory of classical (point) mechanics?

Now, it is often stated that such formulations are equivalent to one another and to Newton's laws, but this is not true unless a number of crucial and challengeable background assumptions are made. Without going into too much detail, the equivalence claim rests on the availability of a translation scheme between the structures of the two formulations, where that of the Hamiltonian is configuration space with a (Riemannian) metric structure and associated distance measure, whereas that of the Lagrangian is phase space with a symplectic structure and associated volume element. Since metric structure determines, or presupposes, a volume structure, but not vice versa, the latter adds another level of structure to what's needed to express the relevant equations of motion. Thus, North has argued that the Hamiltonian formulation should be preferred over the Lagrangian on the grounds that it involves less structure (North 2009). Curiel, on the other hand, argues that, given a plausible characterization of 'classical system', classical systems 'evince' (his word) the Lagrangian structure and not that of the Hamiltonian formulation (Curiel 2014). The argument depends on a crucial distinction between equations of motion and kinematic constraints: when a classical system interacts with another, only certain quantities are directly 'pushed around' and the equations governing quantities that cannot be 'pushed around' in this way should be thought of as kinematical constraints rather than equations of motion per se. However, according to Curiel not only does the Hamiltonian formulation of a system not allow one to express such constraints, it allows solutions to the equations of motion that violate them. Hence it does not offer a 'natural' formulation of these classical systems.

The force of this argument might be resisted by rejecting the distinction between those quantities that can be 'pushed around' and those that cannot, perhaps on the grounds of begging the question in some sense. However, Curiel himself explicitly addresses this, insisting that he is trying to set things up in as neutral and non-question-begging way as possible. And after all, if these two formulations are to be understood as formulations *of* classical mechanics, it may seem plausible to claim that some basic characterization of what it is they are supposed to be formulations of needs to be given. But then we could deny this and argue that in the present context—concerned with what we take classical mechanics to be, as a theory—it is *this* claim that is question-begging, particularly given the role of what a theory is putatively 'about' in delineating it. This would amount to insisting that the formulations should be regarded as different theories, and from the perspective of the Hamiltonian theory, the above 'neutral' characterization is in fact nothing of the sort, being a theoretically informed representation of empirical phenomena, interpreted in terms of fundamental quantities that the advocate of the Hamiltonian theory would reject.

'Theory' Equivalence

Of course, you might feel that despite the above arguments, the two formulations should be regarded as physically equivalent by virtue of yielding the same solutions to the equations of motion. Unfortunately, this collapses into an empiricist notion of physical equivalence that is widely rejected (see Sklar 1982). How then should equivalence be determined? You could try to appeal to the formal relations between the relevant structures (Barrett and Halvorson 2016). However, it's not clear that such relations on their own can account for two important features of our intuitions about and judgements of theoretical equivalence in certain exemplary cases (Coffey 2014). The first is that given that the name of the game is to be able to separate such formulations into those that are theoretically equivalent and those that are not, there is 'surprisingly little agreement' as to which are and which aren't (ibid., p. 824). Thus, when it comes to space-time theories, for example, some take Poincaré variants in which the effects of space-time curvature are mimicked by non-standard forces to be theoretically equivalent to 'standard' general relativity, whereas others take them to be different; likewise, some take Newton-Cartan 'geometrized' Newtonian gravity to be equivalent to the traditional form, others do not (for further discussion of this particular example, see Knox 2014). As Coffey says, 'If we can't agree on those formulations that really are theoretically equivalent, how can we provide an account of theoretical equivalence?' (Coffey 2014, p. 825).[26]

The second feature has to do with the fact that certain judgements of theoretical equivalence exhibit a kind of asymmetry, but if equivalence is taken to mean identity then such judgements should be symmetrical. Thus consider again the Lagrangian formulation touched on above (ibid., pp. 829–32). Although, as we have noted, this is widely taken to be equivalent to the standard Newtonian account, it is the latter that is generally taken to be fundamental in physics *practice* at least, whereas the Lagrangian formulation is understood to be derivative (of course, this may not reflect the attitudes of certain philosophers of science!). As Coffey puts it, 'Our [philosophical] discourse about theoretical equivalence may suggest a symmetric relationship, but there's an implicit asymmetry in how the theoretically equivalent frameworks, or the formulations constructed within those frameworks, are taken to relate [in the practice of physics]' (ibid., p. 831; my additions).

Again we see the significance of the relevant *practices* in this context. The issue then is how to accommodate such features. Let's recall the Syntactic Approach, from Chapter 1: a theory is (typically) identified with or represented by some set

[26] As Coffey goes on to note, one of the issues in play here has to do with the lack of a clear account of theory identity. Of course, I would suggest that these disagreements indicate that that no such account can be given.

of sentences or propositions and a particular formulation can then be identified with a certain *axiomatization* of that set, so that different formulations correspond to different axiomatizations (see Quine 1975). Alternatively, we could insist that theoretically equivalent formulations should endorse the same logical inferences. Following the Semantic Approach, on the other hand, we might take equivalent formulations to have the same *models* in common (Glymour 2013; for a summary of these approaches, see Coffey 2014, pp. 832–3 and also Barrett and Halvorson 2016).

However, it seems prima facie unlikely that such approaches will be able to capture the first feature above, simply because the relevant formal relationships, however expressed, will hold independently of divergent judgements regarding the equivalence or otherwise of the formulations involved. Thus, in the context of the Syntactic Approach, certain theories of physical geometry, touched on above, must be regarded as theoretically equivalent, but as Coffey (2014, p. 834) notes, that does not mesh with everyone's judgement when it comes to the relevant practices. One response would be to go normative, but those are dangerous waters!

Furthermore, as has already been noted, these approaches cannot capture the second feature sketched above: '[t]he relevant formal relations identified by a particular analysis might be either symmetric or asymmetric (or some combination thereof), but it is difficult to see how their application might generate symmetric results in some cases and asymmetric results in others—a sometimes symmetric, sometimes asymmetric, relation of reformulation' (ibid.).

What to do? We could take two putative formulations to be theoretically equivalent if *they say the same thing about what the physical world is like* (ibid., pp. 834–5). Now, of course, as we have seen, there are issues about what it is to 'say the same thing about what the world is like', but assuming we have the relevant semantic tools, the above features can be accommodated, since formulations themselves do not stand in relations of equivalence on this view: '[t]here is no fact of the matter as to whether textbook presentations of (say) Newtonian gravitation theory and Cartan Theory are variant formulations of a single underlying theoretical picture' (ibid., p. 835). According to one interpretation they are equivalent, according to another they are not and the divergence simply mirrors that of our (whether scientists or philosophers of science) interpretive practices.

Likewise, any asymmetry can be accounted for in terms of an interpretive decision to take a certain formulation as primary and the other as secondary or derivative. Thus, we might have good, and pretty obvious, reasons, for taking the point masses and forces associated with Newton's laws as basic (but see Wilson's commentary above!)[27] and the actions, potentials, etc. of the Hamiltonian and Lagrangian formulations as secondary. And with regard to the supposed

[27] As Coffey recognizes (2014, p. 835, n. 33), determining which formulation should be taken as basic may be a very complicated and tricky issue.

equivalence, or not, between the latter, one can understand the dispute between North and Curiel as arising from a difference in *their* interpretive practices, where those practices will also include, on the one hand, methodological elements having to do with surplus structure and, on the other, preliminary decisions as to what is to count as a 'classical system'. I'll come back to such philosophical practices in Chapter 9.

Now, there's an obvious objection to this approach: consider the Heisenberg and Schrödinger formulations of quantum mechanics discussed previously. These are generally accepted as equivalent but, crucially, in the absence of any agreed interpretation and on the basis of the *formal* result partly established by Schrödinger and fully so by von Neumann, as already noted (see, again Muller 1997 who shows how this formal equivalence is nicely captured within the Semantic Approach). Of course, this might be seen as an exceptional case, insofar as it has yet to be established what the world is like according to quantum mechanics. Certainly, the above approach does not rule out such results as *irrelevant* to claims of equivalence (Coffey 2014, p. 837). Indeed, these formal relations may help to shape the relevant interpretive judgements and, in particular, may *motivate* certain interpretive approaches without necessarily specifying a particular interpretation. So, we recall that the equivalence between the Schrödinger and Heisenberg formulations is established by showing that they are merely different representations on an underlying vector space (Hilbert space). This motivates the suggestion that the appropriate ontology should be conceptualized in terms of the *structure* presented by that space (Ladyman 1998). Nevertheless, as much an advocate of such structuralist moves as I am, I can certainly appreciate how some will balk at this apparent consequence of adopting the above approach to theoretical equivalence.[28] To those folk, I can only say, 'all the more reason to come over to the Dark Side and drop the presumption that theories are the sorts of things that can be determined to be equivalent (or not) to begin with!'

Not surprisingly perhaps, this 'consider what the world is like' approach can also be extended beyond cases of different but inter-translatable formulations to well-known examples of the underdetermination of theories by experimental

[28] Furthermore, as Coffey acknowledges (2014, p. 837, n. 35) such a move conflicts with his insistence that one such formulation should be taken as ontologically primary and he suggests that the Schrödinger formulation should be regarded as the more basic one, since it represents the state of a quantum system as dynamically evolving over time and that is what we would expect an interpretation to tell us. But of course, Heisenberg's matrix mechanics does not eschew dynamics—the difference is that whereas in Schrödinger's formulation the wave function, or state vector, changes with time, in Heisenberg's it is the observables, represented by the matrices, that are time dependent, but the state vector is not. And if one did want to play the 'one formulation is more basic than the other' game one could counter Coffey's claim about the dynamics with the methodological points that the Heisenberg formulation both meshes with (certain features of) classical physics insofar as it involves treating the Poisson bracket of the latter as plastic (recalling Saunders's phrase) and with relativistic physics insofar as it does not privilege time (as opposed to space) with regard to the representation of the state via the state vector.

data, so beloved of the antirealist critics of scientific realism. Thus, let us consider again the case of Lorentz's aether theory and Einstein's special theory of relativity, where we have observational equivalence and theoretical similarity but no inter-translatability, notably because Lorentz's notion of an absolute rest frame simply has no counterpart in Einstein's theory. Here Norton has argued that the 'aether structure' of the former is physically superfluous (cf. again North 2009) and hence Lorentz's theory is just an overwrought formulation of Einstein's (Norton 2008).[29] Note that this provides a further argument against the claims of multiple discovery which we looked at previously—Lorentz may have obtained the same core equations as Einstein (the clue is in the name!) but this does not amount to discovering the same *theory* as the interpretations are so different.

Of course, appealing to interpretive practices implies many more cases of underdetermination than realists have typically appreciated, due to the inherent ambiguity in the notion of 'interpretation' and the multiplicity of roles that it plays. So, it is not just a matter of supplying an interpretation of some underlying mathematical formalism with the associated question of what features of that formalism refer to or represent features of the world or not, since a given formulation can be interpreted differently, with the associated question being what kind of physical thing is being represented (Coffey 2014, p. 840). Again, the most obvious example is that of quantum mechanics and one would have to conclude, on this view, that what are often taken to be different interpretations must actually be counted as different theories (see also Acuña 2019)—and so the mill grinds on![30]

But even in the case of Newtonian mechanics, and leaving aside Wilson's worries, force for example, can be interpreted as ontologically primitive or as a causal disposition (ibid., p. 841) and different choices will lead to different accounts of what the world is like. We can go even further and consider, again, how laws in general are interpreted, as either Humean regularities or as the outcome of relational dispositions, (or as primitive; French 2014) so that what may appear to be the same set of laws also in fact supports different theories. Now, at this point the reader may feel that we have now gone too far and across such thin ice that a kind of *reductio* has been reached—if this approach implies that a formulation *with the same laws* can be understood as generating two different theories, then surely any epistemic grounds for such a claim are fast disappearing in the rearview mirror! The point is, trying to establish what a theory formulation says the world is like, in order to determine whether two such formulations are

[29] Lorentz of course famously insisted that Einstein had simply postulated what he had deduced from the equations of electromagnetism.

[30] Coffey also suggests that '"quantum mechanics" doesn't pick out a theory *per se*, but rather a framework for theory construction' (Coffey 2014, p. 841, n. 38; cf. Wallace 2020). That doesn't necessarily conflict with my view, as long as we understand that what is being constructed is not some kind of entity, abstract, artefactual, or otherwise. What the framework does is constrain certain practices—verbal, written, presentational, and so on—in certain ways.

equivalent or not, involves appealing to a range of further considerations, some of which may be regarded as heavily metaphysical. What we get, as a result, is a plethora of putative theories, depending on how deep we go into the metaphysics and what metaphysical tools and devices we choose to deploy. To think these are all no more than different formulations 'of' a core theory is a mistake (a question-begging one at that) and generates all the problems with equivalence and under-determination that have been used as weapons against realism by its critics. Best, again, to dispense with that thought and take them to be different constructions, dependent on our interpretive practices.[31]

Conclusion: There Are No Theories Out There

The aim of this chapter was to cast doubt on the idea that scientific practices, as represented in the history of science, or textbooks and the like, or scientists' reminiscences...demonstrate or indicate in some way that there *really* are theories 'out there' and as a result, following this train of thought, that philosophers of science should, indeed, come up with an ontology of theories that reflects these practices. As I've tried to suggest, these practices are complex, overlapping, and, in some cases, entangled. It is simply not clear how we should delineate classical physics from quantum physics, for example, or what counts as the relevant theory in either case. We can shift our terminology to 'theory facades' or 'frameworks' but those terms obscure the diversity and the complexity of what scientists do and come up with. Best, I would suggest, to drop the idea that they come up with some*thing*, that then lives in some Popperian realm, say, and accept that what we are presented with in the histories and the textbooks and the reminiscences is no more than a construction for which certain features of the relevant practices have been emphasized and highlighted for all sorts of different purposes.[32] Again, it is the latter that come to the fore and which serve as truth-makers for the claims which we then make about putative theories.

This suggestion is further reinforced, I would argue, by thinking about the problem of establishing theoretical equivalence. As we just saw, appealing to formal relationships alone seems too 'thin' a measure, whereas considerations of what

[31] A reader has pointed out that this simply replaces one form of underdetermination with another: that centred on whether theories exist. But of course, insofar as we are talking about some 'thing' that goes beyond what we can observe—namely, scientific practices—that form of underdetermination is unavoidable! Nevertheless, if we can avoid the former kind, progress of a sort has been made.

[32] It has also been suggested that in this chapter I am effectively arguing that no vague or ambiguous term has a referent, so that we should conclude that games, customs, and nation states should also be eliminated. As welcome as such a conclusion might be, at least with respect to that third example, I would resist such a generalization of what I argue for here. There is of course a rich literature on how to deal with vague terms and eliminativism is only one option, to be wielded in specific circumstances and in a context-dependent manner.

different formulations say the world is like invite the concern that it all ends up, metaphysically, too 'thick'. However, eschewing the presumption that there is something to which different formulations may or may not be equivalent raises obvious realist concerns in this context. It is to these that I will now turn, and in particular, the issue of accommodating scientific realism within an eliminativist framework.

9

Theories in the Realism Debate

Introduction

Let us take stock: I have tried to articulate an argument for an eliminativist stance that can be summed up with the slogan: there are no such *things* as theories. It is a stance that involves multiple import/export moves: from metaphysics into the philosophy of music, and from there into the philosophy of science. It is also a stance that, I claim, allows us to make sense of what we and, particularly, scientists say or write about theories, as such talk is made true by the relevant practices—the writing and publication of papers, the seminar presentations, the experimental work, and so on. Thus, as we saw, claims such as 'quantum mechanics is empirically adequate' or 'quantum mechanics is a beautiful theory', can still be taken to be true, even though there is no such *thing*—whether abstract artefact or whatever—as quantum mechanics.

Now, again, I must emphasize that with all these statements, I am not talking about the truth or falsity of claims the theory makes about the *world*, but rather the truth or falsity of claims made about the *theory*. Nevertheless, there are statements that some of us, at least, make about theories that relate to such claims that the theory makes about the world, namely the kinds of statements that are made in the context of the debate over scientific realism. How might these claims be accommodated? Furthermore, how are we to make sense of the idea that theories and models *represent*, in whatever sense and of the various accounts of representation, as discussed in Chapter 3, if there is no*thing* that is doing the representing?! Let's consider each of these questions in turn—as we'll see they push us to both expand the relevant set of truth-makers and reflect on what we, philosophers of science, are doing when we assert that theories represent.

Realism without Theories

Let's begin by considering statements like 'theory T is true (or approximately so)'. These are statements that are typically made by scientific realist philosophers, or scientists in realist mode. On the face of it they appear to be *about* the theory concerned and, indeed, attribute a property to it, namely *truth*. What would be the truth-makers that make such claims true? For the (standard) realist, what makes it true that 'T is true' is that T stands in a certain relationship with the

There Are No Such Things as Theories. Steven French, Oxford University Press (2020). © Steven French.
DOI: 10.1093/oso/9780198848158.001.0001

system concerned, or the world more generally, where that relationship is captured via the notion of 'correspondence', if the realist adopts Tarski's theory of truth, say. Immediately we can see that this takes us beyond the set of what might be called 'purely' scientific practices, insofar as the establishment of such a relationship is not a part of such practices—indeed, if it were, antirealists who want to take scientists' talk and such *literally* would be in trouble! I'll come back to this shortly.

Furthermore, talk of *T* standing in a certain *relationship* might encourage an object-oriented metaphysics, of course, with the encouragement running along the lines of 'to say there's a relationship suggests there's a relation, but relations need relata, hence there has to be a theory, qua object, as relatum here'. An obvious move would be to appeal to Musgrave's response to Fine's fretting over correspondence truth, in the context of the latter's attempt to break the realism–antirealism impasse with his 'Natural Ontological Attitude', namely that it presupposes a relation that all truths bear to reality, in terms of which they can be formed into a natural kind, about which Fine is, naturally, sceptical. Musgave shares this scepticism but insists that the correspondence theory, as formulated by Tarski, makes no such presupposition: 'there is no more to the "correspondence relation" than Tarski gives us', via his 'Convention T' (Musgrave 1989, p. 387).[1] And with no relation, as such, there are no relata and the above encouragement doesn't even get off the ground.[2]

Now, as I just indicated, granted that for the realist the truth-maker of '*T* is true' is that *T* 'corresponds' in the Tarskian sense to the relevant part of the world, this is only the case for said realist. What about antirealists who also want to be eliminativists about theories? What would count as the relevant truth-makers of such claims for them?

Let us consider what the old-school instrumentalist and the cutting-edge constructive empiricist would say. The former would insist that statements that mention unobservable entities are not 'truth-apt', since taken as they stand, the relevant terms are strictly meaningless. As is well known, in order to accommodate the use of such statements in practice, the instrumentalist has to adopt a non-literal semantics by which meaning is conferred upon those terms via their association with observable terms or the appropriate laboratory procedures etc. And it was the difficulties in cashing out that semantics that contributed to the demise of this view. That point aside, for the instrumentalist, then, the claim that '*T* is true', whether uttered by a scientist or another philosopher, is simply not a

[1] An instance of which is 'The statement "electrons have spin ½" is true if and only if electrons have spin ½'.

[2] In response to the concern that the advocate of the Semantic Approach cannot avail herself of the Tarskian framework (Chakravartty 2001), we can again suggest the 'Suppesian switch' to the 'internal' perspective, from which one adopts a characterization in terms of statements or propositions (French 2008; remember: this move carries no ontological weight!).

candidate for either truth or falsity. In effect, the eliminativist instrumentalist excludes such claims from consideration by not taking literally those aspects of scientific practice involving them.

The constructive empiricist, on the other hand, famously insists on adopting a literal understanding of statements involving theoretical terms. Thus, she takes such statements and the claim '*T* is true' to be truth-apt. However, given her empiricist account of knowledge, she would insist that we cannot *know* whether *T* is true. At best we can only assert that *T* is empirically adequate. Thus, for the constructive empiricist, there *is* a potential truth-maker for the claim '*T* is true' and, depending on how she understands truth, it might be correspondence à la Tarski, just as for the realist, but we can never know whether the claim is actually made true or not.

Generalizing, then, both the realist and the constructive empiricist will agree about scientific practices, at least at the level of what scientists actually do, in the sense of conducting certain operations in the laboratory or presenting certain sets of equations on a whiteboard, or whatever. As a result, the relevant practice that justifies the assertion of the above claim can't be *scientific* practice alone but must include those inferential moves that realists typically appeal to in order to justify their claims. Hence both realists and constructive empiricists will agree that 'theory *T* is true' is made true by certain truth-makers such as 'correspondence' but will differ insofar as the realist's justification for such a claim must draw on *philosophical* practice, where such practice will incorporate these inferential moves deployed by the realist (whether philosopher of science or appropriately reflective scientist). Indeed, realists and constructive empiricists, or antirealists in general, differ over the elements that can be taken to be included in such practices to such a degree that they can be regarded as adopting entirely different 'stances' that cannot be straightforwardly adjudicated between (van Fraassen 2002; also see Rowbottom 2011).[3] And of course, non-standard forms of realism that commit to different accounts of what it is for a theory to be true will entail appeal to truth-makers other than 'correspondence' when it comes to the above claim.

Even among 'standard' realists who all take theories to be truth-apt, with truth understood in the usual correspondence sense, and agree that we can know whether a theory is true, or approximately so, there may be differences as to what constitutes the relevant set of philosophical practices. So, for example, consider the recent shift away from what has been called 'over-generalizing wholesale arguments' in the realism debate to those that support forms of 'local' realism (Saatsi 2012) The former are associated with 'global' defences of scientific realism

[3] Of course, as a reader has pointed out, the constructive empiricist does not eschew truth entirely since the empirical adequacy of a theory is cashed out in terms of the truth of the relevant empirical sub-structures. However, to attribute truth to these sub-structures involves a different set of inferential moves and hence different practices than in the case of the theoretical 'super'-structures.

and include the 'No Miracles Argument', for example (for a useful overview, see the relevant section in Chakravartty 2014). There are different ways of framing this argument but the following is sufficient for my purposes: our best theories—such as quantum mechanics—are massively successful in empirical terms (that is, in terms of novel—and confirmed—predictions, explanations of phenomena, and so on); the best explanation of such success is that these theories appropriately represent the relevant features of the world, since the alternative is to take this success to be nothing short of miraculous (and all things considered we should prefer any explanation other than an appeal to a miracle); and the most appropriate way of cashing out 'appropriately represent' is in terms of (correspondence) truth; hence, the best explanation of the empirical success of our best theories is that they are true.

Now, this argument has been subjected to numerous challenges and counters (again see Chakravartty ibid.). However, I'm not going to get into the the ins and outs of the debate here: my point is simply is that if we were to accept such a global move, this argument would be part of the inferential practice of the realist which can serve to justify the claim that 'theory *T* is true'. Putting it another way, when we ask of the realist, 'on what basis do you claim that quantum mechanics, say, is true?', if she is operating in global mode, she will respond by appealing to something like the No Miracles Argument, where that appeal is manifested in certain concrete (philosophical) practices, whether oral, written, through the medium of dance, or whatever.

Furthermore, as presented above the No Miracles Argument incorporates a well-known inferential move, namely 'Inference to the Best Explanation' (IBE). Some realists have then (meta-)justified their deployment of NMA on the grounds that scientists themselves also use IBE to justify acceptance of their favoured theories as true; hence, the claim is, realism can be justified on exactly the same grounds, using the same inferential means, as scientific theories themselves, and thus is naturalistically acceptable. In this sense, the practices of scientists and (realist) philosophers overlap. However, this move has been famously dismissed as question-begging, since many antirealists will of course reject claims that scientists should or even do accept theories as true. Of course, the realist can drop this justification for NMA, accepting, for example, that realism itself is a different fish in a different kettle from scientific theories (thus, again, one might regard it as a 'stance', rather than a theory per se, philosophical or otherwise) and I'll come back to this shortly (see again Saatsi 2012 for more on the arguments justifying or rejecting appeals to NMA), but for now I just want to use this issue as a means of highlighting possible differences between scientific and philosophical practices.

So, one might either insist that NMA should not be seen as or cast in the form of an IBE or that it can be but the latter is in fact not part of scientific practice, in which case(s) we basically have something that philosophers do—some inferential move they appeal to—that scientists do not. In that case, obviously, the set of practices would be fundamentally different in kind and this difference might then be linked to the shift from the 'object' level of scientific practices to the 'meta-'level of those in the philosophy of science; in effect, the shift gets demarcated by means of the kind of inferential move deployed. Or one might argue that NMA *is* a form of IBE and that IBE is used by scientists themselves, or indeed, even further, that IBE in general is nothing more than a form of abductive inference found in everyday reasoning as well as the scientific and philosophical forms (Douven 2011). In this case, there is no real difference between the core inferential move made at the level of science and at the level of the philosophy of science, so there is no sharp distinction between the relevant kinds of practices and likewise no shift in 'levels'. Thus, just to emphasize: whether one sees those practices as, in some sense, contiguous, or not, is going to depend on all sorts of other factors regarding how one views or frames NMA, IBE, abduction, and so on.

More radically, however, someone might eschew NMA entirely as the appropriate way of justifying or supporting a realist stance. Thus, as just noted, there have been recent moves toward a more 'local' approach that draws on the claim that different scientific domains may exhibit different explanatory and inferential practices (Saatsi 2012, p. 13). In neuroscience, say, accounts of explanation that emphasize causal or mechanical features appear to be more appropriate, whereas in modern physics, non-causal forms of explanation may be deployed (ibid., p. 14; on the latter kind of explanation, see in particular, Saatsi, forthcoming and French and Saatsi 2018). More generally, the 'scientific enterprise on the whole exhibits variation in many respects relevant for the realism debate'. Not surprisingly, perhaps, given these differences, there may also be differences in the strength of the (abductive) connection to truth: the non-causal explanations of quantum mechanics, for example, may be deemed to be less reliable in tracking the truth than the mechanical explanations of neuroscience (Saatsi 2012, p. 14).

In that case, the realism we adopt, and arguments in its favour, are going to be tied to specific domains and perhaps, even further, to specific theories. The idea, then, would be that instead of advocating and defending scientific realism *tout court*, we should do so with regard to specific and localized realist stances about particular theories. Thus, the practices that justify the realist's assertion of the claim that 'theory *T* is true' would be not a set of practices that justify any such claim, such as those associated with NMA, but those tied to or associated with *T* alone. In other words, the practices on the basis of which we accept 'quantum mechanics is true' would be different from those that support the claim 'cognitive neuroscience is true'.

Now, of course you might simply balk at the extension of this whole framework of eliminativism and truth-makers to the truth of claims about the truth of theories! After all, how is it appropriate to take a philosophical position such as realism to be itself true—at least in the standard, correspondence sense—for what would this position *correspond* to? Or, thinking of representation again, if theories are said to be true, or approximately so, by virtue of the fact that they represent certain features of the world, what is it that realism, as a philosophical stance, could itself be said to *represent* in this way?

Truth and *Truth*

One option here might be to take a pluralist view of truth itself, with truth-as-correspondence appropriate for certain domains of discourse but not others (see Wright 1999; or for a general introduction see Pedersen and Wright, 2016). In that domain of discourse that concerns the realism–antirealism debate, in which realist claims about the truth of theories are themselves taken to be true (by realists) by virtue of certain elements of philosophical practice, perhaps the appropriate notion of truth is something much more deflationary, like warranted assertibility, according to which proposition *p* is true if we are warranted in asserting *p*. In our context, we could then say that the statement 'theory *T* is true' is true not by virtue of corresponding to anything, and that relatedly, realism is true, not by virtue of the position itself representing anything, but in both cases by virtue of our being warranted in asserting that statement or more generally realist claims respectively. And that we have such warrant is precisely guaranteed by the relevant practices, of course.

If you feel queasy about this kind of pluralism as applied to truth, you might prefer some form of functionalism, according to which we should consider truth in terms of the function it plays in the relevant discourse (Lynch 2009). As such a concept, truth can be realized in different ways (and one can find this multiple realizability in other domains, and applicable to other concepts, such as mental concepts for example); so, when it comes to physical objects, truth is realized in terms of correspondence between propositions and the relevant states of affairs but when it comes to moral statements, truth is realized as, again, warranted assertibility (ibid.). Similarly, when it comes to statements about theories themselves, truth is manifested likewise as warranted assertibility, which is justified, of course, since here, on the eliminativist view, at least, we are not concerned with claims about that which might exist (presumably if one believed theories to be abstract entities one would have to take claims about them to be true in some sort of correspondence sense). This functionalist perspective might be seen as 'pluralism-light': truth is identified with one property—that of playing a certain role—but that role is multiply realizable. Thus, you can be a realist about the world, and in particular, its unobservable features, where that is cashed out via correspondence truth in the usual way, but be an antirealist or, in particular, an eliminativist *about theories themselves*, and take statements *about them* to be true, but in a different sense, such as that of warranted assertibility.

The eliminativist has options, when it comes to truth, then. But what about representation, as covered in Chapter 3? How we should understand that from this eliminativist perspective?

Representation without the Representation

Here, it would seem, we can make similar moves: statements such as those we looked at in Chapter 3, namely 'Model *M* is a good/accurate/faithful/whatever representation of target system *S*', are made true by a certain relationship holding between the model and the system, where this relationship can be cashed out in terms of *M* being similar, in certain respects and to certain degrees, to *S* and then formally characterized in terms of partial isomorphism, say (where *S* is understood to instantiate an appropriate structure, of course).[4] The claim that '*M* is a good/accurate/faithful/whatever representation of *S*' is then justified by certain features of the relevant practices, where again, these will cover both scientific and philosophical practices, insofar as 'good/accurate/faithful/whatever', for example, goes beyond empirical adequacy. Just as we might enquire about the practices—the inferences, judgements, moves in general—that would lead us (or at least, the realists among us) to conclude that a theory should be regarded as true, or approximately true, so we can ask about the similar moves that are made in determining whether a representation is faithful or not.

But of course, determining that a model is similar to the system is akin to the determination that a theory is true of the world and likewise—when it comes to unobservable features, anyway—an epistemic leap must take place (one that the antirealist will resist making, of course). And as with theories, this leap will be motivated by explanatory and predictive success. Thus, consider the liquid drop model of the nucleus, which represents the atomic nucleus as a semi-classical liquid (strictly a Fermi fluid), on the basis of the (approximately) constant density of nuclei yielding a (positive) analogy between such nuclei and liquid drops (see https://en.wikipedia.org/wiki/Semi-empirical_mass_formula). Famously, this model explained nuclear fission in terms of the splitting of the drop (see Bohr 1939) and informed by the relevant experimental results yields the expression for nuclear binding energy known as the Weiszacker formula or semi-empirical mass formula. In updated form, as the 'finite range liquid drop model' it yields accurate descriptions/predictions of the height of fission barriers and of nuclear masses (Litvinov and Sobiczewski 2012). Such success underpins the claim that this model is a reasonably or partially faithful representation of atomic nuclei.

[4] Complete faithfulness would be satisfied by an isomorphism holding between *M* and *S* and degrees of faithfulness could then be captured by the framework of partial isomorphisms, with a particular metric adopted to capture the gradability of the faithfulness (da Costa and French 1990).

Of course, the model incorporates crucial idealizations and misrepresents in certain respects features of the target system.[5] However, it goes too far to dismiss truth entirely in such cases; rather one can argue that insofar as this model is successful, even if only minimally, it should be regarded as partially or pragmatically true (da Costa and French 1990; da Costa and French 2003). Again we can call on a functional conception of truth to accommodate the partial or pragmatic truth of models themselves and the warranted assertibility of our (meta-level) claims about the faithfulness of such modes qua representations. And again, the *determination* of that faithfulness and, particularly, the level or grade of faithfulness occurs at this meta-level of reflection on the representational relationship, whether by philosophers of science or appropriately reflective scientists.

Now, as with my account of theories in general, it is important to distinguish the above view of representation from recent kinds of 'deflationary' approaches, where the latter take the relevant practices to be *constitutive* of representation. Such deflationary views can be divided into the following approaches:

'no-theory'—representation neither possesses nor requires necessary or sufficient conditions

(because it is not a genuine explanatory or substantive concept; Suárez 2015, p. 39);

'abstract minimalism'—although one can give an account of the various platitudes that characterize the abstract concept of 'representation', there is no general account of its use across different domains and thus that there is no universal pattern to its application;

'use-based'—'representation' is a relation that is susceptible of philosophical analysis, but maintains that it is entirely so analysed in terms of its use (or the use of the corresponding predicate) in our linguistic practices.

So the DDI account, as discussed in Chapter 3, eschews necessary and sufficient conditions (Hughes 1997, p. 329) and so appears appropriately deflationary, although the role of denotation, as standardly understood, appears to introduce a substantive element (Suárez 2015, p. 41).[6] One way round this would be to replace denotation with 'denotative function', understood as a feature of use and understand the relevant sense of mapping as not involving an independently existing relation but as simply relating certain claims regarding the source with certain

[5] We recall the point that successful representation involves 'correct surrogative reasoning' (see Swoyer 1991) by which one uses the vehicle of representation—the model, say—to learn about the target but the conclusions of this surrogative reasoning do not need to be true; rather, the model is said to be more or less 'faithful' according to how many of the inferences are sound (and if all are, the model is said to be completely faithful).

[6] Indeed, as Suárez notes, the account is a hybrid, since denotation is a relation, as normally conceived, and demonstration and interpretation can be understood as activities on the part of the user (Suárez 2015, p. 43).

claims regarding the target. By extending the DDI account in this way, all the relevant features of representation can be understood as features of representational *practice* (ibid., p. 45).

Alternatively, according to the inferential account, representation is a 'two vector' notion, requiring on the one hand 'representational force' and on the other hand 'inferential capacities'. The former is a feature of the practice by which the intended target is determined; the latter concerns the capacity of the model to be used by an 'informed and competent user' to draw valid inferences regarding the target, where such capacity is embodied in the internal structure of the model (Suárez 2010). This account is straightforwardly deflationary in any of the above three senses, each of which highlights different features of it. The upshot then is that 'it is impossible, on a deflationary account, for the concept of representation in any area of science to be at variance with the norms that govern representational practice in that area...representation in that area, if anything at all, is *nothing but* that practice' (Suárez 2015, p. 38; see also Knuuttila 2005). This is the core point: on this sort of account, the relevant practices are *constitutive* of the concept of representation and thus any enquiry into the nature of scientific representation reduces to an examination of the relevant forms of practice.[7]

Now, I would argue that this last feature is true of my approach also, insofar as a consideration of the justification of certain claims about particular models qua representations will require the examination of the relevant practices. However, there is a concern about the above deflationary analysis that has to do with the different levels of consideration involved. At the 'object' level, of course, we have the various practices of representation themselves, which might be organized or conceptualized into levels, or distinct domains or whatever; at the meta-level, we have the philosophical (or philosophically reflective, whether by philosophers, scientists, or 'lay' folk) representation of those representational practices, via isomorphism, homomorphism, partial isomorphism, denotation, exemplification, or whatever. By collapsing these two levels, via the constitutive move, the deflationist purchases a focus on the relevant practices—indeed it becomes the sole focus—at the expense of denying philosophers, and like-minded scientists and lay people, the opportunity to reflect on those practices via appropriate devices, formal and otherwise.

Indeed, one might compare this deflationist approach to representation with the 'natural ontological attitude' (Fine 1986), that we touched on above. NOA rejects any meta-level narrative about science by either realists or antirealists, comparing this to the notes one (presumably) receives at the opera, telling one

[7] Thus, 'the study of scientific representation must always be carried out in explicit or implicit reference to a particular scientific practice' (Suárez 2015, p. 47) and as Suárez acknowledges, this meshes with recent moves towards conceptualizing the philosophy of science as 'philosophy of science in practice'.

what is going on. Instead, we should simply focus our collective attention on the practices—on the opera itself, as it were—and simply follow science in accepting what is real and what is not. In response, we might just suggest that urging philosophers to throw away their narratives and their metaphorical opera notes would gut the field of its content, reducing it to mere reportage of scientists' moves and opinions.[8]

More recently, in discussions over the nature of structure invoked by structural realists, Brading and Landry have argued for a 'minimalist' line, according to which we should focus on those structures that are doing all the 'work' at the level of science itself, and not introduce further (and primarily, set-theoretic) structure at the meta-level of the realism–antirealism debate (Brading and Landry 2007). In response, I have tried to draw a distinction between structure as presented at the level of the science itself—such as group theoretic structure in the case of modern physics—and the structures (set-theoretic, typically) that are used at the level of the philosophy of science to *represent* those former structures and associated practices (French 2014, ch. 5). As discussed here, we can be pluralists about these meta-level representations, suggesting that certain of them are more useful for some philosophical purposes than others, but ultimately we need something, some form of meta-representational narrative, if we are to do philosophy of science in this context at all. Indeed, I suggested that if we were to adopt the above sort of approach, we would be reduced to advocating a form of meta-positivism, in which the defence of a structuralist account amounted to little more than a recitation of the various kinds of structures presented by scientists themselves.

Now, that's perhaps a little dramatic, but I feel that the same can be said of deflationist accounts of representation: philosophically reflecting on, talking about, and generally discussing scientific representation would be reduced to little more than a re-presentation of the relevant practices: in this domain, scientists use these forms of representation, in that context, they use those others; here they do this, there they do that; and nothing more. The philosophy of science would become a 'readers' digest' of scientists' sayings and doings and you do not have to be a gung ho advocate of formal approaches to appreciate that there are advantages to couching such sayings and doings within some overarching framework, such as the Semantic Approach, even one that is domain or context dependent. Note, in particular, that by adopting such an approach I am not insisting that partial isomorphisms, say, should be conceived of as independently existing relations—all they are, are devices that we may use at the level of the philosophy of science to represent the 'actual' representational relations found at the level of science. In *that* sense, I am uncomfortable with the contrast with deflationary approaches—I would argue that the object/meta-level distinction can be kept

[8] But for a counterweight, see Ruetsche (2015) who reinterprets Arthur Fine's NOA in line with a piecemeal approach of local sense-making projects that she calls 'Locavoracity'.

without having to reify the relations of the meta-level as 'independently existing', 'substantive', or 'actual' in any sense. And as I've already said, we can still retain the emphasis on practices at the object level via their role as truth-makers and justifiers of claims made at the meta-level.

Still, there is yet a further concern that we have already touched on in the context of the correspondence theory of truth: talk of representation suggests, at least, there is a *that* which is doing the representing; that is, some*thing* that stands at one end of the representational relationship. And eliminativism about theories denies this.

What Is Doing the Representing?

One option—the hard line, perhaps—would be to insist that's just a bullet we have to bite down on. There is strictly *nothing* standing in the representational relationship—theories are *not* like paintings, or depictive art more generally, from which so many examples and counter-examples have been drawn and representation is an inappropriate device for capturing the kinds of claims we want to make. This would indeed be a hard line to take as it would eliminate not only theories, as things, and not only representation as an undeflated relation, but all meta-level talk and consideration of such. Perhaps this could be made to mesh with the deflationary account above—or even be seen as the extension, beyond *reductio*, of it: all we would be entitled to do is present the 'facts' of practice and when scientists use the term 'representation' in the context of theories and models purportedly representing systems or phenomena, we flag or index that use in some way, as an example of 'folk-talk' that is strictly without content, as there is nothing standing in that relationship and thus no relationship to be stood in! As I said, that's a hard line to pursue.

An alternative, softer, response would be to suggest that what is standing in that relationship is a *meta-level construction* that we, as philosophers of science (or scientists when they are in philosophical mode), deploy when we endeavour to make sense of scientific practice and its implications for our understanding of the world. The idea then is that when we, philosophers (or again, scientists or others thinking philosophically), talk (seriously) about theories or models representing some target system, we have in mind, if perhaps only implicitly, some way of 'representing' theories themselves and these systems. If you are an adherent of the Syntactic Approach, then the theory can be said to be 'represented' logico-linguistically and a referential relationship with the relevant elements of reality established in those terms. If you are an advocate of the Semantic Approach, then, likewise, you 'represent' the theory at the meta-level of the philosophy of science in terms of set theory and also 'represent' the way the theory latches onto the world via the formal notion of (partial) isomorphism, which of course then relates to the formal representation of the target system (in order to sidestep protests that set theoretical notions cannot relate mathematical structures and physical ones).

However, I've put the word *represent* in scare quotes because there is nothing for either the logico-lingustic set of propositions or set theoretical structure to actually (meta-)represent. Instead, these devices should be understood as constructions that we philosophers of science introduce and use to do our work, just as the scientists' use of 'theory' is a construction that they introduce and invoke to make the distinctions they are concerned with and to bring together and throw light on the practices they deem to be important, as we have seen. And that work we do is different from scientists' work—we can give all sorts of appropriately nuanced distinctions between the two but put crudely, theirs is to make sense of the world (in either its observable or unobservable aspects or both), whereas ours is to make sense of science itself, that is, to make sense of that making sense of the world. And to repeat, that making sense involves more than simply reading off or reciting the relevant practices. Thus, the object-level relation of representation does not hold between the relevant features of the world and theory, per se, but between those features and our logico-linguistic or set-theoretic or whatever, *construction*. Furthermore, statements about *that* representation are made true by the appropriate features of the practice of the history and philosophy of science.

When we use the Semantic Approach to capture, at the meta-level, the equivalence of wave and matrix mechanics (Muller 1997), then, we are not representing some*thing* at the object level, namely quantum mechanics as formulated in terms of Hilbert space; rather we are presenting a construction that enables us, as philosophers of science, to make sense of certain features of scientific practice. And different sets of features can be captured in different ways, either within the Semantic Approach or using other formal devices entirely, without any commitment to one or other such device uniquely representing 'the' theory. Indeed, extending what Wilson says about classical mechanics, the relevant scientific practices are so diverse and heterogenous that reconceiving what we do as philosophers in terms of meta-level constructions that enable us to make (partial) sense of such practices surely seems the way to go!

Conclusion: What Are We Doing as Philosophers of Science?

As philosophers of science, we deploy various resources, such as the Semantic Approach for example, in order to do what we want to do as philosophers and thus to achieve our aims. These resources come with various benefits and disadvantages and one can argue for a pluralistic view that suggests we deploy whatever available resources are most appropriate for the job at hand. When it comes to capturing inter-theoretic or theory-data relations, like many people I think the Semantic

Approach offers the best set of tools for the job. But it would be a mistake to move from that view to the claim that theories *are* set-theoretic in some sense.

Of course, new developments in the philosophy of science put pressure on those resources. However, there are ways of resisting or, at least, sidestepping that pressure via the appropriation of further resources, including some that are metaphysical or methodological. Thus, we can cut through concerns about what sorts of things theories *are* and end that particular debate by importing a form of eliminativism and associated truth-maker theory from metaphysics. This allows us to continue talking about theories without being committed to them as abstract entities or whatever. From this perspective the Semantic Approach can be seen as *no more than a construction*, deployed by us as philosophers of science to achieve our own ends and certainly not as constitutive (either of theories or the structure of the world).

The focus then shifts back, as it should and as the deflationist demands, to the practices that ground the truth of the claims we make about theories, rather than on the ontological nature and identity conditions of the latter. Just as, it is claimed, new forms of realism consistent with modern physics should free us from a dependence on outdated metaphysics, so this view of theories themselves should free us from relying on sometimes inappropriate comparisons with artworks and encourage us to pay more attention to the particularities of the practice of science itself.

Here we might recall Thomasson's approach to fictions, and her insistence that the role of the philosophy of fiction must be 'to extract and make explicit the principles that are embodied in our practices, assess what these commit us to (or what principles they tacitly presuppose), and how we can best make a consistent, coherent theory that accommodates them' (Thomasson 2003, pp. 146–7). Any philosophy that is so revisionary that it conflicts with those practices must then be regarded as at best a view of what the relevant fictional characters might or should be (which may still be illuminating) rather than how they are. And it is these practices that determine the existence conditions for such fictional characters, such that having pinned down such conditions there is nothing further to be said on the matter of the character's existence. And this, of course, can be straightforwardly repurposed for eliminativism.

Recall also Cameron's statement that,

> one of the benefits of truthmaker theory is to allow that <*x* exists> might be made true by something other than *x*, and hence that '*a* exists' might be true according to some theory without *a* being an ontological commitment of that theory. (Cameron 2008, p. 4)

Thus, as I've tried to suggest, we can still maintain that 'quantum mechanics exists' but with the twist that what makes that claim true is not that there is, in some realm, an abstract entity called 'quantum mechanics', but rather a set of practices 'arranged' across both time and geography. But beyond this, we can perhaps appropriate Thomasson's general approach towards fictions along the following lines, namely that the role of the philosophy of science is to extract and make explicit the principles that are embodied in scientific practices, both current and as presented through the history of science, assess what these commit us to (or what principles they tacitly presuppose), and how we can best make a consistent, coherent (philosophical) theory that accommodates them.

In the context of the extended debate over how to conceive of the relationship between the history of science and the philosophy of science (see for example Dear 2012), this may seem naively inductivist but I have a more iterative conception in mind.

Think of 'local' realism again. Here the idea is that we should eschew general recipes and instead focus on the particular reasons for believing in certain theoretical features that are appropriate to that particular case; or in other words, pay attention to the specific practices associated with that case. It is not being claimed that we inductively infer realism or antirealism from reflection on such practices but rather that having adopted such a stance—whether by virtue of importing certain principles from elsewhere in philosophy or under the influence of certain episodes in the history of philosophy or whatever—we then extract from the relevant practices the 'principles' and reasons that allow us to make sense of realism or antirealism in that specific case. Of course, we may then discover that applying the same move to other cases we come to believe in the same kinds of features (such as certain structures, say!) and hence arrive at what may appear to be a generic recipe; or we may conclude that other cases are not relevantly similar, or that we have good reasons for distinguishing one historical episode or era from others, or whatever. The point is that we should extract the relevant reasons and principles, assess what they commit us to and make a consistent, coherent form of realism that accommodates them.

And we can say something similar when it comes to representation. We don't proceed 'inductively' by examining case after case of purported representation in science; rather, we import, either directly or indirectly, from the philosophy of art some more or less appropriate examples of representation or, perhaps, even a full-blown account and then we examine the relevant scientific practices to identify both examples and 'principles' appropriate to such cases and again construct a coherent account of representation suitable for science, perhaps again importing further features from the philosophy of art, and almost certainly moving to and from that meta-level of account construction to the 'object' level of the practices. But of course, in importing such devices from art and philosophical reflection on art, we obviously need to take care not to allow them to motivate or shape what is ultimately a spurious ontology at that meta-level.

And of course, in moving back and forth in this iterative manner, the relevant differences between the two fields need to be acknowledged. So, coming back to one of our central questions: are theories more like paintings or pieces of music or works of literature? The answer should be that the question is not well-posed as it stands—'theory' is a term that scientists and philosophers of science introduce and throw around for various purposes, in the course of which various meanings and understandings accrete, like barnacles on the hull of a boat. In this case, however, when the barnacles are scraped away, we find there is no boat there! Freeing ourselves from this illusory ontology opens up space for a richer understanding of both the relationship between philosophy of science and philosophy of art and between the former and scientific practice and, more generally, of our *own* practices as philosophers of science.

Bibliography

Achinstein, P. (2002). 'Is There a Valid Experimental Argument for Scientific Realism?'. *Journal of Philosophy* 99(9): 470–95.

Acuña, P. (2019). 'Charting the Landscape of Interpretation, Theory Rivalry, and Underdetermination in Quantum Mechanics'. *Synthese*. doi.org/10.1007/s11229-019-02159-z.

Adlam, E. (2011). 'Poincaré and Special Relativity'. http://arxiv.org/pdf/1112.3175v1.pdf.

Aerts, D., Durt, T., and Czachor, M. (2002). *Probing the Structure of Quantum Mechanics*. Singapore: World Scientific.

Andreas, H. (2014). 'Carnapian Structuralism'. *Erkenntnis* 79(8): 1373–91.

Andreas, H. and Zenker, F. (eds) (2014). 'Perspectives on Structuralism'. *Erkenntnis* 79(8): 1365–1561.

Antoine, J. P., Bishop, R. C., Bohm, A., and Wickramasekara, S. (2009). 'Rigged Hilbert Spaces in Quantum Physics'. *Compendium of Quantum Physics*. Berlin: Springer, pp. 640–50.

Ariew, R. (2018). 'Pierre Duhem', in E. N. Zalta (ed.), *The Stanford Encyclopedia of Philosophy*. https://plato.s tanford.edu/archives/fall2018/entries/duhem/.

Armstrong, D. M. (2004). *Truth and Truthmakers*. Cambridge: Cambridge University Press.

Azzouni, J. (2007). 'The Inconsistency of Natural Languages: How We Live with It'. *Inquiry* 50: 590–605.

Bain, J. (2013). 'Category-theoretic Structure and Radical Ontic Structural Realism'. *Synthese* 190: 1621–35.

Baird, D. (2004). *Thing Knowledge: A Philosophy of Scientific Instruments*. Berkeley, CA: University of California Press.

Baker, L. (2000). *Persons and Bodies: a Constitution View*. Cambridge: Cambridge University Press.

Baker, A. and Colyvan, M. (2011). 'Indexing and Mathematical Explanation'. *Philosophia Mathematica* 19: 323–34.

Baker, L. (2000). *Persons and Bodies: A Constitution View*. Cambridge: Cambridge University Press.

Balaguer, M. (1998). 'Attitudes without Propositions'. *Philosophy and Phenomenological Research* 58: 805–26.

Balaguer, M. (2015). 'Fictionalism in the Philosophy of Mathematics', in E. N. Zalta (ed.), *The Stanford Encyclopedia of Philosophy*. https://plato.stanford.edu/archives/sum2015/entries/fictionalism-mathematics/.

Barthes, R. (1967). 'The Death of the Author'. Aspen issue 5–6; repr. in *Art and Interpretation: An Anthology of Readings in Aesthetics and the Philosophy of Art*, ed. E. Dayton. Peterborough, Ont.: Broadview, pp. 383–6.

Barrett, T. W. and Halvorson, H. (2016). 'Glymour and Quine on Theoretical Equivalence'. *Journal of Philosophical Logic* 45: 467–83.

Bätschmann, O. and Griener, P. (1997). *Hans Holbein*. London: Reaktion Books.

Batterman, R. (2010). 'On the Explanatory Role of Mathematics in Empirical Science'. *The British Journal for the Philosophy of Science* 61: 1–25.

Bell, J. and Machover, M. (1977). *A Course in Mathematical Logic*. Amsterdam: Elsevier.

Beller, M. (1999). *Quantum Dialogue: The Making of a Revolution*. University of Chicago Press.

Benson, P. (2015). 'The Ontology of Photography: From Analogue to Digitial'. *Philosophy Now* 95. https://philosophynow.org/issues/95/The_Ontology_of_Photography_From_Analogue_To_Digital.

Billson, A. (2005). *Buffy the Vampire Slayer*. British Film Institute. London: Bloomsbury, pp. 24–5.

Bitbol, M. (2011). *Schrödinger's Philosophy of Quantum Mechanics*. Dordrecht: Kluwer.

Bloor, D. (1974). 'Essay Review: Popper's Mystification of Objective Knowledge'. *Science Studies* 4: 65–76.

Bogdan, R. J. (1986). 'The Importance of Belief', in R. J. Bogdan (ed.), *Belief: Form, Content and Function*. Oxford: Oxford University Press, pp. 1–16.

Boesch, B. (2015). 'Scientific Representation'. *Internet Encyclopaedia of Philosophy*. http://www.iep.utm.edu/sci-repr/.

Bogen, J. and Woodward, J. (1988). 'Saving the Phenomena'. *Philosophical Review* 97: 303–52.

Bohr, N. (1939). 'Disintegration of Heavy Nuclei'. *Nature* 143: 330.

Bohr, N. (1981). *Neils Bohr Collected Works*, Vol. 2, ed. U. Hoyer. Dordrecht: North-Holland.

Bolinska, A. (2013). 'Epistemic Representation, Informativeness and the Aim of Faithful Representation'. *Synthese* 190: 219–34.

Bourbaki, N. (1968). *Theory of Sets*. Paris: Hermann.

Bowler, P. J. (1976). 'Alfred Russel Wallace's Concepts of Variation'. *Journal of History of Medicine* 31: 17–29.

Born, H., Born, M., and Einstein, A. (1971). *The Born-Einstein Letters*. London: Macmillan.

Brading, K. and Landry, E. (2007). 'Shared Structure and Scientific Structuralism'. *Philosophy of Science* (Proceedings) 73: 571–81.

Braithwaite, R. B. (1962). 'Models in the Empirical Sciences', in E. Nagel, P. Suppes, and A. Tarski (eds), *Logic Methodology and Philosophy of Science*. Stanford: Stanford University Press, pp. 224–31.

Brannigan, A. (1981). *The Social Basis of Scientific Discoveries*. Cambridge: Cambridge University Press.

Brown, H. (2005). *Physical Relativity*. Oxford: Oxford University Press.

Brown, J. (2004). 'Why Thought Experiments Transcend Experience', in C. Hitchcock (ed.), *Contemporary Debates in Philosophy of Science*. Oxford: Blackwell, pp. 23–43.

Brown, L. M. (1996). 'Some QED Myths-in-the-Making'. *Studies in History and Philosophy of Modern Physics* 21: 81–90.

Brown, L. M. (2006). 'Paul A. M. Dirac's *The Principles of Quantum Mechanics*'. *Physics in Perspective* 8: 381–407.

Budd, M. (1993). 'How Pictures Look', in D. Knowles and J. Skorupski (eds), *Virtue and Taste*. Oxford: Blackwell, pp. 154–75.

Bueno, O. (1997). 'Empirical Adequacy: A Partial Structures Approach'. *Studies in History and Philosophy of Science* 28: 585–610.

Bueno, O. (2000). 'Empiricism, Mathematical Change and Scientific Change'. *Studies in History and Philosophy of Science* 31: 269–96.

Bueno, O. (2008). 'Visual Evidence at the Nanoscale'. *Spontaneous Generations* 2: 132–8.

Bueno, O. (2014). 'Nominalism in the Philosophy of Mathematics', in E. N. Zalta (ed.), *The Stanford Encyclopedia of Philosophy*. https://plato.stanford.edu/archives/spr2014/entries/nominalism-mathematics/.

Bueno, O. and Colyvan, M. (2011). 'An Inferential Conception of the Application of Mathematics'. *Noûs* 45: 345–74.

Bueno, O. and French, S. (2011). 'How Theories Represent'. *The British Journal for the Philosophy of Science* 62: 857–94.

Bueno, O. and French, S. (2018). *The Applicability of Mathematics: Immersion, Inference and Interpretation*. Oxford: Oxford University Press.

Bueno, O. and Rowbottom, D. (2011). 'Stance and Rationality: A New Perspective', in D. P. Rowbottom and O. Bueno (eds), *Synthese* 178: 1–5.

Bueno, O., French, S., and Ladyman, J. (2002). 'On Representing the Relationship between the Mathematical and the Empirical'. *Philosophy of Science* 69: 452–73.

Bueno, O., French, S., and Ladyman, J. (2012). 'Empirical Factors and Structural Transference: Returning to the London Account'. *Studies in History and Philosophy of Modern Physics* 43: 95–103.

Bulmer, M. (2005). 'The Theory of Natural Selection of Alfred Russel Wallace FRS'. *Notes Rec. R. Soc.* 59: 125–36.

Burgos, J. E. 2007. 'The Theory Debate in Psychology'. *Behavior and Philosophy* 35: 149–83.

Callender, C. (2011). 'Philosophy of Science and Metaphysics', in S. French and J. Saatsi (eds), *The Continuum Companion to the Philosophy of Science*. London: Continuum, pp. 33–54.

Callender, C. and Cohen, J. (2006). 'There Is No Special Problem about Scientific Representation'. *Theoria* 55: 7–25.

Cameron, R. (2008a). 'There Are No Things That Are Musical Works'. *British Journal of Aesthetics* 48: 295–314.

Cameron, R. (2008b). 'Truthmakers and Ontological Commitment'. *Philosophical Studies* 140: 1–18.

Cameron, R. (2010). 'How to Have a Radically Minimal Ontology'. *Philosophical Studies* 151: 249–64.

Camilleri, K. (2009a). *Heisenberg and the Interpretation of Quantum Mechanics*. Cambridge: Cambridge University Press.

Camilleri, K. (2009b). 'Constructing the Myth of the Copenhagen Interpretation'. *Perspectives on Science* 17: 26–57.

Carlyle, T. (1840). *On Heroes, Hero-Worship and the Heroic in History*. London: Chapman and Hall.

Carnap, R. (1939). *Foundations of Logic and Mathematics*. Foundations of the Unity of Science: Toward an International Encyclopedia of Unified Science, vol. 1(3). Chicago: University of Chicago Press.

Carnap, R. (1956). 'The Methodological Character of Theoretical Concepts', in H. Feigl and M. Scriven (eds). *Minnesota Studies in the Philosophy of Science*, vol. 1. Minneapolis: University of Minnesota Press, pp. 38–76.

Carnap, R. (1966). *Philosophical Foundations of Physics: An Introduction to the Philosophy of Science*, ed. M. Gardner. New York: Basic.

Cartwright, N. (1983). *How the Laws of Physics Lie*. Cambridge: Cambridge University Press.

Cartwright, N. (1990). 'Philosophical Problems of Quantum Theory: The Response of American Physicists'. *The Probabilistic Revolution*, vol. 2. Cambridge, MA: MIT Press, pp. 417–36.

Cartwright, N., Shomar, T., and Suárez, M. (1995). 'The Tool Box of Science (Tools for Building of Models with a Superconductivity Example)', in W. E. Herfel et al. (eds). *Theories and Models in Scientific Processes*. Amsterdam: Editions Rodopi, pp. 137–49.

Cassirer, E. (1956). *Determinism and Indeterminism in Modern Physics*. New Haven, CT: Yale University Press (1936).

Castelvecchi, D. (2014). 'Einstein's Lost Theory Uncovered'. *Nature* 506: 418–19.

Cei, A. and French, S. (2006). 'Looking for Structure in All the Wrong Places: Ramsey Sentences, Multiple Realizability, and Structure'. *Studies in History and Philosophy of Science* 37: 633–55.

Chakravartty, A. (2001). 'The Semantic or Model-theoretic View of Theories and Scientific Realism'. *Synthese* 127: 325–45.

Chakravartty, A. (2007). *A Metaphysics for Scientific Realism: Knowing the Unobservable*. Cambridge: Cambridge University Press.

Chakravartty, A. (2014). 'Scientific Realism', in E. N. Zalta (ed.), *The Stanford Encyclopedia of Philosophy*. http://plato.stanford.edu/archives/spr2014/entries/scientific-realism/.

Chakravartty, A. (2017). *Scientific Ontology*. Oxford: Oxford University Press.

Chalmers, D. (2002). 'Does Conceivability Entail Possibility', in J. Hawthorne and T. Gendler (eds), *Conceivability and Possibility*. Oxford: Oxford University Press, pp. 145–200.

Church, R. (1984). 'Popper's World 3 and the Problem of the Printed Line'. *Australasian Journal of Philosophy* 62: 378–91.

Churchland, P. (1989). *A Neurocomputational Perspective: The Nature of Mind and the Structure of Science*. Cambridge, MA: MIT Press.

Clark, A. and Chalmers, D. J. (1998). 'The Extended Mind'. *Analysis* 58: 7–19.

Clark, W. A. V. and Fossett, M. (2008). 'Understanding the Social Context of the Schelling Segregation Model'. *Proceedings of the National Academy of Sciences* 105: 4109–14.

Coffey, K. (2014). 'Theoretical Equivalence as Interpretative Equivalence'. *British Journal for the Philosophy of Science* 65: 821–44.

Coleman, A. J. (1997). 'Groups and Physics: Dogmatic Opinions of a Senior Citizen'. *Notices of the American Mathematical Society* 44: 8–17.

Collier, J. D. (1992). 'Critical Notice: Paul Thompson: The Structure of Biological Theories'. *Canadian Journal of Philosophy* 22: 287–98.

Collingwood, R. G. (1938/71). *The Principles of Art*. Oxford: Clarendon Press.

Collins, D. (1992). 'Anamorphism and the Eccentric Observer'. *Leonardo* 25(1–2). http://www.public.asu.edu/~dan53/Anamorph.html.

Colyvan, M. (2010). 'There's No Easy Road to Nominalism'. *Mind* 119: 285–306.

Contessa, G. (2007). 'Scientific Representation, Interpretation, and Surrogative Reasoning'. *Philosophy of Science* 74: 48–68.

Contessa, G. (2010). 'Scientific Models and Fictional Objects'. *Synthese* 172: 215–29.

Contessa, G. (2011). 'Scientific Models and Representation', in J. Saatsi and S. French (eds), *The Continuum Companion to the Philosophy of Science*. London: Bloomsbury, pp. 120–37.

Cook, N. D. (1990). 'The Attenuation of the Periodic Table'. *Modern Physics Letters A* 5(17): 1321.

Cook, N. D. (2010). *Models of the Atomic Nucleus: Unification through a Lattice of Nucleons*. Berlin: Springer.

Corry, L. (1992). 'Nicolas Bourbaki and the Concept of Mathematical Structure Author(s): Leo Corry'. *Synthese* 92(3): 315–48.

Craig, W. (1953). 'On Axiomatizability within a System'. *Journal of Symbolic Logic* 18(1): 30–2.

Crawford, S. (2005). 'Propositions', in K. Brown (ed.). *The Encyclopaedia of Language and Linguistics*. 2nd edn. Amsterdam: Elsevier.

Creath, R. (2014). 'Logical Empiricism', in E. N. Zalta (ed.), *The Stanford Encyclopedia of Philosophy*. http://plato.stanford.edu/archives/spr2014/entries/logical-empiricism/.

Crick, F. (1988). *What Mad Pursuit*. New York: Basic Books.

Curiel, E. (2014). 'Classical Mechanics Is Lagrangian; It Is Not Hamiltonian'. *British Journal for the Philosophy of Science* 65: 269–321.

Currie, A. (2017). 'Hot-Blooded Gluttons: Dependence, Method and Coherence in the Historical Sciences'. *The British Journal for the Philosophy of Science* 68: 929–52.

Currie, G. (1990). *The Nature of Fiction*. Cambridge: Cambridge University Press.

Cushing, J. T. (1993). 'Underdetermination, Conventionalism and Realism: The Copenhagen vs. the Bohm Interpretation of Quantum Mechanics', in S. French and H. Kamminga (eds), *Correspondence, Invariance and Heuristics*. Dordrecht: Kluwer. 261–78.

Cushing, J. T. (1994). *Quantum Mechanics: Historical Contingency and the Copenhagen Hegemony*. Chicago: University of Chicago Press.

da Costa, N. C. A., and Chuaqui, R. (1988). 'On Suppes' Set-Theoretical Predicates. *Erkenntnis* 29: 95–112.

da Costa, N. C. A. and French, S. (1990). 'The Model-Theoretic Approach in the Philosophy of Science'. *Philosophy of Science* 57: 248–65.

da Costa, N. C. A. and French, S. (2003). *Science and Partial Truth*. Oxford: Oxford University Press.

Darby, G., Pickup, M., and Robson, J. (2017). 'Deep Indeterminacy in Physics and Fiction', in O. Bueno et al. (eds), *Thinking about Science, Reflecting on Art*. London: Routledge, pp. 99–118.

Darwin, C. G. (1931). *The New Conceptions of Matter*. London: Bell and Sons, Ltd.

Davies, D. (2004). *Art as Performance*. Malden, MA: Blackwell.

Davies, D. (2008). 'Collingwood's "Performance" Theory of Art'. *British Journal of Aesthetics* 48(2): 162–74.

Dear, P. (2012). 'Philosophy of Science and Its Historical Reconstructions', in S. Mauskopf and T. Schmaltz (eds), *Integrating History and Philosophy of Science: Problems and Prospects*. Boston Studies in the Philosophy of Science. London: Springer.

del Campo, J., Goold, J., and Paternostro, M. (2014). 'More Bang for Your Buck: Super-adiabatic Quantum Engines'. *Scientific Reports* 4(6208). doi:10.1038/srep06208.

Demopoulos, W. (2007). 'Carnap and the Reconstruction of Scientific Theories', in R. Creath and M. Friedman (eds), *The Cambridge Companion to Carnap*. Cambridge: Cambridge University Press, pp. 248–72.

Devitt, M. (1981). *Designation*. New York: Columbia University Press.

Di Lazarro, P., Murra, D., and Vitelli, P. (2019). 'The Interdisciplinary Nature of Anamorphic Images in a Journey through Art, History and Geometry'. *Journal of Mathematics and the Arts*. doi:10.1080/17513472.2018.1506627.

Dirac, P. (1930). *The Principles of Quantum Mechanics*. Oxford: Clarendon Press.

Distler, J. (2015). 'Elegant Bach Toccatas'. *Classics Today*. http://www.classicstoday.com/review/elegant-bach-toccatas/.

Dodd, J. (2000). 'Musical Works as Eternal Types'. *British Journal of Aesthetics* 40: 424–40.

Dodd, J. (2002). 'Defending Musical Platonism'. *British Journal of Aesthetics* 42: 380–402.

Dolnick, E. (2008). *The Forger's Spell: A True Story of Vermeer, Nazis, and the Greatest Art Hoax of the Twentieth Century*. New York: Harper Collins.

Dorr, C. (2007). 'There Are No Abstract Objects', in J. Hawthorne, T. Sider, and D. Zimmerman (eds), *Contemporary Debates in Metaphysics*. Oxford: Blackwell, pp. 32–64.

Douven, I. (2011). 'Abduction', in E. N. Zalta (ed.), *The Stanford Encyclopedia of Philosophy*. http://plato.stanford.edu/archives/spr2011/entries/abduction/.

Downes, S. M. (1992). 'The Importance of Models in Theorizing: A Deflationary Semantic View'. *Philosophy of Science Association* 1: 142–53.

Downes, S. M. (2009). 'Models, Pictures, and Unified Accounts of Representation: Lessons from Aesthetics for Philosophy of Science'. *Perspectives on Science* 17: 417–28.

Eckert, M. (2014). 'How Sommerfeld Extended Bohr's Model of the Atom (1913–1916)'. *The European Physical Journal H* 39: 141–56.

Eddington, A. (1928). *The Nature of the Physical World*. Gifford Lectures, Edinburgh University, 1927. Cambridge: Cambridge University Press.

Einstein, A. (1905a). 'On the Electrodynamics of Moving Bodies'. *Annalen der Physik* 17: 891–921. Eng. trans. http://einsteinpapers.press.princeton.edu/vol2-trans/154.

Einstein, A. (1905b). 'Ist die Trägheit eines Körpers von seinem Energieinhalt abhängig?'. *Annalen der Physik* 18: 639–43. Eng. trans. http://www.fourmilab.ch/etexts/einstein/E_mc2/www/.

Eklund, M. (2011). 'Fictionalism', in E. N. Zalta (ed.), *The Stanford Encyclopedia of Philosophy*. http://plato.stanford.edu/archives/fall2011/entries/fictionalism/.

Elgin, C. (1983). *With Reference to Reference*. Indianapolis: Hackett.

Esfeld, M., Hubert, M., Lazarovici, D., and Dürr, D. (2017). 'The Ontology of Bohmian Mechanics'. *British Journal for the Philosophy of Science* 65: 773–96.

Feynman, R. (1985). *Surely You're Joking, Mr. Feynman*. New York: W. W. Norton.

Field, H. (1980). *Science without Numbers*. Princeton, NJ: Princeton University Press.

Fine, A. (1986). 'The Natural Ontological Attitude', in *The Shaky Game: Einstein, Realism, and the Quantum Theory*. Chicago: Chicago University Press.

Fish, S. (1980). *Is There a Text in This Class*. Cambridge, MA: Harvard University Press.

Forster, M. and Sober, E. (1994). 'How to Tell When Simpler, More Unified, or Less *Ad Hoc* Theories Will Provide More Accurate Predictions'. *British Journal for the Philosophy of Science* 45: 1–35.

Frankel, H. (1987). 'The Continental Drift Debate', in H. T. Engelhardt Jr and A. L. Caplan (eds), *Scientific Controversies: Case Solutions in the Resolution and Closure of Disputes in Science and Technology*. Cambridge: Cambridge University Press.

Fraser, D. (2009). 'Quantum Field Theory: Underdetermination, Inconsistency, and Idealization'. *Philosophy of Science* 76: 536–67.

Fraser, J. (2016). *What Is Quantum Field Theory?* PhD Thesis, University of Leeds.

Fraser, J. (forthcoming). 'The Real Problem with Perturbative Quantum Field Theory'. *British Journal for the Philosophy of Science*.

French, S. (1999). 'Models and Mathematics in Physics: The Role of Group Theory', in J. Butterfield and C. Pagonis (eds), *From Physics to Philosophy*. Cambridge: Cambridge University Press, pp. 187–207.

French, S. (2002). 'A Model-Theoretic Account of Representation (or, I Don't Know Much About Art... But I Know it Involves Isomorphism'. *Philosophy of Science* 70: 1472–83.

French, S. (2008). 'Symmetry, Invariance and Reference', in M. Frauchiger and W. K. Essler (eds), *Representation, Evidence, and Justification: Themes from Suppes*. Lauener Library of Analytical Philosophy, vol. 1. Frankfurt: Ontos Verlag, pp. 127–56.

French, S. (2010). 'Keeping Quiet on the Ontology of Models'. *Synthese* 172: 231–49.

French, S. (2014). *The Structure of the World*. Oxford: Oxford University Press.

French, S. (2017). 'Identity Conditions, Idealisations and Isomorphisms: A Defence of the Semantic Approach'. *Synthese* 194: 3311–26. doi:10.1007/s11229-017-1564-z.

French, S. (2018). 'Realism and Metaphysics', in J. Saatsi (ed.), *The Routledge Handbook of Realism*. London: Routledge, pp. 395–406.

French, S. (forthcominga). 'Doing Away with Dispositions', in A. Meinke (ed.), *Dispositionalism: Perspectives from Metaphysics and the Philosophy of Science*. Springer Synthese Library. Berlin: Springer.

French, S. (forthcomingb), 'Imagination in Science and Art'.

French, S. and Krause, D. (2006). *Identity in Physics*. Oxford: Oxford University Press.

French, S. and Ladyman, J. (1997). 'Superconductivity and Structures: Revisiting the London Account'. *Studies in History and Philosophy of Modern Physics* 28: 363–93.

French, S. and Ladyman, J. (1998). 'A Semantic Perspective on Idealisation in Quantum Mechanics', in N. Shanks (ed.), *Idealization IX: Idealization in Contemporary Physics*. Pozna´n Studies in the Philosophy of the Sciences and the Humanities. Amsterdam: Rodopi, pp. 51–73.

French, S. and Ladyman, J. (1999). 'Reinflating the Model-Theoretic Approach'. *International Studies in the Philosophy of Science* 13: 99–117.

French, S. and Ladyman, J. (2011). 'In Defence of Ontic Structural Realism', in A. Bokulich and P. Bokulich (eds), *Scientific Structuralism*. Boston Studies in the Philosophy of Science. Berlin: Springer, pp. 25–42.

French, S. and McKenzie, K. (2012). 'Thinking Outside the (Tool)Box: Towards a More Productive Engagement between Metaphysics and Philosophy of Physics'. *The European Journal of Analytic Philosophy* 8: 42–59.

French, S. and McKenzie, K. (2015). 'Rethinking Outside the Toolbox: Reflecting Again on the Relationship between Philosophy of Science and Metaphysics', in T. Bigaj and C. Wuthrich (eds), *Metaphysics in Contemporary Physics*. Poznan Studies in the Philosophy of the Sciences and the Humanities. Amsterdam: Rodopi, pp. 145–74.

French, S. and Saatsi, J. (2006). 'Realism about Structure: The Semantic View and Non-linguistic Representations'. *Philosophy of Science* (Proceedings) 78: 548–59.

French, S. and Saatsi, J. (2018). 'Symmetries and Explanatory Dependencies in Physics', in A. Reutlinger and J. Saatsi (eds), *Explanation beyond Causation: Philosophical Perspectives on Non-Causal Explanations*. Oxford: Oxford University Press.

French, S. and Vickers, P. (2011). 'Are There No Such Things as Theories?'. *The British Journal for the Philosophy of Science* 62: 771–804.

Friederich, S. (2015). *Interpreting Quantum Theory: A Therapeutic Approach*. London: Palgrave Macmillan.

Friedman, M. (1982). 'Review of *The Scientific Image*'. *Journal of Philosophy* 79: 274–83.

Frigg, R. (2006). 'Scientific Representation and the Semantic View of Theories'. *Theoria* 21: 49–65.

Frigg, R. (2010a). 'Models and Fictions'. *Synthese* 172: 251–68.

Frigg, R. (2010b). 'Fiction and Scientific Representation', in R. Frigg and M. Hunter (eds), *Beyond Mimesis and Nominalism: Representation in Art and Science*. Berlin and New York: Springer, pp. 97–138.

Frigg, R. and Hartmann, S. (2018). 'Models in Science', in E. N. Zalta (ed.), *The Stanford Encyclopedia of Philosophy*. http://plato.stanford.edu/archives/sum2018/entries/models-science/.

Frigg, R. and Nguyen, J. (2017a). 'The Turn of the Valve: Representing with Material Models'. *European Journal for Philosophy of Science*. 8: 205–224.

Frigg, R. and Nguyen, J. (2017b). 'Of Barrels and Pipes: Representation-as in Art and Science', in O. Bueno, G. Darby, S. French, and D. Rickles (eds), *Thinking About Science, Reflecting on Art*. London: Routledge, pp. 41–61.

Frost-Arnold, G. (2010). 'The No-Miracles Argument for Realism: Inference to an Unacceptable Explanation'. *Philosophy of Science* 77(1): 35–58.

Gale, G. (2017). 'Cosmology: Methodological Debates in the 1930s and 1940s', in E. N. Zalta (ed.), *The Stanford Encyclopedia of Philosophy*. https://plato.stanford.edu/archives/sum2017/entries/cosmology-30s/.

Galison, P. (2003). *Einstein's Clocks, Poincaré's Maps: Empires of Time*. New York: W. W. Norton.

Gaudet, E. (2006). *Quine on Meaning: The Indeterminacy of Translation*. (Continuum) London: Bloomsbury.

Gay, H. (1978). 'The Asymmetric Carbon Atom: (a) A Case Study of Independent Discovery; (b) An Inductivist Model for Scientific Method'. *Studies in History and Philosophy of Science* 9: 207–38.

Gelfert, A. (2016). *How to Do Science with Models: A Philosophical Primer*. London: Springer.

Gendler, T. (2013). 'Imagination', in E. N. Zalta (ed.), *The Stanford Encyclopedia of Philosophy*. http://plato.stanford.edu/archives/fall2013/entries/imagination/.

Ghirardi, G. C., Rimini, A., and Weber, T. (1986). 'Unified Dynamics for Microscopic and Macroscopic Systems'. *Physical Review D* 34: 470–91.

Gibbard, A. and Varian, H. R. (1978). 'Model Experiments and Models in Experiments'. *The Journal of Philosophy* 75: 664–77.

Giere, R. (1988). *Explaining Science: A Cognitive Approach*. Chicago: University of Chicago Press.

Giere, R. (2002). 'Scientific Cognition as Distributed Cognition', in P. Carruthers, S. Stich, and M. Siegal (eds), *The Cognitive Basis of Science*. Cambridge: Cambridge University Press, pp. 285–99.

Giere, R. (2004). 'How Models Are Used to Represent Reality'. *Philosophy of Science* 71: 742–52.

Giere, R. (2008a). 'Why Scientific Models Should Not be Regarded as Works of Fiction', in M. Suárez (ed.), *Fictions in Science: Philosophical Essays on Modeling and Idealization*. London: Taylor and Francis, pp. 248–58.

Giere, R. (2008b). 'Models, Metaphysics and Methodology', in S. Hartmann, C. Hoefer, and L. Bovens (eds), *Nancy Cartwright's Philosophy of Science*. London: Routledge, pp. 123–33.

Giovannelli, A. (2010). 'Goodman's Aesthetics', in E. N. Zalta (ed.), *The Stanford Encyclopedia of Philosophy*. http://plato.stanford.edu/archives/sum2010/entries/goodman-aesthetics/.

Glymour, C. (2013). 'Theoretical Equivalence and the Semantic View of Theories Author(s): Clark Glymour'. *Philosophy of Science* 80(2): 286–97.

Godfrey-Smith, P. (2006). 'The Strategy of Model-Based Science', *Biology and Philosophy* 21: 725–40.

Godfrey-Smith, P. (2009). 'Models and Fictions in Science' *Philosophical Studies* 143:101–16.

Goldstein, S. (2013), 'Bohmian Mechanics', in E. N. Zalta (ed.), *The Stanford Encyclopedia of Philosophy*. http://plato.stanford.edu/archives/spr2013/entries/qm-bohm/.

Gombrich, E. (1995). 'Introduction: On Art and Artists', in *The Story of Art*, 16th edn. London: Phaidon Press.

Gooday, G. and Mitchell, D. (2013). 'Rethinking "Classical Physics"', in J. Z. Buchwald and R. Fox (eds), *The Oxford Handbook of the History of Physics*. Oxford: Oxford University Press, pp. 721–64. Goodman, N. (1968).

Goodman, N. (1976). *Languages of Art*. Indianapolis: Hackett.

Gratzer, W. (2001). *The Undergrowth of Science: Delusion, Self-Deception, and Human Frailty*. Oxford: Oxford University Press.

Griesemer, J. R. (1990). 'Modeling in the Museum: On the Role of Remnant Models in the Work of Joseph Grinnell'. *Biology and Philosophy* 5: 3–36.

Grobler, A. (1995). 'The Representational and the Non-representational in Models of Scientific Theories', in W. E. Herfel, W. Krajewski, I. Niiniluoto, and R. Wójcicki (eds), *Theories and Models in Science*. Pozna´n Studies in the Philosophy of the Sciences and the Humanities. Rodopi, pp. 37–48.

Hacking, I. (1975). 'The Identity of Indiscernibles'. *Journal of Philosophy* 72: 249–56.

Hacking, I. (1983). *Representing and Intervening*. Cambridge: Cambridge University Press.

Halvorson, H. (2012). 'What Scientific Theories Could Not Be'. *Philosophy of Science* 79: 183–206.

Halvorson, H. (2013). 'The Semantic View, if Plausible, Is Syntactic'. *Philosophy of Science* 80(3): 475–8.

Hanson, N. R. (1958). *Patterns of Discovery*. Cambridge: Cambridge University Press.

Hardwig, J. (1985). 'Epistemic Dependence'. *Journal of Philosophy* 82: 335–49.

Harman, P. M. (1982b). *Metaphysics and Natural Philosophy: The Problem of Substance in Classical Physics*. Sussex: Harvester Press.

Hecht, E. (2011). 'How Einstein Confirmed E=mc2'. *American Journal of Physics* 79: 591–600.

Heinzmann, G. and Stump, D. (2014). 'Henri Poincaré', in E. N. Zalta (ed.), *The Stanford Encyclopedia of Philosophy*. http://plato.stanford.edu/archives/spr2014/entries/poincare/.

Heisenberg, W. (1930). *The Physical Principles of Quantum Theory*. Chicago: University of Chicago Press.

Hendry, R. and Psillos, S. (2007). 'How to Do Things with Theories: An Interactive View of Language and Models in Science', in J. Brzeziński, A. Klawiter, T. A. F. Kuipers, K. Łastowski, K. Paprzycka, and P. Przybysz (eds), *The Courage of Doing Philosophy*. Amsterdam/New York, NY: Rodopi, pp. 59–115.

Hick, D. H. (2011). 'Toward an Ontology of Authored Works'. *British Journal of Aesthetics* 51: 185–99.

Hick, D. H. (2012). 'Aesthetic Supervenience Revisited'. *British Journal of Aesthetics* 52: 301–16.

Hirsch, E. (2008). 'Language, Ontology, and Structure'. *Noûs* 42: 509–28.

Hodge, J. and Radick, G. (2009). 'The Place of Darwin's Theories in the Intellectual Long Run', in Hodge and Radick (eds), *The Cambridge Companion to Darwin*. Cambridge: Cambridge University Press, pp. 246–74.

Hopkins, A. (1977). *The Nine Symphonies of Beethoven*. London: Scolar Press.

Hopkins, R. (1998). *Picture, Image, and Experience*. Cambridge: Cambridge University Press.

Hopkins, R. (2005). 'The Speaking Image: Visual Communication and the Nature of Depiction', in M. Kieran (ed.), *Contemporary Debates in Aesthetics and the Philosophy of Art*. New York: Blackwell, pp. 145–59.

Hopwood, N. (2002). *Embryos in Wax: Models from the Ziegler Studio*. Cambridge: Cambridge University Press.

Horsten, L. (2014), 'Philosophy of Mathematics', in E. N. Zalta (ed.), *The Stanford Encyclopedia of Philosophy*. http://plato.stanford.edu/archives/fall2014/entries/philosophy-mathematics/.

Howard, D. (2017). 'Einstein's Philosophy of Science', in E. N. Zalta (ed.), *The Stanford Encyclopedia of Philosophy*. https://plato.stanford.edu/archives/fall2017/entries/einstein-philscience/.

Hudetz, L. (2017). 'The Semantic View of Theories and Higher-order Languages'. *Synthese* 196: 1131–49.

Hughes, R. I. G. (1997). 'Models and Representation'. *Philosophy of Science* 64: S325–S336.

Hughes, R. I. G. (2010). *The Theoretical Practices of Physics: Philosophical Essays*. Oxford: Oxford University Press.

Huron, D. B. (2006). *Sweet Anticipation: Music and the Psychology of Expectation*. Cambridge, MA: MIT Press.

Hylton, P. (2016). 'Willard van Orman Quine', in E. N. Zalta (ed.), *The Stanford Encyclopedia of Philosophy*. https://plato.stanford.edu/archives/win2016/entries/quine/.

Irvin, S. (2005). 'Appropriation and Authorship in Contemporary Art'. *British Journal of Aesthetics* 45: 123–37.

Ivanova, M. (2017). 'Aesthetic Values in Science'. *Philosophy Compass*. 12. doi: 10.1111/phc3.12433.

Janssen, M. (forthcoming). 'Arches and Sacaffolds: Bridging Continuity and Discontinuity in Theory Change', in A. Love and W. Wimsatt (eds). *Beyond the Meme. Articulating Dynamic Structures in Cultural Evolution*. Minneapolis: University of Minnesota Press.

Jannsen, M., Mecklenburg, M., Hendricks, V. F., et al., eds. (2007). 'From Classical to Relativistic Mechanics: Electromagnetic Models of the Electron'. *Interactions: Mathematics, Physics and Philosophy*. Dordrecht: Springer, pp. 65–134.

Jansson, L. and Tallant, J. (2017). 'Quantitative Parsimony: Probably for the Better'. *British Journal for the Philosophy of Science* 68: 781–803.

Joas, C., Lehner, C., and Renn, J. (eds) (2007). *HQ-1: Conference on the History of Quantum Physics (Berlin, 2–6 July)*. Preprint 350. Berlin: MPI History of Science.

Jones, R. (1991). 'Realism about What?'. *Philosophy of Science* 58: 185–202.

Judge, J. (2018). 'The Surprising Thing about Musical Surprise'. *Analysis* 78: 225–34.

Kaiser, D. (2013). 'Epilogue: Textbooks and the Emergence of a Conceptual Trajectory', in M. Badino and J. Navarro (eds), *Research and Pedagogy: A History of Quantum Physics through Its Textbooks*. http://edition-open-access.de/studies/2/.

Kania, A. (2008). 'The Methodology of Musical Ontology: Descriptivism and its Implications'. *British Journal of Aesthetics* 48: 426–44.

Kania, A. (2014). 'The Philosophy of Music', in E. N. Zalta (ed.), *The Stanford Encyclopedia of Philosophy*. http://plato.stanford.edu/archives/spr2014/entries/music/.

Kasner, E. and Newman, J. (2001). *Mathematics and the Imagination*. Dover.

Katzir, S. (2005). 'Poincaré's Relativistic Physics: Its Origins and Nature'. *Physics in Perspective* 7: 268–92.

Kelly, T. (2014). 'Evidence', in E. N. Zalta (ed.), *The Stanford Encyclopedia of Philosophy*. http://plato.stanford.edu/archives/fall2014/entries/evidence/.

Kenaan, H. (2002). 'The "Unusual Character" of Holbein's *Ambassadors*'. *Artibus et Historiae* 23(46). http://www.tau.ac.il/~kenaan/readers/holbein.pdf.

Ketland, J. (unpublished). 'Craig's Theorem'. https://www.princeton.edu/~hhalvors/teaching/phi520_f2012/ketland-craig.pdf.

Kieran, M. (2005). *Revealing Art*. London: Routledge.

Kieran, M. (2010). 'Aesthetic Knowledge', in S. Bernecker and D. Pritchard (eds), *Routledge Companion to Epistemology*. London: Taylor and Francis, pp. 369–79.

Knox, E. (2014). 'Newtonian Spacetime Structure in Light of the Equivalence Principle'. *British Journal for the Philosophy of Science* 65: 863–80.

Knuuttila, T. (2005). 'Models, Representation, and Mediation'. *Philosophy of Science* 72: 1260–71.

Koestler, A. (1959). *The Sleepwalkers*. London: Hutchinson.

Koller, P. (2014). 'On the Nature of Norms'. *Ratio Juris* 27: 155–75.

Korman, D. Z. (2014). 'Ordinary Objects', in E. N. Zalta (ed.), *The Stanford Encyclopedia of Philosophy*. http://plato.stanford.edu/archives/spr2014/entries/ordinary-objects/.

Kragh, H. (1990). *Dirac: A Scientific Biography*. Cambridge: Cambridge University Press.

Kragh, H. (2013). 'Paul Dirac and *The Principles of Quantum Mechanics*'. http://edition-open-access.de/studies/2/11/.

Kraugerud, H. A. and Ramberg, B. T. (2010). 'The New Loud: Richard Rorty, Quietist?'. *Common Knowledge* 16: 48–65.

Kripke, S. (1980). *Naming and Necessity*. Cambridge, MA: Harvard University Press.

Kronz, F. and Lupher, T. (2012). 'Quantum Theory: von Neumann vs. Dirac', in E. N. Zalta (ed.), *The Stanford Encyclopedia of Philosophy*. http://plato.stanford.edu/archives/sum2012/entries/qt-nvd/.

Kuhn, T. S. (1970). *The Structure of Scientific Revolutions*. Chicago: University of Chicago Press (1962).

Kuhn, T. S. (1978). *Black Body Theory and the Quantum Discontinuity*. Oxford: Oxford University Press.

Kulvicki, J. (2006). *On Images*. Oxford: Oxford University Press.

Kusch, M. (2015). 'Microscopes and the Theory-Ladenness of Experience in Bas van Fraassen's Recent Work'. *Journal for General Philosophy of Science* 46: 167–82.

Ladyman, J. (1998). 'What Is Structural Realism?'. *Studies in History and Philosophy of Science* 29: 409–24.

Ladyman, J. and Ross, D. (2007). *Everything Must Go*. Oxford: Oxford University Press.

Lam, L. and Wuthrich, C. (2015). 'No Categorial Support for Radical Ontic Structural Realism'. *British Journal for the Philosophy of Science* 66: 605–34.

Landry, E. (2007). 'Shared Structure Need Not Be Shared Set-structure'. *Synthese* 158: 1–17.

Laubichler, M. D. and Müller, G. B. (2007). *Modelling Biology*. Cambridge, MA: MIT Press.

Laudan, L. (1984). *Science and Values: The Aims of Science and Their Role in Scientific Debate*. Berkeley: University of California Press.

Laudan, L. (1990). 'Normative Naturalism'. *Philosophy of Science* 57: 44–59.

Laudan, L. (2007). 'The Epistemic, the Cognitive and the Social', in P. Machamer and G. Wolters (eds), *Science, Objectivity and Values*. Pittsburgh: University of Pittsburgh Press, pp. 14–23.

Laymon, R. (1991). 'Thought Experiments by Stevin, Mach and Gouy: Thought Experiments as Ideal Limits and Semantic Domains', in T. Horowitz and G. Massey (eds), *Thought Experiments in Science and Philosophy*. Lanham, MD: Rowman and Littlefield, 167–91.

LeBihan, S. (2012). 'Defending the Semantic View: What It Takes'. *European Journal for the Philosophy of Science* 2, 249–74.

Leike, A. (2002). 'Demonstration of the Exponential Decay Law Using Beer Froth'. *European Journal of Physics* 23: 21. Bibcode:2002EJPh...23...21L. doi:10.1088/0143-0807/23/1/304.

Leiter, B. (2006). *The Future for Philosophy*. Oxford: Oxford University Press.

Lenain, T. (2011). *Art Forgery: The History of a Modern Obsession*. London: Reaktion Books.

Liao, S., Strohminger, N., and Sripada, C. S. (2014). 'Empirically Investigating Imaginative Resistance'. *British Journal of Aesthetics* 54: 339–55.

Liebesman, D. (2014). 'Necessarily, Sherlock Holmes Is Not a Person'. *Analytic Philosophy* 55: 306–18.

Litvinov, Y. A. and Sobiczewski, A. (2012). 'Theoretical Description of Nuclear Masses'. *EPJ Web of Conferences* 38(05001). doi:10.1051/epjconf/20123805001.

Livingston, P. (2013). 'History of the Ontology of Art', in E. N. Zalta (ed.), *The Stanford Encyclopedia of Philosophy*. http://plato.stanford.edu/archives/sum2013/entries/art-ontology-history/.

Lloyd, E. A. (1994). *The Structure and Confirmation of Evolutionary Theory*. Princeton, NJ: Princeton University Press.

London, F. and London, H. (1935). 'The Electromagnetic Equations of the Supraconductor'. *Proceedings of the Royal Society (London)*, A149: 71–88.

Lopes, D. M. (1996). *Understanding Pictures*. Oxford: Oxford University Press.

Lopes, D. (2005). 'The Domain of Depiction', in M. Kieran (ed.), *Contemporary Debates in Aesthetics and the Philosophy of Art*. New York: Blackwell, 160–74.

Lutz, S. (2012). 'On a Straw Man in the Philosophy of Science: A Defense of the Received View'. *HOPOS: The Journal of the International Society for the History of Philosophy of Science* 2(1): 77–120.

Lutz, S. (2014). 'What's Right with a Syntactic Approach to Theories and Models?'. *Erkenntnis* 79: 1475–92.

Lutz, S. (2017). 'What Was the Syntax-Semantics Debate in the Philosophy of Science?'. *Philosophy and Phenomenological Research* 95: 319–52.

Lynch, M. (2009). *Truth as One and Many*. Oxford: Oxford University Press.

MacBride, F. (2014). 'Truthmakers', in E. N. Zalta (ed.), *The Stanford Encyclopedia of Philosophy*. http://plato.stanford.edu/archives/spr2014/entries/truthmakers/.

McGrath, M. (2014). 'Propositions', in E. N. Zalta (ed.), *The Stanford Encyclopedia of Philosophy*. http://plato.stanford.edu/archives/spr2014/entries/propositions/.

Mac Lane, S. (1996). 'Structure in Mathematics'. *Philosophia Mathematica* 4: 174–83.

McMullin, E. (1985). 'Galilean Idealisation'. *Studies in History and Philosophy of Science* 16: 247–73.

Magnus, P. D. (2012). *Scientific Enquiry and Natural Kinds: From Planets to Mallards*. London: Palgrave Macmillan.

Massimi, M. (2018). 'Perspectival Modeling'. *Philosophy of Science* 85: 335–59.

Melia, J. (2000). 'Weaseling Away the Indispensability Argument'. *Mind* 109: 455–80.

Meir, T. (forthcoming). 'Structural Realism in Generative Grammar'.

Menzel, C. (2017). 'Possible Worlds', in E. N. Zalta (ed.), *The Stanford Encyclopedia of Philosophy*. https://plato.stanford.edu/archives/win2017/entries/possible-worlds/.

Merton, R. K. (1942). 'A Note on Science and Technology in a Democratic Order'. *Journal of Legal and Political Sociology* 1: 115–26.

Merton, R. K. (1963). 'Resistance to the Systematic Study of Multiple Discoveries in Science'. *European Journal of Sociology* 4: 237–82.

Meskin, A. (2011). 'The Philosophy of Comics'. *Philosophy Compass* 6/12: 854–64.

Meskin, A. (2014). 'The Ontology of Comics', in A. Meskin and R. T. Cook (eds), *The Art of Comics: A Philosophical Approach*. Oxford: Wiley-Blackwell, pp. 31–46.

Meskin, A. and Cook, R. (eds) (2014). *The Art of Comics: A Philosophical Approach*. Oxford: Wiley-Blackwell.

Meskin, A., Phelan, M., Moore, M., and Kieran, M. (2013). 'Mere Exposure to Bad Art'. *British Journal of Aesthetics* 53: 139–64.

Mikenberg, I., da Costa, N. C. A., and Chuaqui, R. (1986). 'Pragmatic Truth and Approximation to Truth'. *Journal of Symbolic Logic* 51: 201–21.

Miller, A. (2014). 'Realism', in E. N. Zalta (ed.), *The Stanford Encyclopedia of Philosophy*. http://plato.stanford.edu/archives/win2014/entries/realism/.

Minkowski, H. (1908). 'Nachrichten von der Gesellschaft der Wissenschaften zu Göttingen, Mathematisch-Physikalische Klasse. 1908: 53–111. Eng. trans. https://en.wikisource.

org/wiki/Translation:The_Fundamental_Equations_for_Electromagnetic_Processes_in_Moving_Bodies#.C2.A7_5._Space-time_Vectors._Of_the_1st_and_2nd_kind.

Minkowski, H. (1909). 'Raum und Zeit (1909)'. Jahresberichte der Deutschen Mathematiker-Vereinigung, 1–14. A Lecture delivered before the Naturforscher Versammlung (Congress of Natural Philosophers) at Cologne—21 September.

Mondrian, P. C. (1917). 'The New Plastic in Painting'. Leiden: De Stijl.

Mondrian, P. C. (1919–20). *Natural Reality and Abstract Reality: A Trialogue (While Strolling from the Country to the City)*, trans. E. M. Beekman. New York: G. Braziller Inc.

Morgan, M. (2005). 'Experiments versus Models: New Phenomena, Inference and Surprise'. *Journal of Economic Methodology* 12: 317–29.

Morgan, M. and Morrison, M. (1993). *Models as Mediators*. Cambridge: Cambridge University Press.

Mormann, T. (2007). 'The Structure of Scientific Theories in Logical Empiricism', in A. Richardson and T. Uebel, *The Cambridge Companion to Logical Empiricism*. Cambridge: Cambridge University Press, pp. 136–62.

Muldoon, R., Smith, T., and Weisberg, M. (2012). 'Segregation That No One Seeks'. *Philosophy of Science* 79: 38–62.

Muller, F. (1997). 'The Equivalence Myth of Quantum Mechanics'. *Studies in History and Philosophy of Modern Physics* 28: 35–61 and 219–47.

Muller, F. (2011). 'Reflections on the Revolution at Stanford'. *Synthese* 183: 87–114.

Muller, F. and Saunders, S. (2008). 'Discerning Fermions'. *British Journal for the Philosophy of Science* 59: 499–548.

Murphy, A. (forthcoming). 'Towards a Pluralist Account of Imagination in Science'.

Musgrave, A. (1989). 'NOA's Ark: Fine for Realism'. *Philosophical Quarterly* 39: 383–98.

Nagel, E. (1961). *The Structure of Science*. New York: Harcourt, Brace, & World.

Navarro, J. (2009). '"A Dedicated Missionary": Charles Galton Darwin and the New Quantum Mechanics in Britain'. *Studies in History and Philosophy of Science Part B: Studies in History and Philosophy of Modern Physics* 40: 316–26.

Nerlich, G. (1994). *The Shape of Space*. 2nd edn. Cambridge: Cambridge University Press.

Newburgh, R. Peidle, J., and Reuckner, W. (2006). 'Einstein, Perrin, and the Reality of Atoms: 1905 Revisited'. *American Journal of Physics* 74: 478–81.

Nicholson, A. J. (1960). 'The Role of Population Dynamics in Natural Selection', in S. Tax (ed.), *Evolution after Darwin*, vol. 1. Chicago: Chicago University Press, pp. 477–522.

Nolan, D. (2016). 'Modal Fictionalism', in E. N. Zalta (ed.), *The Stanford Encyclopedia of Philosophy*. https://plato.stanford.edu/archives/spr2016/entries/fictionalism-modal/.

Norton, J. D. (1984). 'How Einstein Found His Field Equations: 1912–1915'. *Historical Studies in the Physical Sciences* 14: 253–316; repr. in D. Howard and J. Stachel, eds. (1989). *Einstein and the History of General Relativity*. Einstein Studies, vol. 1. Boston: Birkhäuser, pp. 101–59.

Norton, J. D. (1985). 'What Was Einstein's Principle of Equivalence?'. *Studies in History and Philosophy of Science* 16: 203–46; repr. in D. Howard and J. Stachel, eds. (1989). *Einstein and the History of General Relativity*. Einstein Studies, vol. 1. Boston: Birkhäuser, pp. 5–47.

Norton, J. D. (1991). 'Thought Experiments in Einstein's Work', in T. Horowitz and G. Massey (eds), *Thought Experiments in Science and Philosophy*. Lanham, MD: Rowman and Littlefield, pp. 129–48.

Norton, J. D. (2004). 'Why Thought Experiments Do Not Transcend Empiricism', in C. Hitchcock (ed.), *Contemporary Debates in the Philosophy of Science*. Oxford: Blackwell, pp. 44–66.

Norton, J. (2008). 'Must Evidence Underdetermine Theory?', in M. Carrier, D. Howard, and J. Kourany (eds), *The Challenge of the Social and the Pressure of Practice*. Pittsburgh, PA: University of Pittsburgh Press, pp. 17–44.

North, J. (2009). 'The "Structure" of Physics: A Case Study'. *Journal of Philosophy* 106: 57–88.

Norton, J. D. (2014). 'The Hole Argument', in E. N. Zalta (ed.), *The Stanford Encyclopedia of Philosophy*. http://plato.stanford.edu/archives/spr2014/entries/spacetime-holearg/.

Norton, J. D. (forthcoming). *The Material Theory of Induction*. https://www.pitt.edu/~jdnorton/papers/material_theory/material.html.

Pais, A. (1982). *Subtle Is the Lord*. Oxford: Oxford University Press.

Pancs, R. and Vriend, N. J. (2007). 'Schelling's Spatial Proximity Model of Segregation Revisited'. *Journal of Public Economics* 91: 1–24.

Paternotte, C. and Grose, J. (2017). 'Robustness in Evolutionary Explanations: A Positive Account'. *Biology and Philosophy* 32: 73–96.

Pechenkin, A. (2012). 'The Early Statistical Interpretations of Quantum Mechanics in the USA and USSR'. *Studies in History and Philosophy of Modern Physics* 43: 25–34.

Pedersen, N. J. L. L. and Wright, C. (2016). 'Pluralist Theories of Truth', in E. N. Zalta (ed.), *The Stanford Encyclopedia of Philosophy*. https://plato.stanford.edu/archives/spr2016/entries/truth-pluralist/.

Peirce C. S. (1878). 'How to Make Our Ideas Clear'. *Popular Science Monthly* 12: 286–302.

Peirce, C. S. (1940). *The Philosophy of Peirce*, ed. J. Buchler. London: Routledge and Kegan Paul.

Penrose, L. and Penrose, R. (1958). 'Impossible Objects: A Special Type of Visual Illusion'. *British Journal of Psychology* 49: 31–3.

Pincock, C. (2005). 'Overextending Partial Structures: Idealization and Abstraction'. *Philosophy of Science* 72: 1248–59.

Planck, M. (1920). 'Max Planck—Nobel Lecture: The Genesis and Present State of Development of the Quantum Theory'. Nobel Media AB 2014, 4 June 2015. http://www.nobelprize.org/nobel_prizes/physics/laureates/1918/planck-lecture.html.

Poincaré, H. (1904/6). 'The Principles of Mathematical Physics'. *The Foundations of Science (The Value of Science)*. New York: Science Press, pp. 297–320.

Poole, S. (2015). 'The Impossible World of MC Escher'. *The Guardian* 20. http://www.theguardian.com/artanddesign/2015/jun/20/the-impossible-world-of-mc-escher.

Popper, K. R. (1972). *Objective Knowledge: An Evolutionary Approach*. Oxford: Clarendon Press.

Popper, K. R. (1974). 'Replies to My Critics', in P. A. Schilpp (ed.), *The Philosophy of Karl Popper*. La Salle: Open Court, pp. 961–1174.

Popper, K. R. (1978). *Three Worlds: The Tanner Lectures on Human Values*. Utah: University of Utah Press.

Popper, K. R. (1986). *Unended Quest*. La Salle: Open Court.

Popper, K. (2002/1959). *The Logic of Scientific Discovery*. London: Routledge.

Post, H. R. (1960). 'Simplicity in Scientific Theories'. *British Journal for the Philosophy of Science* 11: 32–41.

Post, H. R. (1971 [1993]). 'Correspondence, Invariance and Heuristics'. *Studies in History and Philosophy of Science* 2: 213–55. Repr. in S. French and H. Kamminga, eds. (1993). *Correspondence, Invariance and Heuristics: Essays in Honour Of Heinz Post*. Boston Studies in the Philosophy of Science, vol. 148. Dordrecht: Kluwer Academic Press, pp. 1–44.

Potters, J. (2019). 'Stabilization of Phenomenon and Meaning: On the London & London Episode as a Historical Case in Philosophy of Science'. *European Journal for the Philosophy of Science*. doi:10.1007/s13194-019-0247-7.

Priest, G. (2006). *In Contradiction: A Study of the Transconsistent*. Oxford: Oxford University Press. 2nd edn.

Psillos, S. (1999). *Scientific Realism: How Science Tracks Truth*. London: Routledge.

Psillos, S. (2011). 'Making Contact with Molecules: On Perrin and Achinstein', in G. J. Morgan (ed.), *Philosophy of Science Matters: The Philosophy of Peter Achinstein*. Oxford: Oxford University Press, pp. 177–90.

Putnam, H. (1965). 'Craig's Theorem'. *Journal of Philosophy* 62: 251–60.

Quine, W. V. O. (1957). 'The Scope and Language of Science'. *British Journal for the Philosophy of Science* 8: 1–17.

Quine, W. V. O. (1960). *Word and Object*. Cambridge, MA: MIT Press.

Quine, W. V. O. (1975). 'On Empirically Equivalent Systems of the World'. *Erkenntnis* 9: 313–28.

Reck, E. and Price, M. (2000). 'Structures and Structuralism in Contemporary Philosophy of Mathematics'. *Synthese* 125: 341–87.

Redei, M. (1996). 'Why John von Neumann Did Not Like the Hilbert Space Formalism of Quantum Mechanics (and What He Liked Instead)'. *Studies in History and Philosophy of Modern Physics* 21: 493–510.

Redhead, M. (1975). 'Symmetry in Intertheory Relations'. *Synthese* 32: 77–112.

Redhead, M. L. G. (1980). 'Models in Physics'. *British Journal for the Philosophy of Science* 3: 145–63.

Reichenbach, H. (1038). Experience and Prediction: An Analysis of the Foundations and the Structure of Knowledge. Chicago, IL: University of Chicago Press.

Reiss, J. (2008). 'Social Capacities', in S. Hartmann and L. Bovens (eds), *Nancy Cartwright's Philosophy of Science*. London: Routledge, pp. 265–88.

Rickles, D. (2013). 'Musicology as an Object for HPS? An Exploration'. *Aesthetics On-Line*. http://www.aesthetics-online.org/articles/index.php?articles_id=74.

Rickles, D. (2017). 'Some Philosophical Problems of Music Theory (and Some Music-Theoretic Problems of Philosophy)', in O. Bueno et al. (eds), *Thinking about Science, Reflecting on Art*. London: Routledge, pp. 119–39.

Ridley, A. (1997). 'Not Ideal: Collingwood's Expression Theory'. *Journal of Aesthetics and Art Criticism* 55(3): 263–72.

Ridley, A. (2003). 'Against Musical Ontology'. *Journal of Philosophy* 100: 203–20.

Rohrbaugh, G. (2003). 'Artworks as Historical Individuals'. *European Journal of Philosophy* 11: 181–6.

Rohrbaugh, G. (2005). 'I Could Have Done That'. *British Journal of Aesthetics* 45: 209–28.

Rorty, R. (2006). 'Naturalismo e Quietismo'. *Diánoia* 51: 3–18.

Rosen, G. (2017). 'Abstract Objects', in E. N. Zalta (ed.), *The Stanford Encyclopedia of Philosophy*. https://plato.stanford.edu/archives/win2017/entries/abstract-objects/.

Roush, S. (2005). *Tracking Truth: Knowledge, Evidence, and Science*. Oxford: Oxford University Press.

Rowbottom, D. (2011). 'Stances and Paradigms: A Reflection'. *Synthese* 178: 111–19.

Ruark, A. E. and Urey, H. C. (1930). *Atoms, Molecules and Quanta*. International Series in Physics. New York: McGraw-Hill Book Co.

Rueger, A. (1990). 'Independence from Future Theories: A Research Strategy in Quantum Theory'. *Philosophy of Science Association* 1: 203–11.

Ruetsche, L. (2002). 'Interpreting Quantum Field Theory'. *Philosophy of Science* 69: 348–78.
Ruetsche, L. (2011). *Interpreting Quantum Theories*. Oxford: Oxford University Press.
Ruetsche, L. (2015). 'The Shaky Game +25, or: on Locavoracity'. *Synthese* 192: 3425–42.
Rupert Hall, A. (2002). 'Newton versus Leibniz: From Geometry to Metaphysics', in I. B. Cohen and G. E. Smith (eds), *The Cambridge Companion to Newton*. Cambridge: Cambridge University Press, pp. 431–54.
Saatsi, J. (2011). 'The Enhanced Indispensability Argument: Representational versus Explanatory Role of Mathematics in Science'. *British Journal for the Philosophy of Science* 62: 143–54.
Saatsi, J. (2012). 'Scientific Realism and Historical Evidence: Shortcomings of the Current State of Debate'. de Regt et al. (eds), *EPSA Philosophy of Science: Amsterdam 2009*. Berlin: Springer.
Saatsi, J. (forthcoming). 'On Explanations from 'Geometry of Motion'. *British Journal for the Philosophy of Science*.
Salis, F. (forthcoming). 'Knowledge through the Scientific Imagination'.
Salis, F. and Frigg, R. (forthcoming). 'Capturing the Scientfic Imagination', in P. Godfrey-Smith and A. Levy (eds), *The Scientific Imagination*. Oxford: Oxford University Press.
Sanchez-Dorado, J. (2017). 'Methodological Lessons for the Integration of Philosophy of Science and Aesthetics: The Case of Representation', in O. Bueno et al., *Thinking about Science, Reflecting on Art*. London: Routledge, pp. 10–26.
Sartre, J-P. (1972). *The Psychology of Imagination*. London: Methuen.
Saunders, S. (1993). 'To What Physics Corresponds', in S. French and H. Kaminga, (eds), *Correspondence, Invariance, and Heuristics; Essays in Honour of Heinz Post*. Dordrecht: Kluwer, pp. 295–326.
Saunders, S. (2006b). 'Are Quantum Particles Objects?'. *Analysis* 66: 52–63.
Schaffer, J. (2008). 'Truth-maker Commitments'. *Philosophical Studies* 141: 7–19.
Schaffer, S. (1986). 'Scientific Discoveries and the End of Natural Philosophy'. *Social Studies of Science* 16: 387–420.
Schaffner, K. (1969). 'Correspondence Rules'. *Philosophy of Science* 36: 280–90.
Schatzki, T. (2001). 'Introduction', in T. R. Schatzki, K. Knorr Cetina, and E. von Savigny (eds), *The Practice Turn in Contemporary Theory*. London: Routledge.
Schelling, T. (1978). *Micromotives and Macrobehaviour*. New York: W. W. Norton (repr. 2006).
Schiffer, S. (2003). *The Things We Mean*, Oxford: Oxford University Press.
Schindler, S. (2008). 'Model, Theory, and Evidence in the Discovery of the DNA Structure'. *The British Journal for the Philosophy of Science* 59: 619–58.
Schmidt, H-J. (2014). 'Structuralism in Physics', in E. N. Zalta (ed.), *The Stanford Encyclopedia of Philosophy*. http://plato.stanford.edu/archives/win2014/entries/physics-structuralism/.
Scholz, E. (2008). 'Weyl Entering the "New" Quantum Mechanics Discourse', in C. Joas, C. Lehner, and J. Renn (eds), *HQ-1 : Conference on the History of Quantum Physics*, vol. 2. Berlin: Max-Planck-Institut für Wissenschaftsgeschichte, pp. 253–71.
Schurz, G. (2014). 'Criteria of Theoreticity: Bridging Statement and Non-Statement View'. *Erkenntnis* 79(8): 1521–45.
Schweber, S. S. (1994). *QED and the Men Who Made It: Dyson, Feynman, Schwinger, and Tomonaga*. Princeton, NJ: Princeton University Press.
Shapere, D. (1982). 'The Concept of Observation in Science and Philosophy'. *Philosophy of Science* 49: 485–525.
Sharpe, R. M. (2001). 'Could Beethoven Have "Discovered" the Archduke Trio?'. *British Journal of Aesthetics* 41: 325–7.

Shech, E. (2015). 'Scientific Misrepresentation and Guides to Ontology: The Need for Representational Code and Contents'. *Synthese* 192: 3463–85.

Shech, E. (2016). 'Fiction, Depiction, and the Complementarity Thesis in Art and Science'. *The Monist* 99(3): 311–32.

Sibley, F. (2001). 'Aesthetic and Non-aesthetic', in J. Benson, B. Redfern, and J. R. Cox (eds), *Approach to Aesthetics: Collected Papers on Philosophical Aesthetics*. Oxford: Clarendon Press, pp. 33–51.

Sider, T. (2014). 'Hirsch's Attack on Ontologese'. *Noûs* 48: 565–72.

Sklar, L. (1982). 'Saving the Noumena'. *Philosophical Topics* 13: 89–110.

Sklar, L. (2004). 'Spacetime and Conventionalism'. *Philosophy of Science* 71: 950–9.

Sperber, D. (1975). *Rethinking Symbolism*, trans. A. Morton. Cambridge: Cambridge University Press.

Staley, R. (2008). *Einstein's Generation: The Origins of the Relativity Revolution*. Chicago: University of Chicago Press.

Stan, M. (forthcoming). 'Euler, Newton and Foundations for Mechanics', in C. Smeenck and E. Schliesser (eds), *The Oxford Handbook of Newton*. Oxford: Oxford University Press.

Stanley, J. (2001). 'Hermeneutic Fictionalism', in P. French and H. Wettstein (eds), *Midwest Studies in Philosophy 25: Figurative Language*. Oxford: Blackwell, pp. 36–71.

Stock, K. (2006). 'Sartre, Wittgenstein and Learning from Imagination', in Goldie, ch. 10.

Stock, K. (2017). *Only Imagine*. Oxford: Oxford University Press.

Strauch, D. (2009). *Classical Mechanics: An Introduction*. Berlin: Springer.

Strevens, M. (2008). *Depth: An Account of Scientific Explanation*. Cambridge, MA: Harvard University Press.

Stuart, M. (2017). 'Imagination: A Sine Qua Non of Science'. *Croatian Journal of Philosophy* 49: 9–32.

Stuart, M. (forthcoming). 'The Productive Anarchy of Scientific Imagination'.

Suárez, M. (2003). 'Scientific Representation: Against Similarity and Isomorphism'. *International Studies in the Philosophy of Science* 17: 225–44.

Suárez, M. (2009). 'Scientific Fictions as Rules of Inference', in M. Suarez (ed.), *Fictions in Science: Philosophical Essays on Models and Idealisations*. London: Routledge, pp. 158–78.

Suárez, M. (2010). 'Scientific Representation'. *Philosophy Compass* 5(1): 91–101.

Suárez, M. (2011). 'Review of B. van Fraassen, *Scientific Representation*'. *Metascience* 20: 417–42.

Suárez, M. (2015). 'Deflationary Representation, Inference, and Practice'. *Studies in History and Philosophy of Science Part A* 49: 36–47.

Suárez, M. and Cartwright, N. (2008). 'Theories: Tools versus Models'. *Studies in History and Philosophy of Modern Physics* 39: 62–81.

Suárez, M. and Sole, A. (2006). 'On the Analogy between Cognitive Representation and Truth'. *Theoria* 21: 39–48.

Suppe, F., ed. (1974). *The Structure of Scientific Theories*. Urbana, IL: University of Illinois Press.

Suppe, F. (1989). *Scientific Realism and Semantic Conception of Theories*. Urbana, IL: University of Illinois Press.

Suppe, F. (2000). 'Understanding Scientific Theories: An Assessment of Developments, 1969–1998'. *Philosophy of Science* 67, S102–S115.

Suppe, F. (2001). 'Theory Identity', in W. H. Newton-Smith (ed.), *A Companion to the Philosophy of Science*. Malden, MA: Blackwell.

Suppes, P. (1957). *Introduction to Logic*. New York: Van Nostrand.

Suppes, P. (1961). 'A Comparison of the Meaning and Uses of Models in Mathematics and the Empirical Sciences', in H. Freudenthal (ed.), *The Concept and the Role of the Model in Mathematics and Natural and Social Sciences*. Dordrecht: Reidel, pp. 163–77.

Suppes, P. (1962). 'Models of Data', in E. Nagel, P. Suppes, and A. Tarski (eds), *Logic, Methodology and the Philosophy of Science: Proceedings of the 1960 International Congress*. Stanford: Stanford University Press, pp. 252–67.

Suppes, P. (1967). 'What is a Scientific Theory?', in S. Morgenbesser (ed.), *Philosophy of Science Today*. New York: Basic Books, pp. 55–67.

Suppes, P. (2002). *Representation and Invariance of Scientific Structures*. Stanford: Center for the Study of Language and Information (CSLI) Publications.

Swoyer, C. (1991). 'Structural Representation and Surrogative Reasoning'. *Synthese* 87: 449–508.

Tahko, T. E. (2018). 'Fundamentality and Ontological Minimality', in R. Bliss and G. Priest (eds), *Reality and Its Structure: Essays in Fundamentality*. Oxford: Oxford University Press, pp. 237–253.

Teller, P. (2001). 'Twilight of the Perfect Model Model'. *Erkenntnis* 55: 393–415.

Thomas, W. (2017). 'Scientists' Imagined Pasts and Historians' Appreciation of Scientific Thought'. *Isis* 108: 830–5.

Thomasson, A. (1999). *Fiction and Metaphysics*. Cambridge: Cambridge University Press.

Thomasson, A. (2003). 'Fictional Characters and Literary Practices'. *British Journal of Aesthetics* 43: 138–57.

Thomasson, A. (2004). 'The Ontology of Art', in P. Kivy (ed.), *The Blackwell Guide to Aesthetics*. Oxford: Blackwell, pp. 78–92.

Thomasson, A. (2006). 'Debates about the Ontology of Art: What Are We Doing Here?'. *Philosophy Compass* 1: 245–55.

Thomson-Jones, M. (2006). 'Models and the Semantic View'. *Philosophy of Science* 73: 524–35.

Thomson-Jones, M. (2010). 'Missing Systems and the Face Value Practice'. *Synthese* 172: 283–99.

Tolkien, J. R. R. (1968). *The Lord of the Rings*. London: Allen and Unwin (pbk).

Toon, A. (2012). *Models as Make-Believe*. London: Palgrave Macmillan.

Toulmin, S. (1962). 'Review of *The Sleepwalkers*'. *Journal of Philosophy* 59: 500–3.

Turney, P. (1990). 'Embeddability, Syntax, and Semantics in Accounts of Scientific Theories'. *Journal of Philosophical Logic* 19: 429–51.

van Dongen, J. (2009). 'On the Role of the Michelson-Morley Experiment: Einstein in Chicago'. *Archives for the History of the Exact Sciences* 63: 655–63.

van Dongen, J. (2017). 'The Epistemic Virtues of the Virtuous Theorist: On Albert Einstein and His Autobiography', in J. van Dongen and H. Paul (eds), *Epistemic Virtues in the Sciences and the Humanities*. Boston Studies in the Philosophy and History of Science, vol. 321. Cham: Springer, pp. 63–77.

van Fraassen, B. C. (1976). 'To Save the Phenomena'. *Journal of Philosophy* 73: 623–32 (page references are to repr. in D. Papineau, ed. (1996). *Philosophy of Science*. Oxford: Oxford University Press).

van Fraassen, B. (1980). *The Scientific Image*. Oxford: Oxford University Press.

Van Fraassen, B. (1985a). 'Empiricism in the Philosophy of Science', in P. Churchland and C. Hooker (eds). *Images of Science: Essays on Realism and Empiricism* (with a reply from Bas C. van Fraassen). Chicago: University of Chicago Press, pp. 245–308.

van Fraassen, B. (1985b). 'On the Question of Identification of a Scientific Theory'. *Crítica* 17: 21–5.
van Fraassen, B. (1989). *Laws and Symmetry*. Oxford: Oxford University Press.
van Fraassen, B. (1990). *Quantum Mechanics: An Empiricist Approach*. Oxford: Oxford University Press.
van Fraassen, B. (1994). 'Interpretation of Science: Science as Interpretation', in J. Hilgevoord (ed.), *Physics and Our View of the World*. Cambridge: Cambridge University Press, pp. 169–87.
van Fraassen, B. (2002). *The Empirical Stance*. New Haven, CT: Yale University Press.
van Fraassen, B. (2008). *Scientific Representation: Paradoxes of Perspective*. Oxford: Oxford University Press.
van Fraassen, B. C. (2014). 'One or Two Gentle Remarks about Hans Halvorson's Critique of the Semantic View'. *Philosophy of Science* 81: 276–83.
van Fraassen, B. and Stigman, J. (1993). 'Interpretation in Science and in the Arts', in G. Levine (ed.), *Realism and Representation*. Madison: University of Wisconsin Press, pp. 73–99.
van Whye, J. (2013). *Dispelling the Darkness Voyage in the Malay Archipelago and the Discovery of Evolution by Wallace and Darwin*. Singapore: World Scientific.
Vickers, P. (2013). *Understanding Inconsistent Science*. Oxford: Oxford University Press.
Vickers, P. (2014). 'Scientific Theory Eliminativism'. *Erkenntnis* 79: 111–26.
von Neumann, J. (1955). *Mathematical Foundations of Quantum Mechanics*, trans. Robert T. Geyer. Princeton: Princeton University Press.
Walhout, D. (1986). 'Discovery and Creation in Music'. *The Journal of Aesthetics and Art Criticism* 45: 193–5.
Wallace, D. (2001). 'In Defence of Naiveté: The Conceptual States of Lagrangian Quatum Field Theory'. *Synthese* 151: 33–80.
Wallace, D. (2012). *The Emergent Multiverse: Quantum Theory According to the Everett Interpretation*. Oxford: Oxford University Press.
Wallace, D. (2020). 'On the Plurality of Quantum Theories: Quantum Theory as a Framework, and its Implications for the Quantum Measurement Problem', in S. French and J. Saatsi (eds), *Realism and the Quantum*. Oxford: Oxford University Press.
Walton, K. (1990). *Mimesis as Make-Believe*. Cambridge, MA: Harvard University Press.
Walton K. (1993). 'Metaphor and Prop-Oriented Make Believe'. *European Journal of Philosophy* 51: 499–510.
Wasserman, R. (2015). 'Material Constitution', in E. N. Zalta (ed.), *The Stanford Encyclopedia of Philosophy*. http://plato.stanford.edu/archives/spr2015/entries/material-constitution/.
Watkins, J. W. N. (1974). 'The Unity of Popper's Thought', in P. A. Schilpp (ed.), *The Philosophy of Karl Popper*, vol. 1. La Salle: Open Court, pp. 371–412.
Watson, R. A. (1993). 'Shadow History in Philosophy'. *Journal of the History of Philosophy* 31: 95–109.
Weisberg, M. (2006). 'Robustness Analysis'. *Philosophy of Science* 73: 730–42.
Weisberg, M. (2013). *Simulation and Similarity*. New York: Oxford University Press.
Weyl, H. (1931). *The Theory of Groups and Quantum Mechanics* (trans. from the 2nd rev. German edn by H. P. Robertson. 1st edn 1928). New York: Dover.
White, A. (1990). *The Language of Imagination*. Oxford: Blackwell.
Whittaker, E. (1910). *A History of Theories of the Aether and Electricity*. London: Longmans, Green and Co.

Whittaker, E. (1951/3). *A History of the Theories of Aether and Electricity*. 2nd edn. Vol. 1: *The Classical Theories*/Vol. 2: *The Modern Theories 1900–1926*. London: Nelson.
Wilson, A. (2017a). 'Science's Imagined Pasts'. *Isis* 108: 814–26.
Wilson, A. (2017b). 'Response to Comments'. *Isis* 108: 846–51.
Wilson, M. (1996). 'Classical Mechanics', in E. Craig (ed.), *Routledge Encyclopedia of Philosophy*. London: Routledge.
Wilson, M. (2006). *Wandering Significance*. Oxford: Oxford University Press.
Wilson, M. (2014). 'What Is Classical Mechanics Anyway?'. http://www.philosophy.pitt.edu/sites/default/files/whatisclassicalmechanicsanyway.pdf (an abridged version can be found in R. Batterman, ed. (2014) *The Oxford Handbook in Philosophy of Physics*. Oxford: Oxford University of Press).
Wimsatt, W. and Beardsley, M. (1946). 'The Intentional Fallacy'. *The Sewanee Review* 54: 468–88.
Wittgenstein, L. (1958). *Philosophical Investigations*. Oxford: Blackwell.
Witze, A. (2014). 'Special Relativity Aces Time Trial'. *Nature* 19. http://www.nature.com/news/special-relativity-aces-time-trial-1.15970.
Wolterstoff, N. (1980). *Works and Worlds of Art*. Oxford: Clarendon Press.
Woodridge, J. A. (2006). 'Propositions as Semantic Pretense'. *Language and Communication* 26: 343–55.
Woodward, J. (2006). 'Some Varieties of Robustness'. *Journal of Economic Methodology* 13: 219–40.
Woodward, J. (2014). 'Simplicity in the Best Systems Account of Laws of Nature'. *British Journal for the Philosophy of Science* 65: 91–123.
Worrall, J. (1984). 'The Background to the Forefront'. PSA: Proceedings of the Biennial Meeting of the Philosophy of Science Association 1984, vol. 2. East Lansing: PSA, pp. 672–82.
Wright, C. (1999). 'Truth: A Traditional Debate Reviewed'. *Canadian Journal of Philosophy* 24: 31–74.
Yablo, S. (1993). 'Is Conceivability a Guide to Possibility'. *Philosophy and Phenomenological Research* 53: 1–42.
Yablo, S. (2001). 'Go Figure: A Path through Fictionalism'. *Midwest Studies in Philosophy* 25: 72–102.
Zahar, E. (2001). *Poincaré's Philosophy*. Chicago: Carus Publishing Company.
Ziff, J. (1980). 'Review of M. Butlin and E. Joll, *The Paintings of J. M. W. Turner*'. *The Art Bulletin* 62: 166–71.

Name Index

For the benefit of digital users, indexed terms that span two pages (e.g., 52–53) may, on occasion, appear on only one of those pages.

Subject Index

For the benefit of digital users, indexed terms that span two pages (e.g., 52–53) may, on occasion, appear on only one of those pages.